Mathematics of Financial Obligations

Translations of

MATHEMATICAL MONOGRAPHS

Volume 212

Mathematics of Financial Obligations

A. V. Mel′nikov
S. N. Volkov
M. L. Nechaev

American Mathematical Society
Providence, Rhode Island

А. В. Мельников, М. Л. Нечаев, С. Н. Волков

МАТЕМАТИКА ФИНАНСОВЫХ ОБЯЗАТЕЛЬСТВ

ГУ ВШЭ, Москва, 2001

Translated from the Russian by H. H. McFaden

2000 *Mathematics Subject Classification.* Primary 91–02, 91B24, 91B28, 91B26;
Secondary 91B30, 91B82, 60H30, 60G40, 60G48, 60G44, 60G42, 60P20.

Library of Congress Cataloging-in-Publication Data

Mel′nikov, A. V., 1953–
[Matematika finansovykh obiazatel′stv. English]
Mathematics of financial obligations / A.V. Mel′nikov, S.N. Volkov, M.L. Nechaev.
p. cm. — (Translations of mathematical monographs, ISSN 0065-9282 ; v. 212)
Includes bibliographical references and index.
ISBN 0-8218-2945-9 (alk. paper)
1. Investments—Mathematics. 2. Stochastic analysis. 3. Hedging (Finance)—Mathematical models. 4. Insurance—Mathematics. I. Volkov, S. N. (Sergei Nikolaevich), 1972– II. Nechaev, M. L. (Mikhail Leonidovich), 1972– III. Title. IV. Series.

HG4515.3.M4513 2002
332.6′01′51—dc21 2002074395

∞ The paper used in this book is acid-free and falls within the guidelines
established to ensure permanence and durability.
Visit the AMS home page at http://www.ams.org/

10 9 8 7 6 5 4 3 2 1 07 06 05 04 03 02

Contents

Foreword

Contemporary financial mathematics and the associated theory of actuarial calculations in insurance have reached such a level of mathematical complexity and abstraction that it is impossible to present a maximally rigorous exposition of them in brief form without detriment to either mathematical correctness or breadth of illumination of the subject. However, only the general theory gives rise to a fundamental approach to the problem on which a practically acceptable methodology for solving it can be based. Contact with practitioners of financial and insurance businesses has shown that these pragmatists are interested first and foremost in the practical realization of concrete models.

Thus arose the plan for this book: to present key and mathematically very complex results in the contemporary theory of hedging and investment at the brink of mathematical correctness, and to show as rigorously as possible how this general methodology can be interpreted in concrete models of financial markets.

This approach, which broadens the spectrum of readers beyond the narrow circle of specialists in stochastic analysis, was the basis for the advanced courses of lectures on financial and actuarial mathematics which I delivered at the Mechanics and Mathematics Department of Moscow State University in 1997–2000 and at the Laboratory of Actuarial Mathematics of Copenhagen University in 1998.

The problems touched upon here are being worked on very intensively, as reflected by the appearance of a whole spectrum of publications, of which I single out the 1999 monograph *Essentials of stochastic finance: facts, models, theory* by my teacher A. N. Shiryaev. It is a fundamental encyclopedia in this area.

Nevertheless, the monograph literature does not yet adequately reflect incomplete markets, markets with constraints on the strategy and structure of the model, "imperfect" forms of hedging (quantile and mean-variance), the "convergence" of financial and actuarial mathematics, and so on. This whole area, which forms a methodological basis for modern quantitative calculations in finance and insurance, is represented in the present book, and as a rule the chosen form of presentation of the material generally goes "from the general to the particular." The above as a whole distinguishes the book from other publications in the area.

After writing the first three chapters, I enlisted the services of my students and colleagues S. N. Volkov and M. L. Nechaev for work on the book. They had prepared the first Russian dissertations devoted entirely to contemporary financial mathematics at the Steklov Institute of Mathematics of the Russian Academy of Sciences. Chapters 4–8 were written jointly with them. The exposition of the concluding Chapter 9 follows my paper *On the unity of quantitative methods of calculations in finance and insurance* (preprint no. 5, Actuarial-Financial Center, Moscow, 2000).

I would like to believe that this book will find an audience among both theoreticians and practitioners in finance and insurance and that it can serve as a basis for modern courses in actuarial and financial mathematics and quantitative risk management for university specializations in the direction of mathematical economics.

A. V. Mel'nikov

Main Notation

$(\Omega, \mathcal{F}, \mathbf{P})$ probability space

$\mathbb{F} = (\mathcal{F}_t)_{t\geq 0}$ filtration

$(\Omega, \mathcal{F}, \mathbb{F}, \mathbf{P})$ stochastic base, or probability space with a filtration

$\mathcal{O}$ optional σ-algebra

$\mathcal{P}$ predictable σ-algebra

$\xi, \eta, \dots$ random variables

τ, σ stopping times, or Markov times

$\sigma\{\xi_0, \dots, \xi_t\}$ the σ-algebra generated by the variables $\xi_0, \dots, \xi_t$

$\mathbf{E}\xi$ (mathematical) expectation of a random variable ξ

$\mathbf{D}\xi$ variance of a random variable ξ

$\mathbf{P}(A)$ probability of an event A

$\mathbf{E}(\xi \mid \mathcal{A})$ conditional expectation of a random variable ξ with respect to a σ-algebra $\mathcal{A}$

$\mathcal{M}$ set of uniformly integrable martingales

$\mathcal{M}^2$ set of square-integrable martingales

$\mathcal{M}_{\text{loc}}$ set of local martingales

$\mathcal{M}^2_{\text{loc}}$ set of local square-integrable martingales

$\varphi * M$ stochastic integral of a function φ with respect to a local martingale M

$\varphi \circ A$ Lebesgue–Stieltjes integral of a function φ with respect to a process A with bounded variation

$\mathcal{E}_t(X)$ stochastic exponential of a process X

$\mathcal{M}(X, \mathbf{P})$ set of measures equivalent to $\mathbf{P}$ with respect to which X is a local martingale

$\langle M \rangle$ quadratic characteristic of an $M \in \mathcal{M}^2_{\text{loc}}$

$\langle M, N \rangle$ mutual quadratic characteristic of $M, N \in \mathcal{M}^2_{\text{loc}}$

$[X, Y]$ mutual quadratic characteristic of $X, Y \in \mathcal{M}_{\text{loc}}$

$\mathbf{P}^* \ll \mathbf{P}$ absolute continuity of a measure $\mathbf{P}^*$ with respect to $\mathbf{P}$

$\mathbf{P}^* \sim \mathbf{P}$ equivalence of measures $\mathbf{P}^*$ and $\mathbf{P}$

$\mathbb{R}^d$ d-dimensional Euclidean space

I_A indicator function of an event A

$\varnothing$ empty set

$\Phi(x)$ standard normal distribution

$X_{t-} = \lim_{s\uparrow t} X_s$

$\Delta X_t = X_t - X_{t-}$

$a \wedge b = \min\{a, b\}$

$a \vee b = \max\{a, b\}$

$a^+ = a \vee 0$

$[a]$ integer part of a

$\binom{N}{n}$ number of combinations of N things taken n at a time

X^π value of a strategy π

$X^{(\pi, C)}$ value of a strategy π with comsumption C

SF set of self-financing strategies

$\mathbb{C}(T, f)$ price of contingent claim f with exercise time T

$\mathbb{C}^*(f)$, $\mathbb{C}_*(f)$ ask and bid prices of a claim f

${}_tp_x$ probability of living to age $x + t$ starting from age x

$\varphi(v)$, $\psi(v)$ ruin probability and nonruin probability with initial value v

${}_TU_x$ premium for a pure endowment contract for living to age $x + T$ starting from age x

CHAPTER 1

Financial Systems: Innovations and the Risk Calculus

We present the concept of a financial system as a collection of firms, intermediary structures, and individuals interacting through financial markets. A characterization of financial markets (complete and incomplete) is realized by identifying sets of contingent claims and terminal values of self-financing strategies. The dynamic behavior of a financial system is described as the movement of incomplete markets toward complete markets as the volume of tradable derivative securities (financial innovations) grows. Furthermore, the problem of the risk calculus, both financial and insurance risks, is constantly analyzed.

§ 1.1. Financial systems and their innovation changes

A *financial system* is formed by a space endowed with financial markets through which diverse firms, individuals, and intermediary structures interact (Figure 1.1).

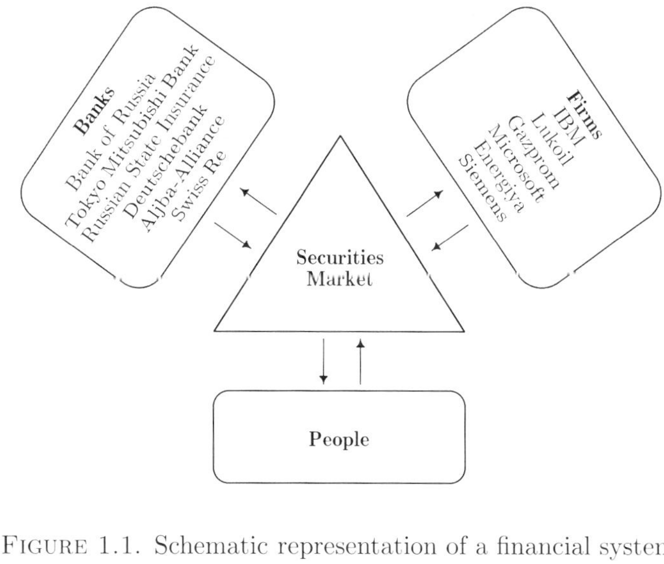

FIGURE 1.1. Schematic representation of a financial system.

The evolution of the international financial system has gone through several basic stages: the gold standard, the Bretton-Woods mechanism, and floating exchange rates. Especially profound changes have taken place over the past two decades, due to the introduction of *new financial instruments* and the process of *computer information technology*. New scales of financial mediation, expanded boundaries of credit mechanisms, and internationalization processes in finance and insurance have become intrinsic features of the modern financial system. To safeguard such vitally important functions of a financial system as control of payment flows and risks, insurance of savings and loans, and so on, it became absolutely necessary to make essential changes in the previously established ways of looking at the operation of financial markets.

Many countries in the West have taken adequate and timely account of this new situation and have created a complete infrastructure for introducing financial innovations, an infrastructure that includes both research centers and companies and a reformed system of actuarial-financial education.

During the Bretton-Woods period (1944–1971) gold prices and exchange rates were rigidly tied (within 1%) to the US dollar in the system of world finance. Therefore, in controlling the financial system *institutional* methods based on the creation of various regulating structures were quite suitable. For example, the International Monetary Fund (IMF) was created with this principal goal.

The abolition of fixed gold prices and the "freeing" of exchange rates and interest rates meant a transition from the Bretton-Woods mechanism to a system of *floating exchange rates*. Interest rates "acquired randomness", and adequate control of a financial system required *functional methods* based on the use of new financial instruments (*financial innovations, derivative securities*): *forward and futures contracts*, *options*, *swaps*, and so on. In financial intermediation there was a shift in the direction of credit mechanisms in which the risks on the capital market are diversified with the help of derivative securities. Further, the problem of *controlling risk* has become more complicated in view of the large risk variability of derivative instruments in comparison with underlying assets (stocks, bonds).

Practitioners of the financial business responded to this in turn by opening specialized exchanges for the trading of derivative securities: the Chicago Board of Options Exchange (CBOE, 1973), the London International Financial Futures Exchange (LIFFE, 1982). Moreover, other exchanges which had opened earlier—the Chicago Board of Trade (CBOT), the American Exchange (AMEX), the New York Stock Exchange (NYSE), and others—were also forced to shift their activities in the direction of trade in these new financial instruments. As a result, the volume of worldwide trade in derivative securities has been constantly growing: from 500 billion US dollars in 1985 to 3500 billion in 1991. This tendency continued also in the 1990s.

Upon considering the innovation processes in finance, which like a diffusion have absorbed one market after another, we should observe certain features of these processes in the "manufacturing industry". In both cases this is a reaction of firms to the demands of the "surrounding medium" with a long-term goal of making a profit. As a rule, innovations are promoted by a high degree of competitiveness of a firm, although a monopoly not particularly disposed toward competition but having sufficient reserves can also stably support the latest research results and their introduction. Industrial know-how can at times lead to cardinal changes in whole branches of the economy. For instance, the development of *computer technology* and *information technology* over the past two decades has led to the present revolution in the financial sector of the economy and in banking service. New information-computer technologies have enabled us to receive, remember, and store huge amounts of information about accounts and transactions, and to use this information in real-time mode. All this has given traders new opportunities for finding arbitrage situations, for implementing continuous monitoring, and for managing a portfolio made up of a complex combination of securities. As a result, such structures of banks and investment companies as front-office, middle-office, and back-office have acquired a completely new aspect, and automated *risk-control systems* have become technologically feasible.

Progress in computer technology has enabled us to penetrate more deeply into the behavior of the markets themselves, since *intraday* information has turned out to be accessible. This is a truly enormous statistical reserve, which has not yet been exploited in full scope. For example, the *ticks*, or moments when prices change, on the foreign exchange market (FX-market) happen roughly 20 times per minute, while the exchange rate between the US dollar and the Deutsche Mark vacillates 18,000 times per day, and this is equivalent to *daily* information for 72 years. All this has led to quantitative changes on a global scale: the daily turnover alone in the FX-market increased by a factor of five during 1986–1995 and constitutes more than a trillion dollars, substantially exceeding the total gold reserves of all the industrially developed countries—the members of the IMF.

Various theories of financial markets have been developed in parallel with the tendencies mentioned. However, it has proved impossible to take into account the above dynamics of a financial system while remaining in the framework of the Markowitz "one-step" theory of diversification and the Schwartz–Lintner Capital Asset Pricing Model (CAPM) created in the period 1950–1960. The foundations for a theory adequate to modern securities markets were laid by Black, Scholes, and Merton. Their *Options Pricing Theory* (OPT) or *Contingent Claim Analysis* (CCA) has provided real possibilities for dynamic hedging and investing in connection with a broad spectrum of financial innovation instruments. It is this theory that one has in view when one speaks of the definitive transition from the *arithmetic of rent payments* to the modern *stochastic mathematics* of finance and insurance. In the next section we give a general description of some key concepts (arbitrage, completeness and incompleteness of markets, and so on), facts, and perspectives of this theory.

In conclusion we mention that an analysis of market information leads to conclusions about the *fractal* structure of price variation, about the "heavy tails" of their distributions, and so on. A whole spectrum of financial models have arisen as a consequence: Autoregressive Conditional Heteroscedasticity (ARCH), Generalized ARCH (GARCH), Fractional Integrated ARCH (FIARCH), FIGARCH, and so on. Finally, we mention another indisputable effect of computer know-how: the widespread use of artificial "*neuron networks*" in financial practice and analysis.

§ 1.2. General statements in the analysis of contingent claims. Models, methods, facts

We consider a *model of a financial market* as a pair of assets: a *nonrisky* asset (bank account) B and a *risky* (stock) asset S, which can be represented by their *prices* B_t and S_t, $t = 0, 1, \dots$ or $t \in \mathbb{R}^1_+$. In this case one speaks of a (B, S)-*market* with discrete or continuous time, respectively. Here the risky component of the (B, S)-market can be multidimensional. The assets B and S will be called *underlying assets* or *underlying securities*.

An important question arises at once: how does one describe the *riskiness* of the asset S? A natural answer is to regard S as a *random process* of the evolution of its prices S_t. Therefore, it seems fully justifiable to assume that a *probability space* $(\Omega, \mathcal{F}, \mathbf{P})$ is given from the start, and that $(S_t)_{t\geqslant 0}$ forms a random process on it. Here the information supplied by the prices of S up to time t is denoted by $\mathcal{F}_t = \mathcal{F}_t^S$, $t \in \mathbb{R}_+^1$.

We fix a time horizon T and define a *contingent claim* $f = f_T$ to be any function determined by the information $\mathcal{F}_T$ (measurable with respect to $\mathcal{F}_T$).

Next, taking the nonrisky asset B_t and the risky asset S_t in the quantities β_t and γ_t, respectively, we form the pair $\pi_t = (\beta_t, \gamma_t)$, called a *portfolio* or (investment) *strategy*. The value of the portfolio π is defined to be the sum

$$X_t^\pi(x) = \beta_t B_t + \gamma_t S_t, \qquad X_0^\pi = x. \tag{1.1}$$

The definition of a (B, S)-market must be completed by indicating what kinds of portfolios can be used. The most important class consists of the *self-financing* portfolios π, defined by saying that $\pi \in SF$ if

$$X_t^\pi - X_{t-1}^\pi \equiv \Delta X_t^\pi = \beta_t \Delta B_t + \gamma_t \Delta S_t, \qquad t = 1, 2, \dots \tag{1.2}$$

(or $dX_t^\pi = \beta_t\, dB_t + \gamma_t\, dS_t$ in the case of continuous time under the assumption that the differentials dB_t and dS_t are well defined).

Arbitrage (at the time T) means the possibility of creating a positive value (with positive probability) at the time T by means of a self-financing strategy with zero initial value $x = 0$.

Any asset given on a (B, S)-market, and hence on the basis of the underlying assets B and S, is called a *derivative security* and is identified with a contingent claim. For example, a *forward contract* (with delivery price F and delivery date T) for delivery of an asset S is equivalent to the claim $f_T = f_T(S_T) = S_T - F$, and an *option to buy* (with exercise price K and exercise date T) is equivalent to the claim $f_T = f_T(S_T) = (S_T - K)^+$. Thus, the set of derivative securities induces the set $\mathbb{CCG}$ of graphs of the corresponding contingent claims (Figure 1.2).

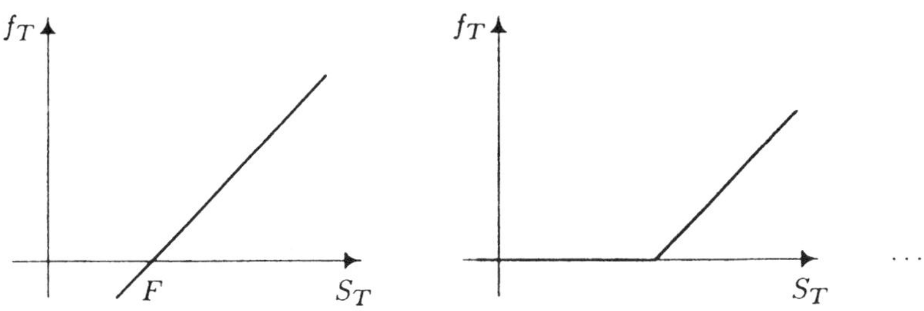

FIGURE 1.2. $\mathbb{CCG}$ (Contingent Claims' Graphs).

On the other hand, the set of self-financing strategies π induces the set $\mathbb{TVG}$ of graphs of the terminal values (Figure 1.3).

A market is said to be *complete* if

$$\mathbb{TVG} = \mathbb{CCG}.$$

Otherwise, it is said to be *incomplete*. In other words, a (B, S)-market is complete if and only if any contingent claim f_T can be *replicated*. Namely, there exist x and $\pi \in SF$ such that $X_T^\pi(x) = f_T$.

Let V_t stand for the *price* at time t of the contingent claim f_T (in other words, the price of a derivative security with payments for it at time T determined by

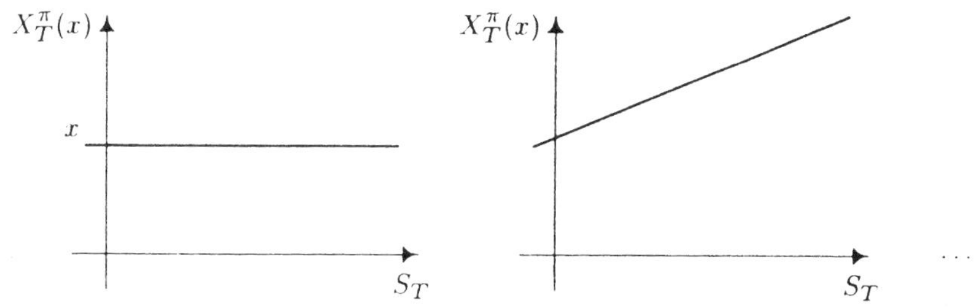

FIGURE 1.3. TVG (Terminal Values' Graphs).

the function f_T). The most important problem is to describe the random process $V = (V_t)_{t \geqslant 0}$ in terms of the (B, S)-market.

The "heuristic" principle for such a description of V_t consists of two ideas. *First*, the amount of the contingent claim f_T must be *discounted* with the help of the nonrisky asset: $B_t B_T^{-1} f_T$. *Second*, consider the averaged quantity $\mathbf{E}(B_t B_T^{-1} f_T \mid \mathcal{F}_t)$ to be a candidate for V_t.

The first idea is absolutely irreproachable, since a measurement of the cost at different moments of time in the same units is achieved with the help of the discounting operation.

The second idea can be questioned: for it does not follow at all that the averaging should be implemented with respect to the initially given "physical" probability $\mathbf{P}$. Moreover, any other probability measure $\widetilde{\mathbf{P}}$ on the space $(\Omega, \mathcal{F})$ determines its own "probability character" of the (B, S)-market. It is clear that the *risk-neutral, stable* character of the chosen probability predetermines the naturalness of the price of the given contingent claim. Consequently, the above "heuristic" principle must be corrected by choosing a more suitable probabilistic character of the market, determined by some measure $\widetilde{\mathbf{P}}$. Here in order to avoid losing essential features of the market ("degenerations" of the character) it is natural to assume that the measures $\mathbf{P}$ and $\widetilde{\mathbf{P}}$ are *equivalent*. These considerations lead to the *no-arbitrage principle* in determining the price V_t. This principle is realized in the following general facts, which represent a somewhat coarsened form of the fundamental theorems of arbitrage and completeness in financial mathematics:

A (B, S)-market does not admit arbitrage opportunities $\Longleftrightarrow$ there is a probability measure $\mathbf{P}^*$ equivalent to $\mathbf{P}$ $(\mathbf{P} \sim \mathbf{P}^*)$ such that the process of discounted prices $B_t^{-1} S_t$ of the risky asset is a martingale with respect to $\mathbf{P}^*$.

Such a measure is called a *martingale* measure. Since a martingale is constant in the mean, the measure $\mathbf{P}^*$ neutralizes, as it were, the riskiness of the asset S. Therefore, $\mathbf{P}^*$ is also called a *risk-neutral* measure of the (B, S)-market. Further:

On a complete no-arbitrage (B, S)-market the price of any contingent claim is uniquely determined $\Longleftrightarrow$ the martingale measure is unique.

Indeed, if there are two such measures $\mathbf{P}_i^*$, $i = 1, 2$, then two prices of the claim f are determined: $V_t^i = \mathbf{E}_i^*(B_t B_T^{-1} f_T \mid \mathcal{F}_t)$. They must coincide, and thus $\mathbf{P}_1^* = \mathbf{P}_2^*$.

Conversely, the price V_t is uniquely determined as $V_t = \mathbf{E}^*(B_t B_T^{-1} f_T \mid \mathcal{F}_t)$ with respect to the unique martingale measure $\mathbf{P}^*$.

As a result we arrive at the following general statement about pricing contingent claims in complete markets:

Let $\mathbf{P}^*$ *be the unique martingale measure for a complete* (B,S)*-market, and let the price* V_t *of the contingent claim* f_T *be defined as* $V_t = \mathbf{E}^*(B_t B_T^{-1} f_T \mid \mathcal{F}_t)$. *Then* (B,S,V) *forms the unique system of prices for which the corresponding extended market does not admit arbitrage opportunities. Moreover, there is a self-financing strategy* π^* *replicating* f_T *such that* $X_t^{\pi^*}(V_0) = V_t$ *for all* $t \in [0,T]$.

This result sums up the *possibility of reducing to zero the risk associated with any contingent claim on a complete market.*

We now present several classical examples of a complete market.

The binomial model/the Cox–Ross–Rubinstein model.

Let $\rho_t = \dfrac{\Delta S_t}{S_{t-1}}$, $t = 1, 2, \dots$, $S_0 > 0$, be the *relative yield* from the risky asset S. We assume that the ρ_t form a sequence of independent random variables taking the two values $b > a$ with probabilities p and $1-p$, $p \in (0,1)$. If $r \equiv \dfrac{\Delta B_t}{B_{t-1}}$ is a nonrisky interest rate such that $-1 < a < r < b$, then $p^* = \dfrac{r-a}{b-a}$ determines a unique martingale measure $\mathbf{P}^*$ on the space $(\Omega, \mathcal{F}, \mathbb{F})$ with $\Omega = \{a,b\}^T$, $\mathcal{F} = \mathcal{F}_T$, $\mathcal{F}_0 = \{\varnothing, \Omega\}$. The price of an option to buy with contingent claim $f_T = (S_T - K)^+$ is given by the *Cox–Ross–Rubinstein formula*:

$$V_0 = \mathbf{E}^*(1+r)^{-T}(S_T - K)^+ = S_0 \mathbb{B}(t_0, T, \widetilde{p}) - K(1+r)^{-T}\mathbb{B}(t_0, T, p^*), \tag{1.3}$$

where

$$t_0 = 1 + \left[\ln \frac{K}{S_0(1+a)^T} \Big/ \ln \frac{1+b}{1+a}\right], \qquad \widetilde{p} = \frac{1+b}{1+r}\, p^*,$$

$$\mathbb{B}(j, T, P) = \sum_{t=j}^{T} \binom{T}{t} p^t (1-p)^{T-t}.$$

The Bachelier model.

This is a purely continuous model of a (B,S)-market, where $B_t \equiv 1$ and the corresponding return from S per unit time is equal to

$$\frac{dS_t}{dt} = \mu + \sigma \dot{w}_t, \qquad S_0 > 0, \tag{1.4}$$

where $\dot{w}_t$ is a Gaussian "white noise", a concept widely used in physics and engineering to simulate chaotic nonregular motions, and $\sigma > 0$ parametrizes the changeability (*volatility*) of the indicated return (Figure 1.4).

In this model the no-arbitrage principle leads to the following price for the option to buy $f_T = (S_T - K)^+$.

Defining the martingale measure

$$d\mathbf{P}_T^* = \exp\left\{-\frac{\mu}{\sigma}\, r\, w_T - \frac{1}{2}\left(\frac{\mu}{\sigma}\right)^2 T\right\} d\mathbf{P},$$

and averaging with respect to it, we come to Bachelier's formula:

$$V_0 = \mathbf{E}^*(S_T - K)^+ = (S_0 - K)\Phi\left(\frac{S_0 - K}{\sigma\sqrt{T}}\right) + \sigma\sqrt{T}\,\phi\left(\frac{S_0 - K}{\sigma\sqrt{T}}\right), \tag{1.5}$$

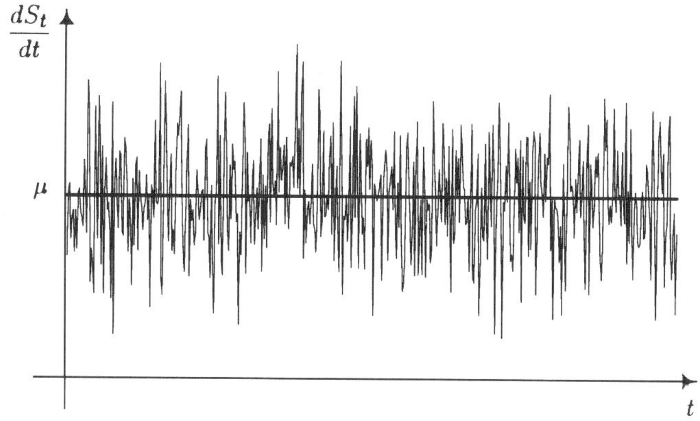

FIGURE 1.4. Evolution of return in the Bachelier model.

where

$$\Phi(x) = \int_{-\infty}^{x} \phi(y)\, dy, \qquad \phi(y) = \frac{1}{\sqrt{2\pi}} e^{-\frac{y^2}{2}}.$$

In particular, for $S_0 = K$

$$V_0 = \sigma \sqrt{\frac{T}{2\pi}}.$$

The Black–Scholes model.

We consider the following *relative yield* per unit of time for B and S:

$$\frac{dB_t}{B_t\, dt} = r, \qquad \frac{dS_t}{S_t\, dt} = \mu + \sigma \dot{w}_t, \qquad S_0 > 0. \tag{1.6}$$

It is clear that the graphical realization of $\dfrac{dS_t}{S_t\, dt}$ will be the same[1] as in Figure 1.4 for $\dfrac{dS_t}{dt}$.

This model can be rewritten in the form of the stochastic differential equations

$$dB_t = rB_t\, dt, \qquad dS_t = S_t(\mu\, dt + \sigma\, dw_t), \qquad S_0 > 0, \tag{1.7}$$

where w_t is a standard Wiener process (Brownian motion).

Another, exponential, form of this model was proposed by Samuelson:

$$B_t = B_0 e^{rt}, \qquad S_t = S_0 e^{\mu t - \frac{\sigma^2 t}{2} + \sigma w_t}. \tag{1.8}$$

The process S_t represented in this form is called a *geometric Brownian motion*.

The unique martingale measure $\mathbf{P}^*$ in this model is determined by the density $Z_T^* = \exp\Big\{-\Big(\frac{\mu - r}{\sigma}\Big) w_T - \frac{1}{2}\Big(\frac{\mu - r}{\sigma}\Big)^2 T\Big\}$. According to the no-arbitrage principle, the price of an option to buy is given by the *Black–Scholes formula*:

$$\begin{aligned} \mathbb{C}_{BS}(S_0, T, r, \sigma) = V_0 &= \mathbf{E}^* e^{-rT}(S_T - K)^+ \\ &= S_0 \Phi(d_+(r)) - K e^{-rT} \Phi(d_-(r)), \end{aligned} \tag{1.9}$$

where $d_\pm(r) = \dfrac{\ln(S_0/K) + T(r \pm \sigma^2/2)}{\sigma\sqrt{T}}$.

[1] The realizations of S_t given by the formulas (1.4) and (1.6) will be different, of course.

It is interesting that the "heuristic" principle for calculating the price gives

$$\mathbb{C}_{\text{heu}}(\mu) = e^{(\mu-r)T} S_0 \Phi(d_+(\mu)) - K e^{-rT} \Phi(d_-(\mu)),$$

which for $\mu = r$ coincides with the Black–Scholes price:

$$\mathbb{C}_{\text{heu}}(r) = \mathbb{C}_{BS}(S_0, T, r, \sigma).$$

For the model (1.6)–(1.8) of a (B, S)-market we can use the *differential equations* method, which was also used by Black and Scholes to obtain their famous formula (1.9).

Suppose that the contingent claim has the form $f_T = g(S_T)$, where g is a nonnegative function. We consider a portfolio π whose value is a smooth function of S_t and $t \leqslant T$: $X_t^\pi = v(S_t, t)$. It is clear that

$$\begin{aligned} v(x, T) &= g(x), \quad x > 0, \\ v(x, t) &\geqslant 0, \quad x > 0, \quad t \leqslant T. \end{aligned} \tag{1.10}$$

Application of the *Kolmogorov–Itô formula* to $v(S_t, t)/B_t$ leads to the relation

$$\frac{v(S_t, t)}{B_t} = \frac{v(S_0, 0)}{B_0} + \int_0^t \frac{\partial v(S_u, u)}{\partial x}\, d\left(\frac{S}{B}\right)_u + \int_0^t \left[\frac{\partial v}{\partial t} + Lv - rv\right] B_u^{-1}\, du, \tag{1.11}$$

where $L = rx\dfrac{\partial}{\partial x} + \dfrac{1}{2}\sigma^2 x^2 \dfrac{\partial^2}{\partial X^2}$ is the generating operator of the diffusion process S_t (see (1.7)).

It follows from (1.11) that

$$\pi \in SF \iff \frac{\partial v}{\partial t} + Lv - rv = 0. \tag{1.12}$$

The equation (1.12) with the boundary conditions (1.10) is called *the fundamental differential equation of Black and Scholes.* We remark that v is a *harmonic* function for the operator $\dfrac{\partial}{\partial t} + L - r$.

If the function $g(x)$ has polynomial growth, then a solution of the problem (1.10)–(1.12) exists and is given by the formula

$$v(x, t) = \int_0^\infty \beta(y, T-t, x) g(y) e^{-r(T-t)}\, dy, \tag{1.13}$$

where $\beta(y, t, x) = \dfrac{1}{y\sigma\sqrt{2\pi t}} \exp\left\{-\dfrac{\ln y - \ln x - (r - \sigma^2/2)}{2\sigma^2 t}\right\}$ is the density of the lognormal distribution.

In the case of complete markets the no-arbitrage principle has led to exhaustive answers with regard to the calculation of prices of contingent claims. How is it realized in *incomplete markets*, when the set $\mathcal{M}(S/B)$ of martingale measures does not reduce to a single measure?

Let us consider a set of strategies richer than *SF*, the so-called *strategies with consumption*:

$$\{(\pi, C) : \pi = (\beta, \gamma),\ C \text{ a nondecreasing process}\}.$$

For such strategies (π, C) the corresponding (discounted) value has the form

$$\frac{X_t^{(\pi,C)}}{B_t} = \frac{X_0^{(\pi,C)}}{B_0} + \int_0^t \gamma_u \, d\left(\frac{S}{B}\right)_u - \int_0^t B_u^{-1} \, dC_u, \tag{1.14}$$

and hence is a positive supermartingale with respect to any measure $\widetilde{\mathbf{P}} \in \mathcal{M}(S/B)$.

On the other hand, a positive process that is a supermartingale with respect to any measure $\widetilde{\mathbf{P}} \in \mathcal{M}(S/B)$ can be represented in the form (1.14), called the *optional decomposition*.

We now consider the process

$$\frac{V_t^*}{B_t} = \operatorname*{ess\,sup}_{\widetilde{\mathbf{P}} \in \mathcal{M}(S/B)} \widetilde{\mathbf{E}}\big(B_T^{-1} f_T \,|\, \mathcal{F}_t\big),$$

which is well known in the theory of optimization as the *Snell envelope*. This is a positive supermartingale (with respect to any measure $\widetilde{\mathbf{P}} \in \mathcal{M}(S/B)$) such that $V_T^* = f_T$ (a.s.).

Suppose that the value of a strategy with consumption (π, C) *hedges* the contingent claim f_T in the following sense:

$$V_T = V_T^{(\pi,C)} \geqslant f_T \quad \text{(a.s.)}. \tag{1.15}$$

According to its supermartingale property we have that for any $\widetilde{\mathbf{P}} \in \mathcal{M}(S/B)$

$$\frac{V_t}{B_t} \geqslant \widetilde{\mathbf{E}}\left(\frac{V_T}{B_T} \,\Big|\, \mathcal{F}_t\right) \geqslant \widetilde{\mathbf{E}}\left(\frac{f_T}{B_T} \,\Big|\, \mathcal{F}_t\right) \quad \text{(a.s.)}.$$

Hence, V_t^* is the minimal value with the property (1.15), and corresponding to the optional decomposition there is a portfolio with consumption (π^*, C^*) called a *minimal hedge*.

Thus,

$$\operatorname*{ess\,sup}_{\widetilde{\mathbf{P}} \in \mathcal{M}(S/B)} \widetilde{\mathbf{E}}\big(B_T^{-1} f_T \,|\, \mathcal{F}_t\big) \tag{1.16}$$

must be the natural price of a contingent claim f_T on an incomplete (B, S)-market.

The approach leading to the price (1.16) in the calculus of contingent claims is called *superhedging*.

We now give an example of an incomplete (B, S)-market whose incompleteness is due to *stochastic volatility*.

A market with stochastic volatility.

Let us consider the (B, S)-market

$$dB_t = B_t r \, dt, \qquad dS_t = S_t(\mu \, dt + \Sigma_t \, dw_t), \tag{1.17}$$

where $r \geqslant 0$, $\Sigma_t^2 = \sigma^2 + (-1)^{N_t} \Delta\sigma^2$, $0 \leqslant \Delta\sigma^2 < \sigma^2$, and N_t is a standard Poisson process independent of w_t.

According to the Kolmogorov–Itô formula, we have

$$d\left(\frac{S}{B}\right)_t = \left(\frac{S}{B}\right)_t \Sigma_t \, dw_t^*,$$

where $dw_t^* = dw_t + \dfrac{\mu - r}{\Sigma_t} \, dt$.

It is clear that $\widetilde{\mathbf{P}} \in \mathcal{M}(S/B) \iff w^*$ is a (local) martingale. Further, according to Girsanov's theorem,

$$\frac{d\widetilde{\mathbf{P}}}{d\mathbf{P}} = \widetilde{Z}_t = \exp\left\{\int_0^t \left(\frac{\mu - r}{\Sigma_t}\,dw_u - \frac{(\mu - r)^2}{2\Sigma_u^2}\,du\right)\right\}\widetilde{N}_t = \overline{Z}_t\widetilde{N}_t,$$

where $\widetilde{N}_t$ is a (local) martingale orthogonal to w with respect to the measure $\mathbf{P}$.

By Novikov's condition

$$\mathbf{E}\exp\left\{\frac{1}{2}\int_0^T \left(\frac{\mu - r}{\Sigma_u}\right)^2 du\right\} = \mathbf{E}\exp\left\{\frac{1}{2}\int_0^T \frac{(\mu - r)^2\,du}{\sigma^2 + \Delta\sigma^2(-1)^{N_u}}\right\} < \infty,$$

the process Z_t is a uniformly integrable martingale, and hence the two martingales

$$\overline{Z}_t^{\pm} = \exp\left\{\int_0^t \frac{r - \mu}{\sqrt{\sigma^2 \pm \Delta\sigma^2}}\,dw_u - \int_0^t \frac{(r - \mu)^2}{2\sigma^2 \pm \Delta\sigma^2}\,du\right\}$$

determine two martingale measures in $\mathcal{M}(S/B)$, which testifies to the incompleteness of the market (1.17).

One more method for getting a reasonable price for a contingent claim is based on *mean-variance hedging.*

We define the risk associated with hedging a contingent claim f_T with the strategy π by the formula

$$R_T^{\pi}(x) = \mathbf{E}\left(f_T - x - \int_0^T \gamma_u\,d\left(\frac{S}{B}\right)_u\right)^2.$$

The corresponding strategy π^* and initial value x^* are chosen to minimize this risk:

$$R_T^{\pi^*}(x^*) = \inf_{\pi, x} R_T^{\pi}(x).$$

Risk can be defined similarly for both complete and incomplete markets. We remark that in the case of the Black–Scholes model the price $\mathbb{C}_{BS}$ of a contingent claim f_T and the strategy $\widehat{\pi}$ hedging this claim coincide respectively with the value x^* and the strategy π^* minimizing $R_T^{\pi}(x)$.

In the case of a discrete model of a (B, S)-market, when the physical measure $\mathbf{P}$ is a martingale, x^* and π^* are given by the formulas

$$x^* = \mathbf{E}f_T, \qquad \gamma_t^* = \frac{\mathbf{E}[f_T\Delta S_t \,|\, \mathcal{F}_{t-1}]}{\mathbf{E}[(\Delta S_t)^2 \,|\, \mathcal{F}_{t-1}]}. \tag{1.18}$$

Accordingly, *for incomplete markets the risk associated with a "not too risky" contingent claim cannot be reduced to zero, but can be minimized.*

§ 1.3. Dynamics of financial markets: from incomplete markets to complete markets through financial innovations

It is necessary to discuss (and now we are in a position to do this) the relationship between complete and incomplete markets.

Let us consider (purely formally) the difference

$$\Delta = \sup_{\widetilde{\mathbf{P}} \in \mathcal{M}(S/B)} B_0\widetilde{\mathbf{E}}B_T^{-1}f_T - \inf_{\widetilde{\mathbf{P}} \in \mathcal{M}(S/B)} B_0\widetilde{\mathbf{E}}B_T^{-1}f_T, \tag{1.19}$$

which can be interpreted as a *measure of the incompleteness* of a (B,S)-market, since $\Delta \equiv 0$ for a complete market. The quantity Δ will be called the *spread*.

We mention also other characteristics associated with incompleteness of a market: *leasing* and *overhead costs*. The leasing of an asset S is usually proportional to its price: $l_t S_t$. And the payment for overhead costs is proportional to its price and to the magnitude of change of the portfolio: $\delta_t \left|\Delta\gamma_t\right| S_t$. The parameters l and δ are called the *leasing* and *overhead coefficients* and serve again as distinctive measures of the incompleteness of the market.

The introduction of new financial products makes the original market "more complete" and correspondingly diminishes Δ, l, and δ. R. Merton, the Nobel laureate for economics in 1997, first defined the movement of markets as a *financial innovation spiral* in a direction toward ideal complete markets to the extent that the incompleteness parameters Δ, l, and δ are reduced toward zero. On this Merton wrote:

> "From the perspective of our theory, these same facts about change are seen as consistent with a real-world dynamic path evolving toward an idealized target of an efficient financial-market and intermediation system. On this premise, these changes can be interpreted as part of a "financial innovation spiral". That is, the proliferation of new trading markets makes feasible the creation of new financial products; to hedge these products, producers trade in these new markets and volume expands; increased volume reduces marginal transaction costs and thereby makes possible further implementation of new products and trading strategies, which in turn leads to still more volume. Success of these trading markets encourages investment in creating additional markets, and so on it goes ..., spiraling toward the theoretically limiting case of zero marginal transaction costs and dynamically complete markets." ([**118**], p. 468)

For this reason we call Figure 1.5 *the financial innovation spiral of Merton*.

Thus, according to Merton all the events in contemporary financial reality form a part of a financial innovation spiral leading to an efficient market and an intermediation system. Furthermore, the intermediation structures, in becoming more and more open to new financial products and services, will also expand their geography, thereby leveling out the geopolitical advantages of various financial institutes. However, it does not follow that this general direction of development of the financial system "spirals" real financial processes in a smooth way. The growth in the size and complexity of financial transactions and the global interconnectedness of markets amplify the total credit risk and create real possibilities for financial crises on a broad scale.

Striking examples of this kind are given by the financial crises of 1987 and 1997–1998. A drop of more than 500 points (22%) in one of the most adequate indexes[2] of the American economy—the Dow Jones Industrial Average index—can lead to very negative consequences, as shown in Figure 1.6.

The figure reflects the dynamics of this index in 1929 and in 1987. Fairly quickly after the crash in 1929 there ensued a period of even steeper drop, now called the

[2] It is made of information about 30 "blue chips".

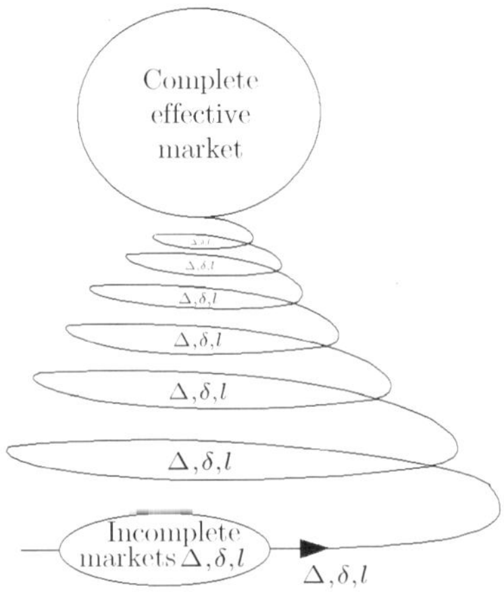

FIGURE 1.5. The financial innovation spiral of Merton: dynamics of incomplete markets toward complete markets as the volumes of financial innovations increase ($\Delta, l, \delta \to 0$).

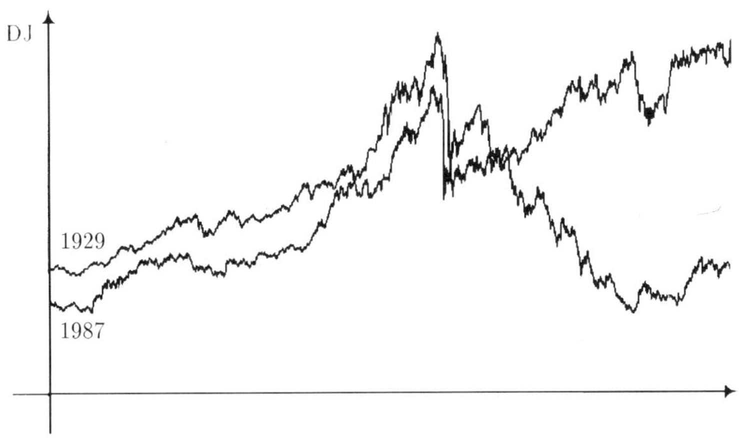

FIGURE 1.6. Dynamics of the DJIA index in 1929 and 1987.

Great Depression of the American economy. On the other hand, after October 19, 1987 it was possible to avoid such negative consequences, and the index stabilized.

We have already put sufficient emphasis on the most important component in the dynamics of a financial system both in a qualitative and a quantitative aspect: financial innovation products. They represent the most efficient tools for investment and risk management. They cost 10–20 times less than underlying assets. All this has created unprecedented opportunities for financial *leverage*, when for investment goals not only one's own means but also borrowed means are used (often simply future rights to dispose over means), thereby increasing the sensitivity of the financial structure to changes in the market. Not taking this circumstance sufficiently into account can have a "lethal" outcome.

How seriously one should take the functioning of derivative securities can be beautifully illustrated by the bankruptcy in 1995 of Barings, the oldest commercial bank in Great Britain, "thanks to" the actions on this market of the head of its Singapore branch, N. Leeson. He used futures to create the following position at a cost (astronomical for that bank) of 27 billion dollars: 7 billion in Japanese equities by means of futures on the NIKKEI 225, and 20 billion in Japanese bonds again through futures on these bonds and "Euroyen" futures. Because of the relative cheapness of entry on the futures market, Leeson did not need a lot of money, and at first he stayed within the established limits, even though the capital of the bank was estimated at only 615 million dollars. However, he was eventually required to show a substantial amount of money to support the declared position, since the prices of the underlying assets had begun to fall steadily, and the variation margin which Leeson was supposed to pay was growing catastrophically. As a result, Barings collapsed, and Leeson was blamed.

As noted by Merton, financial innovations represent a permanent component of the basic flow of development of the global financial system and now play a decisive role in guaranteeing risk management functions, a role which, like classical Newtonian mechanics, has three dimensions.

The *first* is the reduction of risk by selling its source. Here innovation products enable one to decrease the corresponding costs.

The *second* is the reduction of risk by means of diversification, which can be implemented more quickly with the help of derivative instruments.

The *third* is the reduction of risk with the help of insurance (understood in the broad sense) against loss.

Most of the commonly accepted financial-economic theories, including CAPM, the Markowitz theory, and others, "fit" risk management into the plane of the first two dimensions. OPT (CCA) allows risk control to acquire the third (not less significant) dimension.

The necessity of new approaches to practical risk management was clearly realized in the 1980s, and in June 1988 the central banks of the 12 largest industrially developed countries reached an accord about new requirements, weighted with risk taken into account, on the capital of financial structures. This was the so-called *Basel accord*, finally introduced on December 31, 1992, which requires the differentiation of capital with regard to various types of assets, including balance sheet items, treats bank capital in a more restrictive manner, dividing it into internal and supplemental components, and introduces a minimal level of capital as a percent of the weight of risky assets.

§ 1.4. Financial innovations and insurance risks

Any derivative security is characterized by *insurance properties*. For example, forwards, futures, and option contracts for the purchase of an asset make it possible for their holders to avoid substantial losses when the price of the asset increases. *Traditional insurance* focuses almost exclusively on the *estimation of risks*. The circulation of options and other derivative financial instruments is quite similar in many aspects to the insurance sector of the financial system. However, while insurance companies have to hold considerable reserves of capital in relation to their obligations, "option" insurance reduces these requirements of reserves. The principal reason for this situation is apparently that traditional insurance involves selling the risk by the client to a specific insurance company, while insurance on the basis of derivative securities involves selling the same risk on a financial market with the possibility of continuous observation of prices and adequate reaction to changes.

The main object of the *classical theory of risk* is the *value process of an insurance company*

$$R_t = u + ct - \sum_{k=1}^{N_t} \xi_k, \tag{1.20}$$

where u is the initial value of the company, c is the rate of receipt of premiums, the ξ_k are the payment claims (independent identically distributed variables), and N_t is a Poisson process indicating the number of these claims up to the time $t \geqslant 0$.

A crucial characteristic for any insurance company is its *solvency*, which can be characterized as the *positivity* of the risk process R_t in the course of time. However, R_t is a random process, and hence a deterministic characteristic is needed that expresses the property of solvency. This role is traditionally filled by the *ruin probability*

$$\phi(u) = \mathbf{P}\big\{\omega : R_t(\omega) < 0 \text{ for some } t \,|\, R_0 = u\big\}.$$

Under the above classical assumptions about the risk process (1.20), this probability admits the exponential *Cramér–Lundberg upper estimates*

$$\phi(u) \leqslant \exp\{-Ru\}, \tag{1.21}$$

where R is the Lundberg constant.

The typical idea of classical *actuarial mathematics* is the following: a chosen level of risk $\varepsilon > 0$ is compared with the indicated bounds (1.21) for $\phi(u)$, and the level of solvency u_ε is found.

The traditional theory of risk does not take into account the *investment activity* of the company, although the insurance companies of many countries are allowed such activities (on regulated scales). To take this circumstance into account it is necessary first of all to specify a model of a financial market. For example, suppose that the prices of a risky asset S are modeled as a geometric Brownian motion (1.6)–(1.8), and the company invests its capital only in this asset. Then by the Kolmogorov–Itô formula, the risk process evolves according to the equation

$$dR_t = (\mu\, dt + \sigma\, dw_t) R_{t-} + c\, dt - \sum_{k=N_t}^{N_{t+dt}} \xi_k. \tag{1.22}$$

In this case also we can define the ruin probability $\phi(u)$, but the upper bounds for it will not be exponential. Here it is possible to get an *integro-differential equation* for the ruin probability. Solving this equation either exactly or approximately, we then compare the solution with the chosen level of risk $\varepsilon > 0$ and find the corresponding value u_ε of the solvency.

As mentioned earlier, the dynamics of the financial system, with various insurance institutions forming an important element of it, has changed qualitatively from the beginning of the 1970s. The insurance system (occupied with traditional insurance), it seems, did not at once react to these changes and at the beginning of the 1980s went through a fairly difficult period of instability. It was apparently then that the significance of the introduction of the new instruments of insurance "tied to" the basic innovation flow of the financial system was fully realized. Thus, actuarial theory and practice had to be enriched by *stochastic financial mathematics*. This meant taking into account a *new randomness* in actuarial calculations, stemming from the financial *randomness* of insurance guarantees.

For instance, in life insurance there appeared contracts in which the payment by the maturity date is connected with the market value of a portfolio of assets. These are the equity-linked life insurance contracts, that is, insurance linked to risky assets of the financial market. The appearance in actuarial calculations of the Black–Scholes formula so crucial for stochastic financial mathematics in the 1980s and 1990s is therefore not surprising.

Further, the greater part of the insurance industry deals with insurance events that are vitally important for the client but constitute a relatively small part of insurance coverage for the insurance company. These are the so-called "high-frequency, low severity" risks. Nevertheless, at the beginning of the 1990s the insurance business encountered another type of risk: "low-frequency, high severity" risks due to various catastrophes. A catastrophe in the insurance sense is regarded as an event leading to losses exceeding 5 million dollars (25 million since 1997) and affecting a large number of people. As we know, catastrophes happen quite regularly. It suffices to note that in the period 1970–1993 there were 34 catastrophes in the USA, with average yearly losses of 2.5 billion dollars. Using even more representative statistics over the period 1949–1993, one can assert that the absolute majority of such events had losses confined within 250 million dollars. However, the potential losses can really be catastrophic. For example, hurricane Andrew in 1992 brought 16 billion dollars in losses, causing the collapse of 6 large insurance companies. The Northridge earthquake in 1994 brought 12.5 billion dollars in losses.

On the other hand, if the famous San Francisco earthquake of 1906 had happened in 1988, then it would have cost 38 billion dollars, and an earthquake of the same magnitude in Los Angeles would cost 50 billion. For the USA the reality is such that in the period 1989–1995 the total losses of this kind amounted to 35 billion dollars for the insurance business, which is equivalent to 18% of its capital.

So how can events of this kind be dealt with when the insurance and reinsurance business in the USA has capital not exceeding 270 billion dollars? A realistic answer again lies in the plane of the capital market, since the "capacities" of the insurance and financial sectors are roughly in the ratio of 1 : 100. Therefore, finding instruments on the capital market for diversification of insurance risks impossible to cover by traditional methods is one of the realistic ways to solve this actuarial problem. Following these ideas, the Chicago Board of Trade (CBOT) has been trading in "catastrophe" futures and options since 1992. It is clear that calculations of appropriate prices are impossible in the framework of traditional actuarial science and must exploit in full scope the modern stochastic theory of contingent claims analysis.

References for Chapter 1

[2]–**[4]**, **[6]**, **[11]**, **[17]**, **[18]**, **[23]**, **[24]**, **[26]**–**[36]**, **[40]**, **[43]**, **[44]**, **[46]**, **[48]**, **[49]**, **[51]**, **[54]**, **[64]**, **[68]**, **[72]**, **[73]**, **[79]**, **[86]**–**[88]**, **[90]**, **[93]**, **[94]**, **[100]**, **[105]**, **[106]**, **[108]**–**[112]**, **[118]**, **[121]**, **[122]**, **[126]**–**[128]**, **[134]**, **[141]**–**[143]**, **[158]**, **[161]**.

CHAPTER 2

Random Processes and the Stochastic Calculus

The main goal of this chapter is to present the concepts and facts from the theory of random processes needed in what follows, with sufficient rigor and logical veracity. We give facts from the general theory of random processes and information about Wiener, Poisson, and diffusion processes, together with a brief sketch of the theory of the stochastic calculus of semimartingales.

§ 2.1. Random processes and their distributions. The Wiener process

In contemporary stochastic analysis the concept of a *stochastic base* has become firmly established. This is a complete probability space $(\Omega, \mathcal{F}, \mathbf{P})$ endowed with a *filtration* $\mathbb{F} = (\mathcal{F}_t)_{0 \leqslant t}$, which is a nondecreasing right-continuous family of (complete) σ-algebras $\mathcal{F}_t$. This is the framework in which we consider all *random processes* $(X_t)_{0 \leqslant t}$, under the assumption that the random variable X_t is $\mathcal{F}_t$-measurable for each $t \geqslant 0$ (is *adapted to* $\mathbb{F}$).

The *measurability* of X is understood as the measurability of the mapping

$$X \colon \Omega \times \mathbb{R}_+ \to R \quad (\text{usually, } R = \mathbb{R}^d, \quad d \geqslant 1)$$

with respect to the σ-algebras $\mathcal{F} \times \mathcal{B}(\mathbb{R}_+)$ and $\mathcal{B}(R)$, where $\mathcal{B}(\,\cdot\,)$ is the Borel σ-algebra, and the *progressive measurability* of X is understood as the measurability of the mapping

$$X \colon (\Omega \times [0, t], \mathcal{F} \times \mathcal{B}[0, t]) \to (R, \mathcal{B}(R)) \quad \text{for any} \quad t \geqslant 0.$$

It is assumed that for almost all ω the process $X_t(\omega)$ has a *sample path* $X.(\omega)$ that is right continuous and has limits $X_{t-}(\omega)$ from the left.

Two processes X and Y are said to be *modifications* of each other if

$$\mathbf{P}\{X_t = Y_t\} = 1 \quad \text{for any} \quad t \geqslant 0.$$

The stronger concept of *indistinguishability* of X and Y is expressed as

$$\mathbf{P}\{\omega : X_t = Y_t \text{ for all } t \geqslant 0\} = 1.$$

The "exotic" nature of the continuous case ($t \in \mathbb{R}_+$ and not $\mathbb{Z}_+$) is illustrated, for example, by the process

$$X_t(\omega) = \begin{cases} 1, & \omega = t, \\ 0, & \omega \neq t, \end{cases}$$

defined on $\Omega = [0, 1]$, with $\mathcal{F} = \mathcal{B}[0, 1]$ and $\mathbf{P} = \text{meas}$ (Lebesgue measure). It is clear that $Y \equiv 0$ is a modification of X, but they are not *indistinguishable*, because of the obvious equality

$$\mathbf{P}\{\omega : X_t(\omega) = 0 \text{ for all } t \in [0, 1]\} = 0.$$

Most important for our purposes is a *Wiener process* W, or *Brownian motion*, defined by the following three conditions:

1) $W_0 = 0$ ($\mathbf{P}$-a.s.);
2) $W_t - W_s$ does not depend on $\mathcal{F}_s$, $s \leqslant t$;
3) $W_t - W_s$ has a *normal distribution* with zero mean and variance $t - s$, that is, $N(0, t-s)$.

The "discrete" analogue of a Wiener process is obvious: these are sums of standard independent Gaussian variables ξ_k: $W_n = \sum_{k=1}^n \xi_k$, $\mathcal{F}_n = \sigma\{\xi_1, \dots, \xi_n\}$.

For what follows we need to understand what the *distribution* of a random process $(X_t(\omega))_{0 \leqslant t}$ is. Denote by $\mathbb{R}^{[0,\infty)}$ the set of all possible functions $x\colon \mathbb{R}_+ \to \mathbb{R}$. A subset A of $\mathbb{R}^{[0,\infty)}$ is called a *cylindrical* set if there is an n-tuple $t_1 \leqslant \dots \leqslant t_n$ and a set $B_n \in \mathcal{B}(\mathbb{R}^n)$ such that

$$A = \{x : (x_{t_1}, \dots, x_{t_n}) \in B_n\}.$$

Such sets form an algebra, which generates a corresponding smallest σ-algebra $\mathcal{B}^{[0,\infty)}$ containing them.

The *distribution* of a random process X is defined to be the measure μ on $\mathcal{B}^{[0,\infty)}$ such that

$$\mu(C) = \mathbf{P}\{\omega : X.(\omega) \in C\}, \qquad C \in \mathcal{B}^{[0,\infty)}.$$

A *finite-dimensional* distribution of X is a measure $\mu_{t_1,\dots,t_n}$ given by an equality of the form

$$\mu_{t_1,\dots,t_n}(A) = \mathbf{P}\{\omega : (X_{t_1}(\omega), \dots, X_{t_n}(\omega)) \in B_n\}.$$

Furthermore, if a cylindrical set A admits another representation

$$A = \{x : (x_{s_1}, \dots, x_{s_m}) \in B_m\},$$

then the equality (for any A)

$$\mu_{s_1,\dots,s_m}(A) = \mu_{t_1,\dots,t_n}(A)$$

is called the *consistency condition*.

Clearly, the measure μ gives rise to a system of consistent finite-dimensional distributions. *Kolmogorov's theorem* is the converse of this assertion:

For a given system $\{\mu_{t_1,\dots,t_n}\}$ of consistent finite-dimensional distributions there exists on $(\mathbb{R}^{[0,\infty)}, \mathcal{B}^{[0,\infty)})$ a unique distribution μ with the measures $\mu_{t_1,\dots,t_n}$ as its finite-dimensional distributions.

We remark that μ is the distribution of the *coordinate* process $X_t(\omega) = \omega(t)$, $\omega \in \Omega = \mathbb{R}^{[0,\infty)}$.

Suppose that the measures $\mu_{t_1,\dots,t_n}$ are determined by the Gaussian density

$$p_{t_1,\dots,t_n}(x_1, \dots, x_n) = (2\pi t_1)^{-\frac{1}{2}} e^{-\frac{x_1^2}{2t_1}} \cdots (2\pi(t_n - t_{n-1}))^{-\frac{1}{2}} e^{-\frac{(x_n - x_{n-1})^2}{2(t_n - t_{n-1})}}.$$

It can be verified directly that this system is consistent, and hence by Kolmogorov's theorem there exists a unique distribution $\mu = \mu^W = \mathbf{P}^W$, called the *Wiener measure*, and the coordinate process $W_t(\omega) = \omega(t)$ is called a *Wiener* process.

Associated with the distributions μ^X and μ^Y of random processes X and Y are the concepts of their *absolute continuity* ($\mu^Y \ll \mu^X$) and their *equivalence* ($\mu^X \sim \mu^Y \Leftrightarrow \mu^Y \ll \mu^X$ and $\mu^X \ll \mu^Y$). In this case a *density* for the corresponding measures exists (the Radon–Nikodým derivative): $h(x.) = \dfrac{d\mu^Y}{d\mu^X}(x.)$, $x. \in \mathbb{R}^{[0,\infty)}$, and for an arbitrary bounded measurable function $f(x.)$ it satisfies the relation

$$\mathbf{E}f(Y) = \int_{\mathbb{R}^{[0,\infty)}} f(x.)\, d\mu^Y = \int_{\mathbb{R}^{[0,\infty)}} f(x.)h(x.)\, d\mu^X = \mathbf{E}f(X)h(X). \tag{2.1}$$

Denoting $h(x)$ by Z, we get from (2.1) that $\mathbf{E}Z = 1$, and hence Z determines another measure $\widetilde{\mathbf{P}} \ll \mathbf{P}$ on the original probability space.

How does one establish that $\mu^Y \ll \mu^X$? The standard way is first of all to assume that $\mu^Y_{t_1,\dots,t_n} \ll \mu^X_{t_1,\dots,t_n}$.

$$\frac{d\mu^Y_{t_1,\dots,t_n}}{d\mu^X_{t_1,\dots,t_n}}(x.) = h_{t_1,\dots,t_n}(x.) \quad \text{and} \quad h_{t_1,\dots,t_n}(X) = Z_{t_1,\dots,t_n}.$$

Then one can investigate the question of the absolute continuity of the distributions μ^Y and μ^X themselves with the help of $Z^* = \lim_{n\to\infty} Z_{t_1,\dots,t_n}$, where the limit is understood in the sense of convergence in probability, $\mathbf{E}Z^* \leqslant 1$:

$$\mu^Y \ll \mu^X \iff \mathbf{E}Z^* = 1.$$

Singularity $\mu^Y \perp \mu^X$ can be characterized in the same terms as the equality $\mathbf{E}Z^* = 0$. For example, $\mu^W \perp \mu^{2W}$ for a Wiener process.

The construction of a Wiener process was realized in the space $\mathbb{R}^{[0,\infty)}$. However, its sample paths are *continuous* functions, as follows from another *theorem of Kolmogorov*:

Suppose that for a random process $(X_t)_{t\geqslant 0}$ there exist constants $C, \varepsilon, \beta > 0$ such that $\mathbf{E}|X_t - X_s|^\beta \leqslant C|t-s|^{1+\varepsilon}$ for all $t \geqslant s$. Then X has a continuous modification.

In the case $X_t = W_t$ we have $\mathbf{E}|W_t - W_s|^4 = 3|t-s|^2$, and hence the conditions of this theorem hold.

The Wiener processes turn out to be invariant under shifts, inversion, and dilation in time. Namely, the following random processes are also Wiener processes:

$$(W_{t+a} - W_t)_{t\geqslant 0}, \quad a > 0; \qquad (tW_{1/t})_{t\geqslant 0}; \qquad (\lambda^{-1}W_{\lambda^2 t})_{t\geqslant 0}, \quad \lambda \neq 0.$$

The last property is called the *self-similarity* of W.

To conclude our discussion of the "sample path" properties of a Wiener process W we remark that the sample paths $W.$, though continuous, are extremely irregular: they are *nowhere differentiable* and on each finite interval they have *unbounded variation*. It is not hard to convince oneself of this on an intuitive level. Thus, for $h > 0$ the increment $W_{t+h} - W_t$ has normal distribution $N(0,h)$, and hence $h^{-1/2}(W_{t+h} - W_h) \sim N(0,1)$. However, the "pre-limit" expression $h^{-1}(W_{t+h} - W_h)$ for the derivative has variance increasing to infinity:

$$\mathbf{D}h^{-1}(W_{t+h} - W_h) = h^{-2}\mathbf{D}(W_{t+h} - W_h) = h^{-1} \uparrow \infty \quad (h \downarrow 0).$$

We mention that although for any interval $[0,t]$ the first variation

$$\sup_{0=t_0<\dots<t_n=t} \sum_{k=1}^{n} |W_{t_k} - W_{t_{k-1}}|$$

is infinite (**P**-a.s.), the second variation has the following remarkable property:

$$\sum_{k=1}^{n} |W_{t_k} - W_{t_{k-1}}|^2 \xrightarrow[n\to\infty]{} t \qquad (\mathbf{P}\text{-a.s.}). \tag{2.2}$$

Heuristic arguments for (2.2) stem from the following relations. From the independence of the increments $W_{t_k} - W_{t_{k-1}}$ and the properties of the variance we have, first,

$$\mathbf{E}\sum_{k=1}^{n} |W_{t_k} - W_{t_{k-1}}|^2 = \sum_{k=1}^{n} \mathbf{D}(W_{t_k} - W_{t_{k-1}}) = \sum_{k=1}^{n} (t_k - t_{k-1}) = t,$$

and second,

$$\begin{aligned}
\mathbf{D}\sum_{k=1}^{n} (W_{t_k} - W_{t_{k-1}})^2 &= \sum_{k=1}^{n} \mathbf{D}(W_{t_k} - W_{t_{k-1}})^2 \\
&= 2\sum_{k=1}^{n} (t_k - t_{k-1})^2 \\
&\leqslant 2\max_k (t_k - t_{k-1}) \sum_{k=1}^{n} (t_k - t_{k-1}) \\
&= 2t \cdot \max_k (t_k - t_{k-1}) \xrightarrow[n\to\infty]{} 0.
\end{aligned}$$

Another beautiful but implicit construction of a Wiener process W and the Wiener measure is based on the *Prokhorov–Donsker invariance principle*:

Let $\xi = (\xi_n)_{n\geqslant 1}$ be a sequence of independent identically distributed random variables with $\mathbf{E}\xi_n = 0$ and $\mathbf{D}\xi_n = 1$. From the sums $S_n = \sum_{k=1}^{n} \xi_k$ define a sequence of continuous random processes by

$$X_t^n = n^{-1/2}\{S_{[nt]} + (nt - [nt])\xi_{[nt]+1}\}, \quad t \in [0,1].$$

Then there exists a measure μ^W concentrated on $C[0,1]$ such that the distributions of the processes X^n converge weakly to μ^W. Furthermore, the coordinate process $W_t(\omega) = \omega_t$ is a Wiener process.

Thus, on the space Ω of continuous functions there exists a measure **P** with respect to which the coordinate process $W_t = \omega(t)$ is a Wiener process. For each $t \geqslant 0$ we define $\mathcal{F}_t$ as the σ-algebra generated by the coordinate process $\omega(s)$, $s \leqslant t$, and completed by the sets of **P**-measure zero. The stochastic base constructed will be called the *coordinate* or *canonical* base.

§ 2.2. Diffusion processes. The Kolmogorov–Itô formula, Girsanov's theorem, representations of martingales

An important place in what follows is occupied by *diffusion processes*, which we consider at first starting from the general theory of Markov processes.

A random process X is called a *Markov* process if

$$\mathbf{P}\{X_t \in A \,|\, X_u,\ u \leqslant s\} = \mathbf{P}\{X_t \in A \,|\, X_s\}$$

for any $t \geqslant s$ and $A \in \mathcal{B}(\mathbb{R})$.

For $X_s = x$ the last probability is denoted by $\mathbf{P}(s, x, t, A)$ and interpreted as the *transition probability* of the process X from the state x (at the time s) to the set A (at the time t). A Markov process is *homogeneous* if $\mathbf{P}(s, x, t, A) = \mathbf{P}(t-s, x, A)$. The transition probability is a measurable function with respect to x and a probability measure with respect to A, and $\mathbf{P}(s, x, s, A) = I_A(x)$.

Further, $\mathbf{P}(t, x, A)$ determines a semigroup of operators

$$\mathbf{P}^t f(x) = \int_{\mathbb{R}} \mathbf{P}(t, x, dy) f(y)$$

on some supply of functions $f: \mathbb{R} \to \mathbb{R}$. This semigroup has a right-hand derivative at zero on a certain set $\mathbb{D}_A$ of functions:

$$Af = \frac{d^+}{dt} \mathbf{P}^t f \bigg|_{t=0} = \lim_{t \to 0+} t^{-1} [\mathbf{P}^t f - f].$$

This is called the *infinitesimal* operator of the semigroup $(\mathbf{P}^t)$.

A Markov process X is called a *diffusion* process if A is a *differential operator* on the *smooth* functions. More precisely, we denote by $U_\varepsilon(x)$ and $V_\varepsilon(x)$ the neighborhood of radius $\varepsilon > 0$ about a point $x \in \mathbb{R}^d$ ($d \geqslant 1$) and its complement in the whole space.

Suppose that a Markov process X is such that there are continuous functions $a^{ij}(x)$ and $b^i(x)$ satisfying

$$\mathbf{P}(t, x, V_\varepsilon(x)) = o(t),$$

$$\int_{U_\varepsilon(x)} (y^i - x^i) \mathbf{P}(t, x, dy) = b^i(x) t + o(t),$$

$$\int_{U_\varepsilon(x)} (y^i - x^i)(y^j - x^j) \mathbf{P}(t, x, dy) = a^{ij}(x) t + o(t)$$

for $i, j = 1, \dots, d$.

Such a process is called a (homogeneous) *diffusion process* with *drift vector* b and *diffusion matrix* a. On the class of twice continuously differentiable functions f we have

$$Af = Lf = \frac{1}{2} \sum_{i,j=1}^{d} a^{ij} \frac{\partial^2 f}{\partial x^i \partial x^j} + \sum_{i=1}^{d} b^i \frac{\partial f}{\partial x^i},$$

and this differential operator L is called the *generating operator* of the diffusion process.

If the measure $\mathbf{P}(t, x, dy)$ admits a smooth transition density $p(t, x, dy)$, then it satisfies the *forward and backward Kolmogorov equations*:

$$\frac{\partial p}{\partial t} = L^* p \quad \text{and} \quad \frac{\partial p}{\partial t} = Lp,$$

where

$$L^* p = \frac{1}{2} \sum_{i,j=1}^{d} a^{ij} \frac{\partial^2 p}{\partial x^i \partial x^j} - \sum_{i=1}^{d} b^i \frac{\partial p}{\partial x^i}.$$

A Wiener process W is a diffusion process with transition density

$$p(t, x, y) = \frac{1}{\sqrt{2\pi t}} e^{-\frac{(x-y)^2}{2t}}.$$

Furthermore, its *drift* is zero, and its *diffusion* is the identity, and there are reasons to hope that all diffusion processes can be obtained from W.

The appropriate construction can be realized with the help of *stochastic integration.*

Let $W = (W_t, \mathcal{F}_t)_{t\geqslant 0}$ be a Wiener process. For "simple" functions

$$f(t) = \sum_{k=1}^{n} f_{k-1} I_{(t_{k-1}, t_k]}(t),$$

where $0 = t_0 < t_1 < \cdots < t_n = T$, $f_{k-1} \in \mathcal{F}_{t_{k-1}}$, and $t \leqslant T$, we define the *stochastic integral* (with respect to W) by the sum

$$\int_0^t f(s) dW_s = \sum_{k=1}^{n} f_{k-1}(W_{t_k \wedge t} - W_{t_{k-1}\wedge t}).$$

The integral has the following properties:

$$\int_0^t (\alpha f + \beta g)\, dW_s = \alpha \int_0^t f\, dW_s + \beta \int_0^t g\, dW_s, \tag{2.3}$$

$$\mathbf{E}\biggl(\int_0^T f\, dW_s \biggm| \mathcal{F}_t\biggr) = \int_0^t f\, dW_s, \tag{2.4}$$

$$\mathbf{E}\biggl(\int_0^t f\, dW_s \int_0^t g\, dW_s\biggr) = \mathbf{E} \int_0^t gf\, ds. \tag{2.5}$$

This integral can be extended as a linear operator to the set of progressively measurable functions f such that $\mathbf{E}\int_0^T f^2\, ds < \infty$ with preservation of the properties (2.3)–(2.5), and is then denoted by the same symbol (here T can be ∞).

If $b(t,x)$ and $\sigma(t,x)$ are continuous functions with respect to $(t,x) \in \mathbb{R}_+ \times \mathbb{R}$ and satisfy a *Lipschitz condition* with respect to x, that is,

$$|b(t,x) - b(t,y)| + |\sigma(t,x) - \sigma(t,y)| \leqslant L\,|x-y| \quad \text{for all} \quad t > 0, \;\; x, y \in \mathbb{R}$$

(L is a positive constant), then there is a unique continuous process X such that

$$X_t = X_0 + \int_0^t b(s, X_s)\, ds + \int_0^t \sigma(s, X_s)\, dW_s \tag{2.6}$$

for all $t > 0$ ($\mathbf{P}$-a.s.).

The equality (2.6) defines a *stochastic differential equation*, and the process X is called a (strong) solution of it. It turns out that in this case X is a diffusion process with generating operator

$$Lf = \frac{1}{2}\sigma^2 \frac{\partial^2 f}{\partial x^2} + b\frac{\partial f}{\partial x}.$$

Moreover, with the help of a smooth function $F(t,x) \in C^{1,2}$ and X we form the new random process $Y_t = F(t, X_t)$, which is also a diffusion process (Figure 2.1) and satisfies the relation (for all $t > 0$, $\mathbf{P}$-a.s.)

$$F(t, X_t) = F(0, X_0) + \int_0^t \left[\frac{\partial F}{\partial t} + \frac{\partial F}{\partial x} b + \frac{1}{2}\frac{\partial^2 F}{\partial x^2}\sigma^2\right] ds + \int_0^t \sigma \frac{\partial F}{\partial x}\, dW_s. \tag{2.7}$$

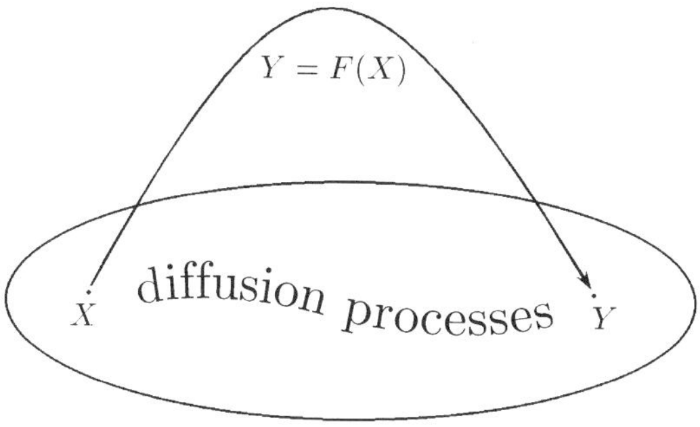

FIGURE 2.1. Closedness of the class of diffusion processes with respect to smooth transformations F.

This formula is called the *Kolmogorov–Itô formula.* Thus, according to (2.7) we have that the square of a Wiener process is equal to

$$W_t^2 = 2\int_0^t W_s\,dW_s + t.$$

In what follows we shall need the concept of a *controlled diffusion process* X_t^α as a solution of a stochastic differential equation

$$X_t = x + \int_0^t b(s,\alpha_s,X_s)\,ds + \int_0^t \sigma(s,\alpha_s,X_s)\,dW_s, \tag{2.8}$$

where the *control process* α_s is determined according to the current information about the state of X_u, $u \leqslant s$, and takes values in some parametric set $\mathbb{A}$ of "controls".

The quality of control of the process X_t is determined by the *loss functional* $\int_0^\infty f^{\alpha_t}(X_t)\,dt$, $f^\alpha(x)$ being a certain function, and by the *payoff function*

$$v(x) = \sup_{\alpha\in\mathbb{A}} \mathbf{E}\int_0^\infty f^{\alpha_t}(X_t)\,dt.$$

A control is found from the *Bellman differential equation*

$$\sup_{\alpha\in\mathbb{A}}\{L^\alpha v + f^\alpha\} = 0, \tag{2.9}$$

where the operator

$$L^\alpha(x) = \frac{1}{2}\sigma^2(\alpha,x)\frac{\partial^2}{\partial x^2} + b(\alpha,x)\frac{\partial}{\partial x}$$

is the generating operator for the diffusion process X^α.

Transformations of probability measures can lead to random processes with "nice" properties of distributions with respect to the new measures. The first theorem of this kind is *Girsanov's theorem*:

Consider a progressively measurable process β_t for which the process (called the Girsanov exponential)

$$Z_t = \exp\left\{\int_0^t \beta_s\,dW_s - \frac{1}{2}\int_0^t \beta_s^2\,ds\right\}, \qquad t \leqslant T,$$

is well defined, where W is a Wiener process. Let $\mathbf{E}Z_T = 1$, $d\widetilde{\mathbf{P}} = Z_T d\mathbf{P}$, and $\widetilde{W}_t = W_t - \int_0^t \beta_s\,ds$. Then the process $\widetilde{W}$ is a Wiener process with respect to $\widetilde{\mathbf{P}}$.

In particular, a diffusion process $X_t = \int_0^t b(X_s)\,ds + W_t$ with bounded measurable drift coefficient is a Wiener process (see Figure 2.2) with respect to the measure

$$d\widetilde{\mathbf{P}} = \exp\left\{-\int_0^T b(X_s)\,dW_s - \frac{1}{2}\int_0^T b^2(X_s)\,ds\right\} d\mathbf{P}.$$

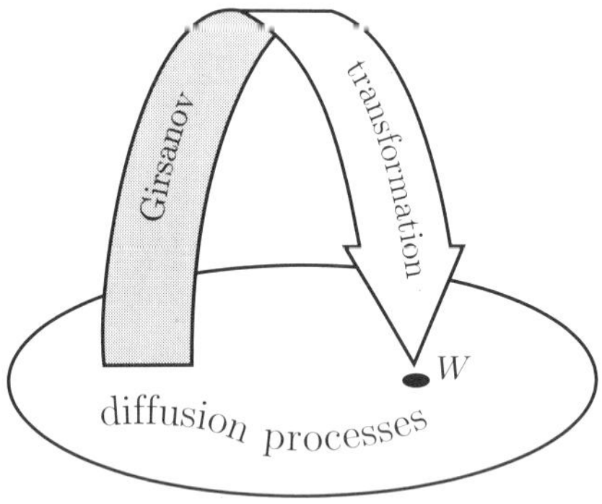

FIGURE 2.2. Transformation of diffusion processes into a Wiener process by a change of measure with the help of the "Girsanov exponential".

Just as the class of diffusion processes is exhausted by the *solutions* of stochastic differential equations, the class of random processes M adapted to the filtration $\mathbb{F} = \mathbb{F}^W$ generated by a Wiener process W, and such that

$$\mathbf{E}\,|M_t| < \infty \quad \text{and} \quad \mathbf{E}\,(M_t \,|\, \mathcal{F}_s) = M_s \;\; (\mathbf{P}\text{-a.s.}) \tag{2.10}$$

is exhausted by the stochastic integrals of the form $\int_0^t \phi_s\,dW_s$. A process M satisfying (2.10) is called a *martingale*. The following *martingale representation theorem* will play a key role in what follows.

Any martingale $M = (M_t, \mathcal{F}_t^W)_{t\geqslant 0}$ can be represented in the form

$$M_t = M_0 + \int_0^t \phi_s\,dW_s, \tag{2.11}$$

where ϕ is a progressively measurable random process.

We remark that the properties (2.4)–(2.5) of stochastic integrals have the *martingale property* built in from the start, and the representation theorem says that there are simply no other martingales (with respect to $\mathbb{F} = \mathbb{F}^W$).

Returning to the example of the process W_t^2, we note that the process $X_t = W_t^2 - \mathbf{E}W_t^2 = W_t^2 - t$ is adapted to the filtration $\mathbb{F}$ and is a martingale with respect to $\mathbf{P}$. Hence, it can be represented in the form

$$X_t = X_0 + \int_0^t \phi_s\,dW_s.$$

An explicit formula for X_t was obtained earlier with the help of the Kolmogorov–Itô formula.

Another basic process is a *Poisson process* $X = (X_t, \mathcal{F}_t)_{t\geqslant 0}$ with intensity $\lambda > 0$, which is defined as follows:

1) $X_0 = 0$ ($\mathbf{P}$-a.s.);
2) $X_t - X_s$ does not depend on $\mathcal{F}_s$, $s \leqslant t$;
3) $X_t - X_s$ has a Poisson distribution with parameter $\lambda(t-s)$.

It is clear that $N_t = X_t - \lambda t$ is a martingale (a Poisson martingale). It turns out that any other martingale $M = (M_t, \mathcal{F}_t^N)_{t\geqslant 0}$ can be represented in the form of an integral with respect to a Poisson martingale:

$$M_t = M_0 + \int_0^t \phi_s\, dN_s, \tag{2.12}$$

where ϕ is a progressively measurable process, and the integral $\int_0^t \phi_s\, dN_s$ can be understood as the difference

$$\int_0^t \phi_s\, dN_s = \int_0^t \phi_s\, dX_s - \int_0^t \phi_s\, d\lambda s$$

of two Lebesgue–Stieltjes integrals, since X_t and λ_t are increasing processes.

In conclusion we give vector forms of the equation (2.6) and the Kolmogorov–Itô formula (2.7). Namely, (2.6) can be understood as a vector equation with $b = (b^1, \dots, b^d)'$, $\sigma = (\sigma^{ij})^{i=1,\dots,d}_{j=1,\dots,d_1}$, and $W = (W^1, \dots, W^{d_1})'$.

In this case the generating operator of a d-dimensional diffusion process X has the form

$$Lf = \sum_{i=1}^d b^i \frac{\partial f}{\partial x_i} + \frac{1}{2} \sum_{i,j=1}^d \frac{\partial^2 f}{\partial x_i \partial x_j} \sum_{k=1}^{d_1} \sigma^{ik}\sigma^{jk},$$

and the formula (2.7) for $F\colon \mathbb{R}_+ \times \mathbb{R}^{d_1} \to \mathbb{R}^1$ is

$$\begin{aligned} F(t, X_t) = F(0, X_0) &+ \int_0^t \left[\frac{\partial F}{\partial s}(X_s) + LF(X_s)\right] ds \\ &+ \int_0^t \sum_{i=1}^d \sum_{j=1}^{d_1} \frac{\partial F}{\partial x_i} \sigma^{ij}\, dW_s^j. \end{aligned} \tag{2.7$'$}$$

The structure of diffusion processes is determined by their *drift* and *diffusion*. The next section shows how the idea of singling out in the structure a *low-frequency* component (a process of bounded variation) and a *high-frequency* component (a martingale) leads to far-reaching and definitive generalizations in the form of processes called *semimartingales*.

§2.3. Semimartingales and the stochastic calculus

In correspondence with the general theory of random processes we introduce in addition to the stochastic base $(\Omega, \mathcal{F}, \mathbb{F}, \mathbf{P})$ the following two σ-algebras on $\mathbb{R}_+ \times \Omega$:

the σ-algebra $\mathcal{P}$ generated by all the $\mathbb{F}$-adapted *continuous* processes;

the σ-algebra $\mathcal{O}$ generated by all the $\mathbb{F}$-adapted *right-continuous* processes.

These are the so-called *predictable* and *optional* σ-algebras. The $\mathcal{P}$-measurable and $\mathcal{O}$-measurable processes are called *predictable* processes and *optional* processes, respectively.

A *stopping time* (with respect to $\mathbb{F}$) is defined to be a random variable $\tau\colon \Omega \to \mathbb{R}_+ \cup \{+\infty\}$ such that for all $t \geqslant 0$

$$\{\omega : \tau(\omega) \leqslant t\} \in \mathcal{F}_t.$$

A martingale satisfying (2.10) is said to be *uniformly integrable* if there is an integrable random variable M_∞ such that (**P**-a.s.)

$$M_t = \mathbf{E}(M_\infty \,|\, \mathcal{F}_t).$$

The martingale M is said to be *square integrable* if $\mathbf{E}\,|M_\infty|^2 < \infty$.

These classes of martingales are denoted by $\mathcal{M}$ and $\mathcal{M}^2$, respectively.

The set of processes having finite (integrable) variation on each finite interval will be denoted by $\mathcal{V}$ ($\mathcal{A}$), and the subset of nondecreasing processes by $\mathcal{V}^+$ ($\mathcal{A}^+$).

A process $M = (M_t)_{t\geqslant 0}$ is a *local martingale* ($M \in \mathcal{M}_{\text{loc}}$) if there is a sequence of stopping times $\tau_n \uparrow \infty$ ($n \uparrow \infty$) such that $M^{\tau_n} = (M_{t\wedge\tau_n})_{t\geqslant 0}$ belongs to $\mathcal{M}$ for each $n \in \mathbb{N}$. The classes $\mathcal{M}^2_{\text{loc}}$, $\mathcal{A}^+_{\text{loc}}$, and $\mathcal{A}_{\text{loc}}$ are defined similarly according to the indicated *localization procedure*, and the corresponding processes are called *locally square-integrable martingales*, and so on.

For a process $\mathbb{A} \in \mathcal{V}^+ \equiv \mathcal{V}^+_{\text{loc}}$ we have a decomposition into the *purely continuous* component A^c and the *purely discontinuous* component A^d:

$$A_t = A^c_t + A^d_t,$$

where $A^d_t = \sum_n \alpha_n I_{\{\tau_n \leqslant t\}}$, and the sequence $\{\tau_n\}$ of stopping times "exhausts" the jumps of the process A.

If in the definition (2.10) the equality is replaced by the inequality

$$\mathbf{E}(M_t \,|\, \mathcal{F}_s) \geqslant M_s \quad (\leqslant M_s),$$

then the process M_t is called a *submartingale* (*supermartingale*).

It turns out that the indicated processes are structurally formed from local martingales (or even from martingales) and increasing predictable processes. For example, an arbitrary nonnegative supermartingale M admits the following (unique) *Doob–Meyer decomposition*:

$$M_t = m_t - A_t, \tag{2.13}$$

where $m \in \mathcal{M}_{\text{loc}}$ and $A \in \mathcal{A}^+_{\text{loc}} \cap \mathcal{P}$. In particular, for $A \in \mathcal{A}^+_{\text{loc}}$ there exists a unique predictable process (the *compensator*) $\widetilde{A} \in \mathcal{A}^+_{\text{loc}} \cap \mathcal{P}$ such that the difference $A - \widetilde{A}$ is in $\mathcal{M}_{\text{loc}}$.

The Doob–Meyer decomposition implies that any $M, N \in \mathcal{M}^2_{\text{loc}}$ have a *mutual quadratic characteristic* $\langle M, N\rangle$, which is a predictable process in $\mathcal{A}_{\text{loc}}$ with the property that it *compensates* MN to form a local martingale: $MN - \langle M, N\rangle \in \mathcal{M}_{\text{loc}}$. In particular, $M^2 - \langle M, M\rangle \in \mathcal{M}_{\text{loc}}$ for $M \equiv N \in \mathcal{M}^2_{\text{loc}}$.

An arbitrary local martingale M does not have such a characterization. Nevertheless, we have the following decomposition into a *continuous* component M^c and a *purely discontinuous* component M^d:

$$M_t = M^c_t + M^d_t,$$

where M^c is a continuous local martingale ($M^c \in \mathcal{M}^c_{\text{loc}}$), and M^d belongs to the class $\mathcal{M}^d_{\text{loc}}$ of *purely discontinuous local martingales*, which is characterized by the condition that $KL \in \mathcal{M}_{\text{loc}}$ for any $K \in \mathcal{M}^c_{\text{loc}}$ and $L \in \mathcal{M}^d_{\text{loc}}$.

For $M, N \in \mathcal{M}_{\mathrm{loc}}$ we define their mutual (nonpredictable in general) quadratic characteristic

$$[M, N]_t = \langle M^c, N^c \rangle_t + \sum_{0<s\leqslant t} \Delta M_s \Delta N_s.$$

In particular, for $M \equiv N \in \mathcal{M}_{\mathrm{loc}}$

$$[M, M]_t - \langle M, M \rangle_t \in \mathcal{M}_{\mathrm{loc}}.$$

Suppose that $M \in \mathcal{M}^2_{\mathrm{loc}}$ and the predictable process ϕ are such that (for almost all ω) the Lebesgue–Stieltjes integral

$$\int_0^t \phi_s^2 \, d\langle M \rangle_s \equiv \phi^2 \circ \langle M, M \rangle_t$$

is defined and is in $\mathcal{A}^+_{\mathrm{loc}}$. Then the stochastic integral

$$(\phi * M)_t = \int_0^t \phi_s \, dM_s$$

is defined as a process in $\mathcal{M}^2_{\mathrm{loc}}$ such that

$$\Delta(\phi * M)_t = \phi_t \Delta M_t, \qquad \langle \phi * M, \phi * M \rangle_t = \phi^2 \circ \langle M, M \rangle_t.$$

In the class $\mathcal{M}^2_{\mathrm{loc}}$ all possible stochastic integrals with respect to a given martingale $M \in \mathcal{M}^2_{\mathrm{loc}}$ form a subspace. Consequently, any other $N \in \mathcal{M}^2_{\mathrm{loc}}$ must have a decomposition into a sum of two projections (on the indicated subspace and on its complement). The realization of this idea is the following (unique) *Kunita–Watanabe decomposition*:

$$N_t = \int_0^t \phi_s \, dM_s + M'_t,$$

where ϕ is a predictable process (the integrand), and $M' \in \mathcal{M}^2_{\mathrm{loc}}$ is *orthogonal* to M in the sense that $\langle M, M' \rangle = 0$.

A *semimartingale* is defined to be a process representable in the form

$$X_t = X_0 + A_t + M_t, \tag{2.14}$$

where $A \in \mathcal{V}$, $M \in \mathcal{M}_{\mathrm{loc}}$, and $X_0 \in \mathcal{F}_0$.

If in this decomposition $A \in \mathcal{P}$, then X is a *special semimartingale*.

Now let $\mathbb{R}_0 = \mathbb{R}^d \setminus \{0\}$ and let $\mathcal{B}(\mathbb{R}_0)$ be the Borel σ-algebra. From the jumps ΔX_t of a d-dimensional semimartingale X we define the *integer-valued random measure*

$$\mu(\omega, B, \Gamma) = \sum_s I_{\{(s, \Delta X_s) \in (B, \Gamma)\}},$$

where $B \in \mathcal{B}(\mathbb{R}_+)$ and $\Gamma \in \mathcal{B}(\mathbb{R}_0)$.

The integer-valued random measure $\nu = \nu(\omega, B, \Gamma)$ is said to be *predictable* if for any $\mathcal{P} \times \mathcal{B}(\mathbb{R}_0)$-measurable nonnegative function ϕ the process

$$\phi \circ \nu_t = \int_0^t \int_{\mathbb{R}_0} \phi(s, x) \, \nu(ds, dx)$$

is predictable. If in this case $\mathbf{E}(\phi \circ \nu)_\infty = \mathbf{E}(\phi \circ \mu)_\infty$, then ν is called the *compensator* of the measure μ.

Let a $\mathcal{P} \times \mathcal{B}(\mathbb{R}_0)$-measurable function ϕ be such that $(|\phi| \circ \nu)_t \in \mathcal{A}^+_{\text{loc}}$. Then the stochastic integral

$$\phi * (\mu - \nu)_t = \int_0^t \int_{\mathbb{R}_0} \phi(s, x)\, d(\mu - \nu) \tag{2.15}$$

with respect to the difference $(\mu - \nu)$ of measures is defined, and it is a purely discontinuous local martingale whose jumps are determined from the relation

$$\Delta\phi * (\mu - \nu)_t = \phi(t, \Delta X_t) I_{\Delta X_t \neq 0} - \int_{\mathbb{R}_0} \phi(t, x)\, \nu(\{t\}, dx).$$

Using the original representation (2.14) and (2.15), we can now write the following *canonical decomposition* of X:

$$X_t = X_0 + B_t + M^c_t + xI_{\{x:|x|\leqslant 1\}} * (\mu - \nu)_t + xI_{\{x:|x|>1\}} \circ \mu_t, \tag{2.16}$$

where $B \in \mathcal{V} \cap \mathcal{P}$ and $M^c \in \mathcal{M}^c_{\text{loc}}$.

The triple $(B, \langle M^c, M^c \rangle, \nu)$ is called the *triplet of predictable characteristics* of the semimartingale X.

It is now easy to define the stochastic integral with respect to a semimartingale X as the sum of the corresponding integrals with respect to the components of the decompositions (2.13) or (2.16).

It turns out that the class of semimartingales, like the class of diffusion processes, is invariant with respect to *smooth* changes of variables. This is what is said by the corresponding *generalization* of the Kolmogorov–Itô formula given by Doleans and Meyer:

Let X be a semimartingale (2.14) *with values in $\mathbb{R}^d$, $d \geqslant 1$, and let $F\colon \mathbb{R}^d \to \mathbb{R}^1$ be a twice continuously differentiable function. Then*

$$\begin{aligned} F(X_t) = F(X_0) &+ \int_0^t \sum_{i=1}^d \frac{\partial F}{\partial x_i}(X_{s-})\, dX^i_s \\ &+ \frac{1}{2} \int_0^t \sum_{i,j=1}^d \frac{\partial^2 F}{\partial x_i \partial x_j}(X_{s-})\, d\langle X^{c,i}, X^{c,j}\rangle_s \\ &+ \sum_{0<s\leqslant t} \left[F(X_s) - F(X_{s-}) - \sum_{i=1}^d \frac{\partial F}{\partial x_i}(X_{s-})(X^i_s - X^i_{s-}) \right], \end{aligned} \tag{2.17}$$

where X^c is the continuous component of a local martingale M.

The existence of stochastic integrals with respect to semimartingales enables us to consider corresponding stochastic equations. Of crucial significance are *linear stochastic equations*

$$Y_t = Y_0 + \int_0^t Y_{s-}\, dX_s, \tag{2.18}$$

where X is a given semimartingale (2.14).

A solution Y is determined with the help of the *stochastic exponential*

$$\mathcal{E}_t(X) = e^{X_t - X_0 - \frac{1}{2}\langle X^c, X^c\rangle_t} \prod_{0<s\leqslant t} (1+\Delta X_s)e^{-\Delta X_s}, \qquad (2.19)$$
$$Y_t = Y_0 \mathcal{E}_t(X).$$

It is clear that $\mathcal{E}_t(X)$ is a solution of (2.18) with $Y_0 = 1$.

We list some important properties of $\mathcal{E}_t(X)$:

1) $\mathcal{E}_t(X)$ is a process of bounded variation or a local martingale if X is;

2) $\mathcal{E}_t^{-1}(X) = \mathcal{E}_t(-X^*)$, where $X_t^* = X_t - \langle X^c, X^c\rangle_t - \sum_{s\leqslant t} \dfrac{(\Delta X_s)^2}{1+\Delta X_s}$;

3) $\mathcal{E}_t(X)\mathcal{E}_t(X') = \mathcal{E}_t(X + X' + [X, X'])$.

The last property (the multiplication rule for stochastic exponentials) is called *Yor's formula.*

Suppose that besides the original measure on the stochastic base $(\Omega, \mathcal{F}, \mathbb{F}, \mathbf{P})$ we are given a measure $\widetilde{\mathbf{P}}$ locally equivalent to it and with local density Z_t. The equivalence implies strict positiveness of Z_t for almost all t ($\mathbf{P}$-a.s.). Hence, we can define a local martingale N (with respect to $\mathbf{P}$: $N \in \mathcal{M}_{\text{loc}}(\mathbf{P})$) as the stochastic integral $N_t = \int_0^t Z_{s-}^{-1}\, dZ_s$, which leads to a representation of Z in the form of the stochastic exponential

$$Z_t = \mathcal{E}_t(N). \qquad (2.20)$$

Just as a diffusion process acquires the properties of a Wiener process under a transformation of measure with the help of the "Girsanov exponential", a given semimartingale R here acquires the properties of a local martingale with respect to a measure $\widetilde{\mathbf{P}}$ with density that is a stochastic exponential (2.20):

$$R \in \mathcal{M}_{\text{loc}}(\widetilde{\mathbf{P}}) \iff ZR \in \mathcal{M}_{\text{loc}}(\mathbf{P}). \qquad (2.21)$$

For what follows, facts such as the representation of a martingale in the form of a stochastic integral with respect to a Wiener process and the Doob–Meyer decomposition with respect to some *family of measures* will be important.

Let X be a given process. We denote by $\mathcal{M}(X, \mathbf{P})$ the set formed by all measures $\widetilde{\mathbf{P}}$ equivalent to $\mathbf{P}$ such that X is a local martingale with respect to $\widetilde{\mathbf{P}}$. Correspondingly, this set will be called the *family of martingale measures.*

The first fact in this direction concerns the structure of nonnegative processes that are local martingales with respect to all $\widetilde{\mathbf{P}} \in \mathcal{M}(X, \mathbf{P})$.

MARTINGALE REPRESENTATION THEOREM. *If $\mathcal{M}(X, \mathbf{P}) \neq \varnothing$, then a nonnegative process M belongs to the class $\mathcal{M}_{\text{loc}}(\widetilde{\mathbf{P}})$ (is a local martingale) for any measure $\widetilde{\mathbf{P}} \in \mathcal{M}(X, \mathbf{P}) \iff$ there exists a predictable process ϕ such that*

$$M_t = M_0 + \int_0^t \phi_s\, dX_s \qquad (2.22)$$

for any $t \geqslant 0$.

The next fact gives a description of the structure of nonnegative processes that are supermartingales with respect to all $\widetilde{\mathbf{P}} \in \mathcal{M}(X, \mathbf{P})$.

OPTIONAL DECOMPOSITION THEOREM. *Suppose that* $\mathcal{M}(X, \mathbf{P}) \neq \varnothing$ *and* V *is a nonnegative process. Then* V *is a supermartingale for any measure* $\widetilde{\mathbf{P}} \in \mathcal{M}(X, \mathbf{P})$ $\Longleftrightarrow$ *there exist a predictable process* ϕ *and a nondecreasing optional process* C *(often called a consumption process in financial mathematics) such that*

$$V_t = V_0 + \int_0^t \phi_s \, dX_s - C_t \tag{2.23}$$

for any $t \geqslant 0$.

Finally, we give the last assertion connected with the family $\mathcal{M}(X, \mathbf{P})$, which in this or that capacity is an indispensable attribute of the theory of optimal stopping rules:

THEOREM. *Let* $\mathcal{M}_t^T$ *be the set of stopping times taking values in the interval* $[t, T]$, *and let* $f = (f_t)_{t \geqslant 0}$ *be a positive optional process such that*

$$\sup_{\widetilde{\mathbf{P}} \in \mathcal{M}(X, \mathbf{P}),\, \tau \in \mathcal{M}_t^T} \widetilde{\mathbf{E}} f_\tau < \infty.$$

Then there exists a random process

$$\operatorname*{ess\,sup}_{\widetilde{\mathbf{P}} \in \mathcal{M}(X, \mathbf{P}),\, \tau \in \mathcal{M}_t^T} \widetilde{\mathbf{E}}[f_\tau \mid \mathcal{F}_t] \tag{2.24}$$

that is a supermartingale with respect to each measure in $\mathcal{M}(X, \mathbf{P})$.

In particular, if a positive random variable f is such that $\sup_{\widetilde{\mathbf{P}} \in \mathcal{M}(X, \mathbf{P})} \widetilde{\mathbf{E}} f_\tau < \infty$, then there exists a random process

$$\operatorname*{ess\,sup}_{\widetilde{\mathbf{P}} \in \mathcal{M}(X, \mathbf{P}),\, \tau \in \mathcal{M}_t^\infty} \widetilde{\mathbf{E}}[f \mid \mathcal{F}_t] \tag{2.25}$$

that is a supermartingale with respect to each measure in $\mathcal{M}(X, \mathbf{P})$.

References for Chapter 2

[**49**], [**55**], [**81**], [**82**], [**90**], [**94**], [**98**], [**103**], [**104**], [**106**]–[**108**], [**141**], [**149**], [**154**], [**158**].

CHAPTER 3

Hedging and Investment in Complete Markets

We consider a general model of a (B, S) market, and for it we give a martingale characterization of strategies and develop methods of perfect and quantile hedging under the assumption that there is a unique martingale measure. A method is presented for finding the martingale measure, with applications to some well-known models. Investment problems are looked at on the basis of utility functions and the associated apparatus. The optimal value and optimal investment strategies are calculated in some concrete models.

§ 3.1. A martingale characterization of strategies and perfect hedging

For a given standard stochastic base $(\Omega, \mathcal{F}, \mathbb{F}, \mathbf{P})$, we define on it a (B, S)-market as a collection of two positive *semimartingales* B and S. The values of B_t and S_t are interpreted as the prices of a *nonrisky* asset B and a *risky* asset S.

A portfolio or investment strategy is defined to be a pair $\pi_t = (\beta_t, \gamma_t)$ of random processes, with predictable second component, that determines a nonnegative *value for the investor* according to the balance equation

$$\begin{aligned} X_t^\pi &= X_t^\pi(x) = \beta_t B_t + \gamma_t S_t, \\ X_0^\pi &= \beta_0 B_0 + \gamma_0 S_0 = x. \end{aligned} \tag{3.1}$$

We remark that the process S, like B, will be assumed to be one-dimensional in what follows, although all the results remain true also for d-dimensional S if $\gamma_t \cdot S_t$ is understood as the *inner product* of two vectors: a d-dimensional asset $S = (S^1, \dots, S^d)$ and a d-dimensional strategy $\gamma = (\gamma^1, \dots, \gamma^d)$.

We write $X_t = S_t/B_t$ and call it the discounted price process (of the risky asset, or stock). Correspondingly, the quotient X_t^π/B_t will be called the *discounted value of the portfolio* π.

Assuming that $\beta, \gamma \in \mathcal{V}$, we now consider the discounted value of the portfolio $\pi = (\beta, \gamma)$. Using the Kolmogorov–Itô formula and (3.1), we get that

$$\begin{aligned} \frac{X_t^\pi}{B_t} &= \frac{X_0^\pi}{B_0} + (\beta_t - \beta_0) + \int_0^t \left(\frac{S}{B}\right)_{u-} d\gamma_u + \int_0^t \gamma_u \, d\left(\frac{S}{B}\right)_u \\ &= \frac{X_0^\pi}{B_0} + \int_0^t d\beta_u + \int_0^t \left(\frac{S}{B}\right)_{u-} d\gamma_u + \int_0^t \gamma_u \, d\left(\frac{S}{B}\right)_u . \end{aligned} \tag{3.2}$$

The economic meaning of the last term in this formula is clear: it describes the dynamics of the discounted value as the discounted prices of the stocks change. It is therefore quite reasonable to distinguish those portfolios π such that

$$\frac{X_t^\pi}{B_t} = \frac{X_0^\pi}{B_0} + \int_0^t \gamma_u \, d\left(\frac{S}{B}\right)_u . \tag{3.3}$$

We call them *self-financing* and denote the corresponding set by SF.

With the discounted stock price process we associate the set $\mathcal{M}(X, \mathbf{P})$ of martingale measures (see § 2.3).

We say that a (B, S)-market admits an *arbitrage opportunity* (admits *arbitrage*) at a time $T > 0$ if there exists a $\pi \in SF$ such that ($\mathbf{P}$-a.s.) $X_0^\pi = 0$, $X_T^\pi \geqslant 0$, and $\mathbf{P}\{X_T^\pi > 0\} > 0$.

We remark that *if $\mathcal{M}(X, \mathbf{P}) \neq \varnothing$, then the (B, S)-market does not admit such an arbitrage opportunity.*

Indeed, let $\mathbf{P}^* \in \mathcal{M}(X, \mathbf{P})$ and let π^* be an arbitrage strategy. Then (3.3) implies that $X_t^{\pi^*}/B_t \in \mathcal{M}_{\mathrm{loc}}(\mathbf{P}^*)$, and hence $\mathbf{E}^*[X_t^{\pi^*}/B_t] = X_0^{\pi^*}/B_0 = 0$. It is clear that $\mathbf{P}^*(X_T^{\pi^*} > 0) = 0$, and since the measures $\mathbf{P}^*$ and $\mathbf{P}$ are equivalent, we get that $\mathbf{P}(X_T^{\pi^*} > 0) = 0$, which contradicts the assumption that π^* is an arbitrage strategy.

In forming a portfolio diverse payments are possible, called *consumption*. It is therefore natural to consider the class of *strategies π with consumption C* as a nonnegative nondecreasing process for which

$$X_t^\pi = X_0^\pi + \int_0^t \beta_u \, dB_u + \int_0^t \gamma_u \, dS_u - C_t, \tag{3.4}$$

or, equivalently,

$$\frac{X_t^\pi}{B_t} = \frac{X_0^\pi}{B_0} + \int_0^t \gamma_u \, dX_u - D_t, \tag{3.4$'$}$$

where $D_t = \int_0^t B_u^{-1} \, dC_u$.

Returning to the formulas (3.2), we see the meaning that can now be given to the sum of the second and third terms: the cumulative discounted consumption D_t.

The classes of strategies introduced admit a very useful martingale characterization.

THEOREM 3.1. *Suppose that $\mathcal{M}(X, \mathbf{P})$ consists of a single measure $\mathbf{P}^*$, and let Y be a nonnegative random process. Then*:

1) *Y is the discounted value of a self-financing strategy $\iff$ Y is a local martingale with respect to $\mathbf{P}^*$;*
2) *Y is the discounted value of a strategy with consumption $\iff$ Y is a supermartingale with respect to $\mathbf{P}^*$.*

In the first case the proof follows directly from the martingale representation theorem and (3.3), and in the second from the optional decomposition of supermartingales and (3.4)–(3.4$'$).

Investor activity on a (B, S)-market leads to the conclusion of agreements "delayed" over a time T into the future, based on contingent claims of the parties. By a *contingent claim* (with exercise date T) we understand any nonnegative $\mathcal{F}_T$-measurable random variable f. We remark that if such an agreement is an option and f represents the corresponding payments, then it is called a *European option*.

After concluding a financial transaction and accepting a contingent claim f, an economic agent tries to choose an attainable investment strategy π with the goal that by the expiration date of the claim the value X_T^π of his chosen strategy *hedges* f in the sense that

$$X_T^\pi \geqslant f \quad (\mathbf{P}\text{-a.s.}). \tag{3.5}$$

In this case π is called a *hedging strategy* (portfolio), or a *hedge*. If in some class of hedging strategies π there is a strategy π^* such that $X_t^{\pi^*} \leqslant X_t^\pi$ ($\mathbf{P}$-a.s.) for all $t \in [0,T]$, then π^* is called a *minimal* (in this class) *hedge*. The theorem below says that such hedges really do exist in the set SF.

In what follows let g denote the discounted value of the contingent claim f.

THEOREM 3.2. *Suppose that $\mathcal{M}(X,\mathbf{P})$ consists of a single measure $\mathbf{P}^*$, and let g be a discounted contingent claim such that*

$$\mathbf{E}^* g < \infty. \tag{3.6}$$

Then in the class of self-financing strategies there is a minimal hedge π^*, and its discounted value $X_t^{\pi^*}/B_t$ and "risky" component γ^* are determined by the formulas

$$\frac{X_t^{\pi^*}}{B_t} = \mathbf{E}^*(g \mid \mathcal{F}_t), \qquad t \in [0,T], \tag{3.7}$$

$$\frac{X_t^{\pi^*}}{B_t} = \frac{X_0^{\pi^*}}{B_0} + \int_0^t \gamma_u^* \, dX_u, \quad t \in [0,T]. \tag{3.8}$$

PROOF. In view of the condition (3.6) a (local) martingale $(Y_t^*)_{t \geqslant 0}$, $Y_t^* = \mathbf{E}^*(g \mid \mathcal{F}_t)$, is defined such that

$$Y_T^* = g \quad (\mathbf{P}^*\text{-a.s.}). \tag{3.9}$$

On the one hand, there then exists by the martingale representation theorem a predictable process γ^* such that

$$Y_t^* = Y_0^* + \int_0^t \gamma_u^* \, dX_u,$$

and on the other hand, Y^* is the discounted value of some self-financing strategy π^* in view of Theorem 3.1. It has thereby been established by virtue of (3.9) that π^* is a hedge satisfying the relations (3.7)–(3.8). To prove it is minimal we consider any other hedge $\pi \in SF$. By the martingale characterization, X_t^π/B_t is a nonnegative local martingale, and hence a supermartingale (with respect to $\mathbf{P}^*$). Therefore, we get from (3.5) that for all $t \in [0,T]$ (a.s.)

$$\frac{X_t^\pi}{B_t} \geqslant \mathbf{E}^*\left(\frac{X_T^\pi}{B_T} \,\Big|\, \mathcal{F}_t\right) \geqslant \mathbf{E}^*(g \mid \mathcal{F}_t) = Y_t^*.$$

We call a (B,S)-market *complete* if for any bounded discounted claim g there exists a self-financing hedge π^* such that $X_t^{\pi^*}/B_T = g$ (a.s.). In this case one also says that π^* *replicates* the claim, and it is said to be *attainable*.

THEOREM 3.3. *Let $\mathcal{M}(X,\mathbf{P}) \neq \varnothing$. Then the (B,S)-market is complete $\iff$ $\mathcal{M}(X,\mathbf{P})$ consists of a single measure $\mathbf{P}^*$.*

PROOF. The sufficiency follows from the preceding theorem. To prove necessity we assume that there exist two distinct martingale measures $\mathbf{P}_1^*, \mathbf{P}_2^* \in \mathcal{M}(X,\mathbf{P})$. Let $A \in \mathcal{F}_T$ be a set where they do not coincide, and consider the claim $g = I_A$ (that is, $f = I_A \cdot B_t$). The definition of completeness of a (B,S)-market implies the existence of a self-financing strategy π^* that replicates g: $Y_T^* = X_T^{\pi^*}/B_T = g = I_A$ ($\mathbf{P}$-a.s.).

Hence, $\mathbf{P}_1^*(A) = \mathbf{E}_1^* I_A = \mathbf{E}_1^* Y_T = \mathbf{E}_2^* Y = \mathbf{E}_2^* I_A = \mathbf{P}_2^*(A)$, and therefore $\mathbf{P}_1^*(A) = \mathbf{P}_2^*(A)$. This contradiction concludes the proof.

Summing up these theorems, we can say that in a complete (B, S)-market *perfect* hedging of any contingent claim is possible, in the sense that there exists a self-financing hedge with minimal value that replicates this claim. Perfect hedging enables us for any $t \in [0, T]$ to uniquely determine the (no-arbitrage, fair) *price of a contingent claim as the value of a minimal hedge* (at the indicated moment of time). This price is fair also because it is convenient both for the buyer of the option (in view of the absence of a cheaper strategy) and for the seller of the option (since at the given price he can "pay off" according to the accepted claim).

All the foregoing speaks of the necessity not only of establishing the nonemptiness itself of the class $\mathcal{M}(X, \mathbf{P})$, but also of being able to find martingale measures. In the next section we present a general method of looking for a (B, S)-market whose prices B_t and S_t satisfy linear stochastic equations in semimartingales. Most financial markets are modeled in just this way.

§ 3.2. A methodology for finding martingale measures and pricing contingent claims for different models of a (B, S)-market

Suppose that a (B, S)-market is determined by the two linear stochastic equations

$$B_t = B_0 + \int_0^t B_{u-}\, dh_u, \quad \Delta h_t > -1,$$
$$S_t = S_0 + \int_0^t S_{u-}\, dH_u, \quad \Delta H_u > -1, \tag{3.10}$$

where h and H are given semimartingales. Then by § 2.3,

$$B_t = B_0\, \mathcal{E}_t(h),$$
$$S_t = S_0\, \mathcal{E}_t(H). \tag{3.11}$$

Our problem here is to find conditions (in terms of the original semimartingales h and H) under which a measure $\mathbf{P}^*$ equivalent to $\mathbf{P}$ takes the process $X = S/B$ into a local martingale

$$\mathbf{P}^* \in \mathcal{M}(X, \mathbf{P}) \iff X = \frac{S}{B} \in \mathcal{M}_{\text{loc}}(\mathbf{P}^*). \tag{3.12}$$

Let us first determine when the original measure $\mathbf{P}$ is a martingale measure on the market (3.10)–(3.11).

According to the properties of stochastic exponentials (see § 2.3), we have

(3.13)
$$\begin{aligned} X_t &= X_0\, \mathcal{E}_t(H)\mathcal{E}_t^{-1}(h) = X_0\, \mathcal{E}_t(H)\mathcal{E}_t(-h^*) \\ &= X_0\, \mathcal{E}_t\biggl(H - h + \langle h^c, h^c\rangle + \sum \frac{(\Delta h)^2}{1+\Delta h} - \langle H^c, h^c\rangle - \sum \frac{\Delta H \Delta h}{1+\Delta h}\biggr) \\ &= X_0\, \mathcal{E}_t\biggl(H - h + \langle h^c, h^c - H^c\rangle + \sum \frac{\Delta h(\Delta h - \Delta H)}{1+\Delta h}\biggr). \end{aligned}$$

We use the notation

$$\Psi_t(h,H) = H_t - h_t + \langle h^c, h^c - H^c\rangle + \sum_{s\leqslant t} \frac{\Delta h_s(\Delta h_s - \Delta H_s)}{1+\Delta h_s}$$

and rewrite (3.13) in the form of the stochastic equation

$$X_t = X_0 + \int_0^t X_{s-}\, d\Psi_s(h,H).$$

Consequently, to verify that $X \in \mathcal{M}_{\text{loc}}(\mathbf{P})$, we must establish that this class contains the process $\Psi(h,H)$.

It is now clear how to find conditions under which the relation (3.12) we need is true. To this end let $Z_t^* = d\mathbf{P}_t^*/d\mathbf{P}_t$ be the local density of the equivalent measures $\mathbf{P}^*$ and $\mathbf{P}$. Then by the generalization (2.21) of Girsanov's theorem,

$$X \in \mathcal{M}_{\text{loc}}(\mathbf{P}^*) \iff ZX \in \mathcal{M}_{\text{loc}}(\mathbf{P}). \tag{3.14}$$

Next, by (2.20),

$$Z_t = \mathcal{E}_t(N), \tag{3.15}$$

where $N_t = \int_0^t Z_{s-}^{-1}\, dZ_s \in \mathcal{M}_{\text{loc}}(\mathbf{P})$. Therefore, we get from (3.14)–(3.15) that

$$X \in \mathcal{M}_{\text{loc}}(\mathbf{P}^*) \iff X\,\mathcal{E}(N) \in \mathcal{M}_{\text{loc}}(\mathbf{P}). \tag{3.16}$$

In view of (3.13)

$$X_t\,\mathcal{E}(N) = X_0\,\mathcal{E}(\Psi(h,H))\,\mathcal{E}_t(N),$$

and by the rules for multiplying stochastic exponents,

$$X_t\,\mathcal{E}_t(N) = X_0\,\mathcal{E}_t(\Psi(h,H,N)), \tag{3.17}$$

where

$$\begin{aligned}\Psi_t(h,H,N) = H_t - h_t + N_t &+ \langle (h-N)^c, (h-H)^c\rangle_t \\ &+ \sum_{s\leqslant t} \frac{(\Delta h_s - \Delta N_s)(\Delta h_s - \Delta H_s)}{1+\Delta h_s}.\end{aligned}$$

It follows from (3.17) that to verify (3.16) we must determine when

$$\Psi(h,H,N) \in \mathcal{M}_{\text{loc}}(\mathbf{P}). \tag{3.18}$$

Let us apply this method to some well-known models of a (B,S)-market.

The Black–Scholes model.

$$\begin{aligned} dB_t &= rB_t\,dt, \quad B_0 = 1, \\ dS_t &= S_t(\mu\,dt + \sigma\,dw_t), \quad S_0 > 0, \end{aligned} \tag{3.19}$$

where $r, \mu \in \mathbb{R}^1_+$, $0 < \sigma$ (the volatility), and $w = (w_t)_{t\geqslant 0}$ is a standard Wiener process generating the filtration $\mathbb{F}$.

In the model (3.19), $h_t = rt$ and $H_t = \mu t + \sigma w_t$, and since w_t is the sole source of randomness in the prices S_t, it is perfectly natural to look for the local martingale N_t in the form $N_t = \varphi w_t$. Using this, we get from (3.17) that

$$\Psi_t(h,H,N) = \mu t + \sigma w_t - rt + \varphi w_t + \varphi\sigma t = (\mu - r + \varphi\sigma)t + (\sigma+\varphi)w_t.$$

Hence, the condition (3.18) will be satisfied (and $\Psi_t(h, H, N)$ will even be a *martingale*) if φ satisfies the equation

$$\mu - r + \varphi\sigma = 0, \quad \text{or} \quad \varphi = -\frac{\mu - r}{\sigma}. \tag{3.20}$$

As a result we get from (3.15) and (3.20) the density of the corresponding martingale measure $\mathbf{P}^*$:

$$Z_t^* = \frac{d\mathbf{P}_t^*}{d\mathbf{P}_t} = \mathcal{E}_t(N) = \exp\biggl\{-\frac{\mu - r}{\sigma}\, w_t - \frac{1}{2}\Bigl(\frac{\mu - r}{\sigma}\Bigr)^2 t\biggr\}, \tag{3.21}$$

and, as we already know from Girsanov's theorem, the process $w_t^* = w_t + \Bigl(\frac{\mu - r}{\sigma}\Bigr) t$ will be a Wiener process with respect to the new measure $\mathbf{P}^*$. It is also clear in view of the uniqueness of the solution of (3.20) that $\mathbf{P}^*$ is the unique martingale measure, and the resulting (B, S)-market (3.19) is complete.

We consider the contingent claim $f = (S_T - K)^+$ corresponding to an option to buy (the asset S at the price K). According to Theorem 3.3, there exists a minimal hedge $\pi^* = (\beta^*, \gamma^*)$ determined by the formulas (3.7) and (3.8), which in this case we should specify more concretely.

Let us first calculate the initial value of the indicated hedge. To this end we let $\varphi(x) = \frac{1}{\sqrt{2\pi}} \exp\{-x^2/2\}$ denote the density of the normal distribution $\Phi(x) = \int_{-\infty}^x \varphi(y)\, dy$, let $\alpha = \Bigl(\frac{\mu - r}{\sigma}\Bigr)\sqrt{\pi}$, and calculate $\mathbf{E}^* e^{-rT}(S_T - K)^+$, making the obvious transformations ($x = z - \alpha$ and so on):

(3.22)

$$\begin{aligned}
\mathbf{E}^* e^{-rT}(S_T - K)^+ &= e^{-rT}\mathbf{E} Z_T^* \bigl(S_0 \cdot e^{(\mu - \frac{\sigma^2}{2})T + \sigma w_T} - K\bigr)^+ \\
&= e^{-rT}\mathbf{E} e^{-(\frac{\mu - r}{\sigma}) w_T - \frac{1}{2}(\frac{\mu - r}{\sigma})^2 T} \bigl(S_0 \cdot e^{(\mu - \frac{\sigma^2}{2})T + \sigma w_T} - K\bigr)^+ \\
&= e^{-rT} \int_{-\infty}^{\infty} e^{-(\frac{\mu - r}{\sigma})\sqrt{T} - \frac{T}{2}(\frac{\mu - r}{\sigma})^2} \\
&\qquad\qquad \times \bigl(S_0 \cdot e^{(\mu - \frac{\sigma^2}{2})T + \sigma x\sqrt{T}} - K\bigr)^+ \varphi(x)\, dx \\
&= e^{-rT} \int_{-\infty}^{\infty} e^{-(\frac{\mu - r}{\sigma})\sqrt{T} - \frac{T}{2}(\frac{\mu - r}{\sigma})^2} \\
&\qquad\qquad \times \bigl(S_0 \cdot e^{(r - \frac{\sigma^2}{2})T + \sigma\sqrt{T}(x + (\frac{\mu - r}{\sigma})\sqrt{T})} - K\bigr)^+ \varphi(x)\, dx \\
&= e^{-rT}(2\pi)^{-\frac{1}{2}} \int_{-\infty}^{\infty} e^{-\frac{\alpha^2}{2} - z\alpha + \alpha - \frac{z^2}{2} + z\alpha - \frac{\alpha^2}{2}} \\
&\qquad\qquad \times \bigl(S_0 \cdot e^{(r - \frac{\sigma^2}{2})T + \sigma\sqrt{T} z} - K\bigr)^+ dz \\
&= e^{-rT}(2\pi)^{-\frac{1}{2}} \int_{\{z : (r - \frac{\sigma^2}{2})T + \sigma\sqrt{T} z > \ln \frac{K}{S_0}\}} e^{-\frac{z^2}{2}} \\
&\qquad\qquad \times \bigl(S_0 \cdot e^{(r - \frac{\sigma^2}{2})T + \sigma\sqrt{T} z} - K\bigr)\, dz \\
&= e^{-rT}(2\pi)^{-\frac{1}{2}} \int_{\{z : z > z_0 = (\ln \frac{K}{S_0} + T(\frac{\sigma^2}{2} - r))/\sigma\sqrt{T}\}} S_0\, e^{-\frac{z^2 - 2\sigma\sqrt{T} z + \sigma^2 T}{2}} e^{rT}\, dz \\
&\quad - K e^{-rT}(1 - \Phi(z_0))
\end{aligned}$$

$$= S_0(2\pi)^{-\frac{1}{2}} \int_{\{z: z > z_0\}} e^{\frac{(z-\sigma\sqrt{T})^2}{2}} \, dz - Ke^{-rT}\Phi(-z_0)$$
$$= S_0\big(1 - \Phi(z_0 - \sigma\sqrt{T})\big) - Ke^{-rT}\Phi(-z_0).$$

Introducing the new notation

$$d_{\pm}(T-t) = \frac{\ln\frac{S_t}{K} + (T-t)(r \pm \frac{\sigma^2}{2})}{\sigma\sqrt{T-t}},$$

we rewrite (3.22) in the form of the formula

$$\mathbb{C}^{BS}(T, S_0, K, r, \sigma) = S_0\Phi(d_+(T)) - Ke^{-rT}\Phi(d_-(T)), \tag{3.23}$$

which is commonly called the *Black–Scholes formula.*

Using the independence and the Gaussian property of the increments of a Wiener process, we can repeat the very same procedure for the conditional expectation $\mathbf{E}^*\big(e^{-rT}(S_T - K)^+ \,|\, \mathcal{F}_t\big)$, which represents the value of the minimal hedge at an arbitrary time $t \leqslant T$, and obtain

(3.24)
$$\begin{aligned}\mathbb{C}^{BS}(T-t, S_t, K, r, \sigma) &= \mathbf{E}^*\big(e^{-rT}(S_T - K)^+ \,|\, \mathcal{F}_t\big)\\ &= S_t\Phi(d_+(T-t)) - Ke^{-r(T-t)}\Phi(d_-(T-t)).\end{aligned}$$

It is clear that $\mathbb{C}^{BS}$ is a smooth function of S and t, and therefore application of the Kolmogorov–Itô formula to $\mathbb{C}^{BS}(T-t, S_t)/B_t$ leads to the relation

(3.25)
$$\begin{aligned}\frac{\mathbb{C}^{BS}(S_t)}{B_t} &= \frac{\mathbb{C}^{BS}(S_0)}{B_0} + \int_0^t \frac{\partial\mathbb{C}^{BS}(S_u)}{\partial x}\, d\Big(\frac{S_u}{B_u}\Big)\\ &\quad + \int_0^t B_u^{-1}\Big[\frac{\partial\mathbb{C}^{BS}}{\partial u} + rS_u\frac{\partial\mathbb{C}^{BS}}{\partial x} + \frac{\sigma^2}{2}\,S_u^2\,\frac{\partial^2\mathbb{C}^{BS}}{\partial x^2} - r\mathbb{C}^{BS}\Big]\, du.\end{aligned}$$

On the other hand, (3.8) holds for the discounted value $\mathbb{C}^{BS}$ of the minimal hedge, and hence we get from (3.25) that

$$\gamma_t^* = \frac{\partial\mathbb{C}^{BS}}{\partial x}(S_t) \tag{3.26}$$

and

$$\frac{\partial\mathbb{C}^{BS}}{\partial t}(x,t) + rS_u\frac{\partial\mathbb{C}^{BS}}{\partial x}(x,t) + \frac{\sigma^2}{2}\,S_u^2\,\frac{\partial^2\mathbb{C}^{BS}}{\partial x^2}(x,t) - r\mathbb{C}^{BS}(x,t) = 0. \tag{3.27}$$

The equation (3.27) is called the *Black–Scholes equation.*

To find the second component π^* it is necessary to use the *balance equation* $X_t^{\pi^*} = \beta_t^* B_t + \gamma_t^* S_t$ and get

$$\beta_t^* = \frac{X_t^{\pi^*} - \gamma_t^* S_t}{B_t}. \tag{3.28}$$

Thus, *in the framework of the Black–Scholes model* (3.19) *the components of the minimal hedge* $\pi^* = (\beta^*, \gamma^*)$ *can be found from the formulas* (3.26) *and* (3.28), *and its value is determined by the formula* (3.24) *and satisfies the equation* (3.27). In particular, the initial value (3.22) of this hedge is called the *fair price* (Black–Scholes price) *for a European option to buy* (call option).

As for the dual option, the *option to sell* (put option) with contingent claim $(K - S_T)^+$, its price $\mathbb{P}^{BS}(T)$ and $\mathbb{C}^{BS}(T)$ are connected by the following useful "parity" relation:

$$\mathbb{P}^{BS}(T) = \mathbb{C}^{BS}(T) - S_0 + Ke^{-rT}. \tag{3.29}$$

The Black–Scholes formula enables us to price contingent claims f that are *smooth functions* of S_T: $f = f(S_T)$.

To do this we use integration by parts and write the following Taylor expansion of the function f: for $x > 0$

$$f(x) = f(0) + xf'(0) + \int_0^\infty (x-y)^+ f''(y)\,dy.$$

We now substitute in this formula $x = S_T$ from (3.19), divide both sides by e^{rT}, and average with respect to $\mathbf{P}^*$. As a result we get the following elegant formula for the initial price $\mathbb{C}_f^{BS}(T)$ of such a contingent claim:

$$\mathbb{C}_f^{BS}(T) = e^{-rT} + S_0 f'(0) + \int_0^\infty \mathbb{C}^{BS}(T,y) f''(y)\,dy. \tag{3.30}$$

We remark that it is possible to arrive at a similar formula also for other models of a financial market. For example, in the Merton model below just replace $\mathbb{C}^{BS}$ in (3.30) by $\mathbb{C}^M$.

The Merton model.

$$\begin{aligned} dB_t &= rB_t\,dt, \quad B_0 = 1,\\ dS_t &= S_{t-}(\mu\,dt - \nu\,d\Pi_t), \end{aligned} \tag{3.31}$$

where $\mu, r \in \mathbb{R}^1_+$, $\nu < 1$, and $\Pi = (\Pi_t)_{t\geqslant 0}$ is a Poisson process with intensity $\lambda > 0$ generating the filtration $\mathbb{F}$.

In the model (3.31), $h_t = rt$ and $H_t = \mu t - \nu\Pi_t$, and the martingale N_t will be found in the form $N_t = \psi(\Pi_t - \lambda t)$. Further,

$$\begin{aligned} \Psi_t(h,H,N) &= \mu t - \nu\Pi_t - rt + \psi(\Pi_t - \lambda t) - \psi\nu\Pi_t \\ &= (\mu - r - \nu\lambda - \psi\nu\lambda)t - \nu(\Pi_t - \lambda t) + \psi(\Pi_t - \lambda t) - \psi\nu(\Pi_t - \lambda t), \end{aligned}$$

and hence the condition (3.18) is satisfied for $\mu - r - \nu\lambda - \psi\nu\lambda = 0$, or

$$\psi = \frac{\mu - r}{\nu\lambda} - 1. \tag{3.32}$$

We note at once the *uniqueness* of the solution of (3.32), and hence the uniqueness of the martingale measure $\mathbf{P}^*$ and the completeness of the (B,S)-market (3.31). Using (3.15) and (3.32), we get that

$$Z_t^* = \frac{d\mathbf{P}_t^*}{d\mathbf{P}_t} = \mathcal{E}_t(N) = \exp\bigl\{(\lambda - \lambda^*)t + (\ln\lambda^* - \ln\lambda)\Pi_t\bigr\}, \tag{3.33}$$

where $\lambda^* = (\mu - r)/\nu$ is the intensity of Π_t with respect to the new measure $\mathbf{P}^*$.

We use the formula (3.33) found for the density to calculate the initial value of the minimal hedge for the contingent claim $f = (S_T - K)^+$.

According to Theorem 3.3, this initial value is

$$\mathbb{C}^M(T) = \mathbf{E}^* e^{-rT}(S_T - K)^+. \tag{3.34}$$

It is clear that

$$(3.35)\qquad \left(\frac{S}{B}\right)_T = \left(\frac{S_0}{B_0}\right)\exp\{-\nu\Pi_T + \nu\lambda^* T\}\prod_{t\leqslant T}(1-\nu\Delta\Pi_t)e^{\nu\Delta\Pi_t} = \left(\frac{S_0}{B_0}\right)\exp\{\Pi_T\ln(1-\nu) + \nu\lambda^* T\}.$$

It follows from (3.34)–(3.35) (and the Poisson property of Π_t with respect to $\mathbf{P}^*$) that

$$(3.36)\qquad \mathbb{C}^M(T) = B_0\,\mathbf{E}^* B_T^{-1}(S_T-K)^+ = B_0\mathbf{E}^*\left(\frac{S_T}{B_T}-\frac{K}{B_T}\right)^+ = B_0\sum_{n=0}^{\infty} e^{-\lambda^* T}\frac{(\lambda^* T)^n}{n!}\left(\frac{S_0}{B_0}e^{n\ln(1-\nu)+\nu\lambda^* T} - B_T^{-1}K\right)^+.$$

Using the notation

$$n_0 = \inf\left\{n : \frac{S_0}{B_0}e^{n\ln(1-\nu)+\nu\lambda^* T} \geqslant B_T^{-1}K\right\} = \left[\frac{\ln\frac{K}{S_0} - \mu T}{\ln(1-\nu)}\right],$$

$$\Psi(x,y) = \sum_{n=x}^{\infty} e^{-y}\frac{y^n}{n!},$$

we get from (3.36) that

$$(3.37)\qquad \mathbb{C}^M(T) = S_0\sum_{n=n_0}^{\infty} e^{-\lambda^* T+\lambda^*\nu T}e^{n\ln(1-\nu)}\frac{(\lambda^* T)^n}{n!} - Ke^{-rT}\Psi(n_0,\lambda^* T) = S_0\sum_{n=n_0}^{\infty} e^{\lambda^*(\nu-1)T}\frac{(\lambda^*(1-\nu)T)^n}{n!} - Ke^{-rT}\Psi(n_0,\lambda^* T) = S_0\Psi(n_0,\lambda^*(1-\nu)T) - Ke^{-rT}\Psi(n_0,\lambda^* T).$$

The formula (3.37) is called *Merton's formula* for a fair price of an option to buy in the market model (3.31).

The jump-diffusion model of a market.

$$(3.38)\qquad \begin{aligned} dB_t &= rB_t\,dt, \quad B_0 = 1, \\ dS_t^i &= S_{t-}^i(\mu^i\,dt + \sigma^i\,dw_t - \nu^i\,d\Pi_t), \quad S_0^i > 0, \ i = 1,2, \end{aligned}$$

where $\mu^i, r \in \mathbb{R}_+^1$, $\sigma^i > 0$, $\nu^i < 1$, $w = (w_t)_{t\geqslant 0}$ is a Wiener process, $\Pi = (\Pi_t)_{t\geqslant 0}$ is a Poisson process (with intensity λ), and w and Π are mutually independent and generate the filtration $\mathbb{F}$.

As in the models (3.19) and (3.31), we calculate the function $\Psi_t(h,H,N)$, and then we look for the martingale N_t in the form $N_t = \varphi w_t + \psi(\Pi_t - \lambda t)$: for $i = 1,2$

$$\begin{aligned} \Psi_t(h,H,N) &= \mu^i t + \sigma^i w_t - \nu^i\Pi_t - rt + \varphi\sigma^i t - \nu^i\psi\Pi_t \\ &= (\mu^i - r - \nu^i\lambda + \varphi\sigma^i - \psi\nu^i\lambda)t + \text{a martingale}. \end{aligned}$$

Hence, φ and ψ satisfy the two equations ($i = 1,2$):

$$(3.39)\qquad \mu^i - r - \nu^i\lambda + \varphi\sigma^i - \psi\nu^i\lambda = 0,$$

or, with the notation $\lambda^* = \lambda(1+\psi)$,

$$\mu^i - r + \varphi\sigma^i - \nu^i\lambda^* = 0. \tag{3.39$'$}$$

Assuming that $\sigma^2\nu^1 - \sigma^1\nu^2 \neq 0$, we write out the (unique) solution of (3.39′) (and hence of (3.39)):

$$\varphi = \frac{(\mu^2 - r)\nu^1 - (\mu^1 - r)\nu^2}{\sigma^2\nu^1 - \sigma^1\nu^2},$$
$$\lambda^* = \frac{(\mu^1 - r)\sigma^2 - (\mu^2 - r)\sigma^1}{\sigma^2\nu^1 - \sigma^1\nu^2}.$$

The density of the original martingale measure $\mathbf{P}^*$ can be uniquely determined from this:

$$\begin{aligned} Z_t^* = \frac{d\mathbf{P}_t^*}{d\mathbf{P}_t} &= \mathcal{E}_t(N) \\ &= \exp\left\{\varphi w_t - \frac{\varphi^2}{2}t + \psi(\Pi_t - \lambda t)\right\}\prod_{s\leqslant t}(1+\psi\Delta\Pi_s)e^{-\psi\Delta\Pi_s} \\ &= \exp\left\{\varphi w_t - \frac{\varphi^2}{2}t + (\lambda - \lambda^*)t + (\ln\lambda^* - \ln\lambda)\Pi_t\right\}. \end{aligned} \tag{3.40}$$

It is clear that with respect to $\mathbf{P}^*$ the process $w_t^* = w_t - \varphi t$ is a Wiener process, while Π is a Poisson process with intensity λ^*.

We remark that the consideration of a (B,S)-market with a *single* risky asset leads (in the search for a martingale measure) only to the *single equation* (3.39) *with two unknowns*, and hence to the possibility of finding a *whole set* of martingale measures, and together with this to a very clear example of an *incomplete market*.

We now show how to price the contingent claim $f = (S_T^1 - K)^+$ on a complete market (3.38) with two risky assets. To this end we first rewrite the discounted price S^1 in the exponential form

(3.41)
$$\begin{aligned} \frac{S_T^1}{B_T} &= \frac{S_0^1}{B_0}\exp\left\{\sigma^1 w_t^* - \nu^1\Pi_T + \nu^1\lambda^* T - \frac{(\sigma^1)^2}{2}T\right\}\prod_{t\leqslant T}(1-\nu^1\Delta\Pi_t)e^{\nu^1\Delta\Pi_t} \\ &= S_0^1\exp\left\{\left(\nu^1\lambda^* - \frac{(\sigma^1)^2}{2}\right)T + \sigma^1 w_T^* - \nu^1\Pi_T\right\}e^{\Pi_T\ln(1-\nu^1)+\nu^1\Pi_T} \\ &= S_0^1\exp\left\{\left(\nu^1\lambda^* - \frac{(\sigma^1)^2}{2}\right)T + \sigma^1 w_T^* + \Pi_T\ln(1-\nu^1)\right\}. \end{aligned}$$

By Theorem 3.3, the (initial) price $\mathbb{C}$ of this claim is in view of (3.41) equal to

(3.42)
$$\begin{aligned} \mathbb{C} &= \mathbf{E}^* e^{-rT}(S_T^1 - K)^+ = \mathbf{E}^* B_T^{-1}(S_T^1 - K)^+ = \mathbf{E}^*\left(\left(\frac{S^1}{B}\right)_T - \frac{K}{B_T}\right)^+ \\ &= \mathbf{E}^*\left(S_0^1 e^{(\nu^1\lambda^* - (\sigma^1)^2/2)T + \sigma^1 w_T^* + \Pi_T\ln(1-\nu^1)} - B_T^{-1}K\right)^+. \end{aligned}$$

Further, using the independence of the Wiener process w^* and the Poisson process Π with respect to $\mathbf{P}^*$, we get from (3.42) that

(3.43)

$$\begin{aligned}
\mathbb{C} &= \mathbf{E}^*\Big(\mathbf{E}^*\big[\big(S_0^1 e^{\left(\nu^1\lambda^* - \frac{(\sigma^1)^2}{2}\right)T + \sigma^1 w_T^* + \Pi_T \ln(1-\nu^1)} - B_T^{-1}K\big)^+ \,|\, \Pi_T\big]\Big) \\
&= \sum_{n=0}^{\infty} \mathbf{E}^*\Big[\big(S_0^1 e^{\left(\nu^1\lambda^* - \frac{(\sigma^1)^2}{2}\right)T + \sigma^1 w_T^* + \Pi_T \ln(1-\nu^1)} - B_T^{-1}K\big)^+ \,|\, \Pi_T = n\Big] \\
&\qquad \times e^{-\lambda^* T}\frac{(\lambda^* T)^n}{n!} \\
&= \mathbf{E}^* \sum_{n=0}^{\infty} \big(S_0^1 (1-\nu^1)^n e^{\left(\nu^1\lambda^* - \frac{(\sigma^1)^2}{2}\right)T + \sigma^1 w_T^*} - B_T^{-1}K\big)^+ e^{-\lambda^* T}\frac{(\lambda^* T)^n}{n!} \\
&= e^{-\lambda^* T}\sum_{n=0}^{\infty}\Big[\frac{(\lambda^* T)^n}{n!}\mathbf{E}^*\big(S_0^1 (1-\nu^1)^n e^{\left(\nu^1\lambda^* - \frac{(\sigma^1)^2}{2}\right)T + \sigma^1 w_T^*} - B_T^{-1}K\big)^+\Big] \\
&= e^{-\lambda^* T}\sum_{n=0}^{\infty}\frac{(\lambda^* T)^n}{n!}\mathbb{C}^{BS}\big(T, S_0^1(1-\nu^1)^n e^{\nu^1\lambda^* T}, r, \sigma^1, K\big),
\end{aligned}$$

where we recall that $\mathbb{C}^{BS}$ is the price determined by the Black–Scholes formula (3.23).

Thus, the price of the contingent claim $(S_T^1 - K)^+$ on this jump-diffusion market is the *Poisson weighting* (3.43) of the *Black–Scholes formula*.

The Cox–Ross–Rubinstein model.

This is a model of a purely discrete market, often called the *binomial* model because the prices of the risky asset $(S_n)_{n=0,1,\dots}$ represent a *binomial random walk*:

$$\begin{aligned}
\Delta B_n &= B_n - B_{n-1} = rB_{n-1}, \quad B_0 > 0, \\
\Delta S_n &= S_n - S_{n-1} = \rho_n S_{n-1}, \quad S_0 > 0,
\end{aligned} \tag{3.44}$$

where $(\rho_n)_{n=1,\dots}$ is a sequence of independent random variables taking the two values $a < b$ with probabilities p and $q = 1 - p$.

It is assumed that the filtration $\mathbb{F}$ is generated by (ρ_n) and that $b > r > a > -1$.

Setting $B_t \equiv B_n$ on $[n, n+1)$ (and doing the same with $S_t, \mathcal{F}_t, \dots$), we "imbed" the discrete model (3.44) *in the standard way* in a general semimartingale model of a (B, S)-market, and hence we can use all the results we need.

In this case $\Delta h_n = r$, $\Delta H_n = \rho_n$, and $\Delta N_n = \psi_n(\rho_n - m)$, $m = \mathbf{E}\rho_n$.

Hence,

$$(1 + r)\Delta\Psi_n(h, H, N) = \delta H_n - \Delta h_n + \Delta N_n + \Delta N_n \Delta H_n,$$

and thus by the martingale property of $\Psi_n(h, H, N)$

$$\mathbf{E}\big(\rho_n - r + \psi_n(\rho_n - m) + \varphi_n(\rho_n - m)\rho_n \,|\, \mathcal{F}_{n-1}\big) = 0.$$

From this, $\psi_n = -\dfrac{m - r}{\sigma^2}$, where $\sigma^2 = \mathbf{E}\rho^2 - m^2 = \mathbf{D}\rho_n$. As a result, the corresponding density of the martingale measure $\mathbf{P}^*$ is

$$Z_n^* = \mathcal{E}_n\Big(-\frac{m - r}{\sigma^2}\sum_{k \leqslant n}(\rho_k - m)\Big). \tag{3.45}$$

We find the initial (fair) price of the claim $f = (S_T - K)^+$. By Theorem 3.3, it is

$$\mathbb{C}_N^{CRR} = \mathbf{E}^*(1+r)^{-N}(S_N - K)^+ = (1+r)^{-N}\mathbf{E}^*(S_N - K)^+. \tag{3.46}$$

By (3.45),

$$\begin{aligned}\mathbf{E}^*(S_N - K)^+ &= \mathbf{E}Z_N^*(S_N - K)^+ \\ &= \mathbf{E}\mathcal{E}_N\Big(-\frac{m-r}{\sigma^2}\sum(\rho_k - m)\Big)(S_N - K)\cdot I_{\{S_N>K\}}.\end{aligned} \tag{3.47}$$

Let $k_0 = \min\{k \leqslant N : S_0(1+b)^k(1+a)^{N-k} > K\}$. We observe that $k_0 \ln(1+b) + (N-k_0)\ln(1+a) > \ln\dfrac{K}{S_0}$, and hence

$$k_0 = \left[\ln\frac{K}{S_0(1+a)^N}\Big/\ln\frac{1+b}{1+a}\right] + 1.$$

We let $p^* = \dfrac{r-a}{b-a}$ and use the binomial property and the relations $m = p(b-a)+a$ and $\sigma^2 = (b-a)^2p(1-p)$ to find that in (3.47)

$$\begin{aligned}&\mathbf{E}\mathcal{E}_N\Big(-\frac{m-r}{\sigma^2}\sum(\rho_k - m)\Big)K\cdot I_{\{S_N>K\}} \\ &= K\sum_{k=k_0}^N\binom{N}{k}\Big(1-\frac{m-r}{\sigma^2}(b-m)\Big)^k p^k\Big(1-\frac{m-r}{\sigma^2}(a-m)\Big)^{N-k}(1-p)^{N-k} \\ &= K\sum_{k=k_0}^N\binom{N}{k}\Big(\frac{p^*}{p}\Big)^k p^k\Big(\frac{1-p^*}{1-p}\Big)^{N-k}(1-p)^{N-k} \\ &= K\sum_{k=k_0}^N\binom{N}{k}(p^*)^k(1-p^*)^{N-k}.\end{aligned} \tag{3.48}$$

Using the representation $S_N = S_0\mathcal{E}_N(\sum\rho_k)$, we get an expression for the first term in (3.47) from the rules for multiplying stochastic exponentials:

$$\begin{aligned}&\mathbf{E}\mathcal{E}_N\Big(-\frac{m-r}{\sigma^2}\sum(\rho_k - m)\Big)S_N\cdot I_{\{S_N>K\}} \\ &= S_0\mathbf{E}\mathcal{E}_N\Big(-\frac{m-r}{\sigma^2}\sum(\rho_k - m) - \frac{m-r}{\sigma^2}\sum(\rho_k - m)\rho_k + \sum\rho_k\Big)I_{\{S_N>K\}} \\ &= S_0\sum_{k=k_0}^N\binom{N}{k}\Big(1-\frac{m-r}{\sigma^2}(b-m) + b - \frac{m-r}{\sigma^2}(b-m)b\Big)^k p^k \\ &\quad\times\Big(1-\frac{m-r}{\sigma^2}(a-m)+a-\frac{m-r}{\sigma^2}(a-m)a\Big)^{N-k}(1-p)^{N-k} \\ &= S_0\sum_{k=k_0}^N\binom{N}{k}\Big(\frac{p^*}{p}(1+b)\Big)^k p^k\Big(\frac{1-p^*}{1-p}(1+a)\Big)^{N-k}(1-p)^{N-k}\end{aligned} \tag{3.49}$$

$$= S_0 \sum_{k=k_0}^{N} \binom{N}{k} \big(p^*(1+b)\big)^k \big((1-p^*)(1+a)\big)^{N-k}$$

$$= S_0 \sum_{k=k_0}^{N} \binom{N}{k} \left(p^* \frac{1+b}{1+r}\right)^k \left((1-p^*)\frac{1+a}{1+r}\right)^{N-k} (1+r)^{-N}.$$

Introducing the additional notation

$$\widetilde{p} = \frac{1+b}{1+r} p^* \quad \text{and} \quad \mathbb{B}(j, N, p) = \sum_{k=j}^{N} \binom{N}{k} p^k (1-p)^{N-k},$$

we get the following expression for $\mathbb{C}_N^{CRR}$ from (3.46)–(3.49):

$$\mathbb{C}_N^{CRR} = S_0 \mathbb{B}(k_0, N, \widetilde{p}) - K(1+r)^{-N} \mathbb{B}(k_0, N, p^*). \tag{3.50}$$

This is called the *Cox–Ross–Rubinstein formula* for the fair price of a European option to buy in the binomial model (3.44) of a (B, S)-market.

§3.3. A methodology for optimal investment and its applications

In essence, all activity of economic agents on a financial market is directed toward *optimal* investment of the capital they have at their disposal. One natural criterion for such optimality can be the strategy of "minimal" hedging. In this section, optimality of an investment is connected with the concept of *utility*.

A *utility function* is defined to be a concave nondecreasing continuously differentiable function $U\colon \mathbb{R}^1_+ \to \mathbb{R}^1$ such that

$$\begin{aligned} U'(0+) &= \lim_{x \downarrow 0} U'(x) = \infty, \\ U'(\infty) &= \lim_{x \uparrow \infty} U'(x) = 0. \end{aligned} \tag{3.51}$$

We introduce the conjugate function

$$V(y) = \sup_{x>0} [U(x) - xy], \qquad y > 0,$$

of U, which is the *Legendre transform* of the function $-U(-x)$. Under the above conditions on U, the function V is strictly decreasing, convex, and continuously differentiable, and it satisfies the conditions

$$\begin{gathered} V'(0) = -\infty, \qquad V'(\infty) = 0, \\ V(0) = U(\infty), \quad V(\infty) = U(0), \\ U(x) = \inf_{y>0} [V(y) + xy], \quad x > 0. \end{gathered}$$

Furthermore,

$$I(x) = (U'(x))^{-1} = -V'(x).$$

On the (B, S)-market introduced in §3.1 we must find a self-financing strategy π^* such that for the given utility function U

$$\mathbf{E}\, U\big(X_T^{\pi^*}(x)\big) = \sup_{\pi \in SF} \mathbf{E}\, U\big(X_T^{\pi}(x)\big) = u(x). \tag{3.52}$$

The function $u(x)$ will be called the *price function* of the optimization problem (3.52).

We established earlier a one-to-one correspondence between the class of self-financing strategies and the class of positive processes of the form

$$Y_t(x) = x + \int_0^t \gamma_s \, dX_s \tag{3.53}$$

that are the discounted values of the strategies in SF.

Let $\mathcal{X}(x)$ denote the class of all such processes.

In view of the above remarks, the original problem (3.52) has been transformed into the problem of finding in the class $\mathcal{X}(x)$ a process $Y^*(x)$ such that

$$\mathbf{E}\, U\big(Y_T^*(x)\big) = \sup_{Y \in \mathcal{X}} \mathbf{E}\, U\big(Y_T(x)\big). \tag{3.52$'$}$$

Denoting by $\mathbf{P}^*$ the unique martingale measure, we let $Z_t^* = d\mathbf{P}_t^*/d\mathbf{P}_t$ and define the differentiable function

$$v(y) = \mathbf{E}\, V(yZ_T^*).$$

We also introduce the notation $y_0 = \inf\{y : v(y) < \infty\}$ and $x_0 = \lim_{y \to y_0}(-v'(y))$.

THEOREM 3.4. *Suppose that a (B, S)-market is complete. Then the following assertions hold.*

1) *The function $u(x) < \infty$ is continuously differentiable for all $x > 0$ and is strictly concave on $(0, x_0)$. The function $v(y) < \infty$ is continuously differentiable and is strictly convex on (y, ∞) for sufficiently large y. Further, u and v are conjugate in the sense that*

$$v(y) = \sup_{x>0}(u(x) - xy), \quad y > 0,$$

$$u(x) = \inf_{y>0}(v(y) + xy), \quad x > 0,$$

$$\textit{and} \quad u'(0) = \infty, \quad v'(\infty) = 0.$$

2) *If $y = u'(x)$ in the domain $x < x_0$ and $y < y_0$, then the optimal terminal value for (3.52′) is an integrable $\mathcal{F}_T$-measurable random variable and is determined by the equality*

$$Y_T^*(x) = I(yZ_T^*). \tag{3.54}$$

We avoid the technical complexity of the proof of this theorem and prove only the *optimality* of $Y_T^*(x)$, which is what is most important for us. Let $Y(x) \in \mathcal{X}(x)$ be an arbitrary process. This is a positive local martingale and hence a supermartingale with respect to $\mathbf{P}^*$. Therefore, $\mathbf{E}^* Y_T(x) \leqslant \mathbf{E}^* Y_0(x) = x$.

Next, in view of the properties of the function U and (3.54) we have

$$\begin{aligned}
\mathbf{E}\, U(Y_T(x)) &= \mathbf{E}\big[U(Y_T^*(x)) + \big(U(Y_T(x)) - U(Y_T^*(x))\big)\big] \\
&\leqslant \mathbf{E}\, U(Y_T^*(x)) + \mathbf{E}\, U'(Y_T^*(x))\big(Y_T(x) - Y_T^*(x)\big) \\
&= \mathbf{E}\, U(Y_T^*(x)) + \mathbf{E}^*(Z_T^*)^{-1} U'(Y_T^*(x))\big(Y_T(x) - Y_T^*(x)\big) \\
&= \mathbf{E}\, U(Y_T^*(x)) + y\, \mathbf{E}^*\big(Y_T(x) - Y_T^*(x)\big) \\
&= \mathbf{E}\, U(Y_T^*(x)) + y\big(\mathbf{E}^* Y_T(x) - x\big) \leqslant \mathbf{E}\, U(Y_T^*(x)).
\end{aligned}$$

We now apply this method to the solution of the investment problem (3.52) for the classical Black–Scholes, Merton, and Cox–Ross–Rubinstein models with the utility function $U(x) = \ln x$.

In this case $I(x) = \dfrac{1}{x}$, and hence the numbers $x > 0$ and $y > 0$ in Theorem 3.4 are connected by the relation

$$y = U'(x) = \frac{1}{x}. \tag{3.55}$$

Since $\dfrac{\partial}{\partial x}(\ln x - xy) = 0$ for $x = \dfrac{1}{y}$, $y > 0$, it follows that

$$V(y) = \sup_{x>0}(\ln x - xy) = \ln\frac{1}{y} - 1 = -\ln y - 1. \tag{3.56}$$

The Black–Scholes model.

Using the form of the density Z_T^* in (3.21), we find from (3.55)–(3.56) that

$$\begin{aligned} v(y) = \mathbf{E}V(yZ_T^*) &= -\mathbf{E}\ln(yZ_T^*) - 1 = -\ln y - \mathbf{E}\ln Z_T^* - 1 \\ &= -\ln y - 1 + \frac{1}{2}\left(\frac{\mu - r}{\sigma}\right)^2 T. \end{aligned}$$

Since $\dfrac{\partial}{\partial y}(v(y) + xy) = 0$ for $y = \dfrac{1}{x}$, it is easy to calculate the *price function* $u(x)$ of the problem (3.52)–(3.52′):

$$\begin{aligned} u(x) = \inf_{y>0}(v(y) + xy) &= \ln x - 1 + \frac{1}{2}\left(\frac{\mu - r}{\sigma^2}\right)T + 1 \\ &= \ln x + \frac{1}{2}\left(\frac{\mu - r}{\sigma^2}\right)T. \end{aligned} \tag{3.57}$$

Further, by (3.53) and (3.21), the discounted optimal terminal value $X_T^*(x)/B_T$ is equal to

$$\begin{aligned} \frac{X_T^*(x)}{B_T} = I(yZ_T^*) &= \frac{1}{yZ_T^*} = \frac{x}{Z_T^*} \\ &= x\exp\left\{\frac{\mu - r}{\sigma}w_T + \frac{1}{2}\left(\frac{\mu - r}{\sigma}\right)^2 T\right\}. \end{aligned} \tag{3.58}$$

We introduce the concept of the *proportion* α_t^* of risky part in the value $X_t^*(x)$:

$$\alpha_t^* = \frac{\gamma_t^* S_t}{X_t^*(x)}, \tag{3.59}$$

where γ^* is the number of stocks in an optimal portfolio $\pi^* = (\beta^*, \gamma^*)$. We shall find $\alpha_t^* \equiv \alpha^*$.

Using the fact that π^* is self-financing, we get that

$$\begin{aligned} dX_t^*(x) &= \beta_t^*\,dB_t + \gamma_t^*\,dS_t \\ &= \beta_t^* rB_t\,dt + \gamma_t^* S_t(\mu\,dt - r\,dt + r\,dt + \sigma\,dw_t) \\ &= (\beta_t^* B_t + \gamma_t^* S_t)r\,dt + \gamma_t^* S_t\big((\mu - r)\,dt + \sigma\,dw_t\big) \\ &= X_t^* r\,dt + \alpha_t^* X_t^*\big((\mu - r)\,dt + \sigma\,dw_t\big). \end{aligned}$$

This leads to the following expression for the discounted value:

$$\frac{X_T^*(x)}{B_T} = x \exp\left\{ \sigma\alpha^* w_T + \alpha^*(\mu - r)T - \frac{1}{2}\sigma^2(\alpha^*)^2 T \right\}.$$

Comparison of the last relation with (3.58) leads to a corresponding formula for α^*:

$$\alpha^* = \frac{\mu - r}{\sigma^2}\,. \tag{3.60}$$

We note that from (3.59)–(3.60) and the balance equation $X_t^*(x) = \beta_t^* B_t + \gamma_t^* S_t$ we can uniquely reconstruct the portfolio π^*:

$$\gamma_t^* = \frac{\alpha^* X_t^*(x)}{S_t}\,, \qquad \beta_t^* = \frac{X_t^*(x) - \gamma_t^* S_t}{B_t^*}\,. \tag{3.61}$$

Thus, *in the problem of maximizing the mean logarithmic utility the price function, the optimal value, and the strategy are given by the formulas* (3.57), (3.58), *and* (3.60)–(3.61) *in the Black–Scholes model.*

The Merton model.

For the *Merton model* (3.31) we use (3.55)–(3.56) and the form of the density Z^* in (3.33) to get that

$$\begin{aligned} v(y) = \mathbf{E}V(yZ_T^*) &= -\ln y - 1 - \mathbf{E}\ln Z_T^* \\ &= -\ln y - 1 - (\lambda - \lambda^*)T - \mathbf{E}(\ln\lambda^* - \ln\lambda)\Pi_T \\ &= -\ln y - 1 - (\lambda - \lambda^*)T - (\ln\lambda^* - \ln\lambda)\lambda T, \end{aligned} \tag{3.62}$$

where $\lambda^* = \dfrac{\mu - r}{\nu}$.

Further, from (3.62) we find the price function of the optimization problem as in the case of (3.57):

$$\begin{aligned} u(x) &= \inf_{y>0}(v(y) + xy) \\ &= -\ln y - 1 - (\lambda - \lambda^*)T - (\ln\lambda^* - \ln\lambda)\lambda T + xy \\ &= \ln x - (\lambda - \lambda^*)T - (\ln\lambda^* - \ln\lambda)\lambda T \\ &= \ln x + \frac{\mu - r}{\nu}T - \lambda T \ln\frac{\mu - r}{\nu\lambda}\,. \end{aligned} \tag{3.63}$$

From (3.54) we get that the discounted optimal value is given by

$$\begin{aligned} \frac{X_T^*(x)}{B_T} = I(yZ_T^*) = \frac{x}{Z_T^*} &= x\exp\{-(\lambda - \lambda^*)T - (\ln\lambda^* - \ln\lambda)\Pi_T\} \\ &= x\exp\left\{\left(\frac{\mu - r}{\nu} - \lambda\right)T - \Pi_T \ln\frac{\mu - r}{\nu\lambda}\right\}. \end{aligned} \tag{3.64}$$

As in the Black–Scholes model, we see using the proportion $\alpha_t^* = \dfrac{\gamma_t^*}{S_{t-}} X_{t-}$ that

$$dX_t^*(x) = rX_{t-}^*\,dt + \alpha_t^* X_{t-}^*\big((\mu - r)\,dt \quad \nu\,d\Pi_t\big).$$

We are looking for $\alpha_t^* = \alpha^*$, and thus from the preceding equation we get that

$$\frac{X_T^*(x)}{B_T} = x\exp\{\alpha^*(\mu - r)T + (1 - \alpha^*\nu)\Pi_T\}. \tag{3.65}$$

Equating (3.64) and (3.65), we arrive at the following expression for the optimal proportion:

$$\alpha^* = \frac{\mu - r - \lambda}{\nu(\mu - r)}. \tag{3.66}$$

Thus, *in the Merton model the investment problem* (3.52)–(3.52′) *with logarithmic utility function admits an optimal solution with price function* (3.63), *optimal value* (3.64), *and optimal proportion* (3.66).

The Cox–Ross–Rubinstein model.

In the framework of the Cox–Ross–Rubinstein model (3.44) the same ideas lead to the following results.

Using the form of the martingale density Z_N^* in (3.45), we find the function

$$\begin{aligned} v(y) &= \mathbf{E}V(yZ_N^*) = -\mathbf{E}\ln(yZ_N^*) - 1 \\ &= -\ln y - 1 - \sum_{k=1}^{N} \mathbf{E}\ln\Big(1 - \frac{m-r}{\sigma^2}(\rho_k - m)\Big) \\ &= -\ln y - 1 - \sum_{k=1}^{N}\Big[p\ln\Big(1 - \frac{m-r}{\sigma^2}(b - m)\Big) \\ &\qquad\qquad + (1-p)\ln\Big(1 - \frac{m-r}{\sigma^2}(a - m)\Big)\Big] \\ &= -\ln y - 1 - N\Big[p\ln\frac{p^*}{p} + (1-p)\ln\frac{1-p^*}{1-p}\Big], \end{aligned}$$

and hence the price function is

$$u(x) = \ln x - \mathbf{E}\ln Z_N^* = \ln x - N\Big[p\ln\frac{p^*}{p} + (1-p)\ln\frac{1-p^*}{1-p}\Big]. \tag{3.67}$$

From (3.53) we get the discounted optimal value

$$\frac{X_N^*(x)}{B_N} = \frac{x}{Z_N^*} = x\,\mathcal{E}_N^{-1}\Big(-\frac{m-r}{\sigma^2}\sum_{k=1}^{N}(\rho_k - m)\Big). \tag{3.68}$$

Again we introduce the proportion of risky value in the portfolio $\pi^* = (\beta^*, \gamma^*)$:

$$\alpha_n^* = \frac{\gamma_n^* S_{n-1}}{X_{n-1}^*}.$$

Since it is self-financing,

$$\Delta\frac{X_n^*}{(1+r)^n} = \frac{\alpha_n^*}{(1+r)^n} X_{n-1}^*(\rho_n - r),$$

and hence

$$\frac{X_N^*(x)}{B_N} = x\,\mathcal{E}_N\Big(\sum_{k=1}^{N}\frac{\alpha_k^*}{1+r}(\rho_k - r)\Big). \tag{3.69}$$

Equating (3.68) and (3.69) gives an equation for finding α_n^*:

$$\mathcal{E}_N\left(-\frac{m-r}{\sigma^2}\sum_{k=1}^{N}(\rho_k-m)\right)\mathcal{E}_N\left(\sum_{k=1}^{N}\frac{\alpha_k^*}{1+r}(\rho_k-r)\right)=1. \tag{3.70}$$

We show that here, as in the preceding two models, the proportion is a constant $\alpha_n^* = \alpha^*$. The method used is mathematical induction. For $N = 1$ the equation (3.70) can be taken to the form

$$\left(1-\frac{m-r}{\sigma^2}(\rho_1-m)\right)\left(1+\frac{\alpha_1^*}{1+r}(\rho_1-r)\right)=1. \tag{3.70$'$}$$

Rewriting (3.70′) on the set $\{\rho_1 = b\}$, we get that

$$\left(1-\frac{m-r}{\sigma^2}(b-m)\right)\left(1+\frac{\alpha_1^*}{1+r}(b-r)\right)=1,$$

or

$$\frac{p^*}{p}\left(1+\frac{\alpha_1^*}{1+r}(b-r)\right)=1.$$

Since $p^* = \dfrac{m-a}{b-a}$, this gives us that

$$\alpha_1^* = \frac{(1+r)(m-r)}{(r-a)(b-r)} = \alpha^*. \tag{3.71}$$

Consideration of the set $\{\rho_1 = a\}$ leads to the same formula for α_1^*.

By the induction hypothesis, all the factors in (3.70) satisfy

$$\left(1-\frac{m-r}{\sigma^2}(\rho_k-m)\right)\left(1+\frac{\alpha^*}{(1+r)^k}(\rho_k-r)\right)=1,$$

$k = 1, \ldots, N-1$. Hence,

$$\left(1-\frac{m-r}{\sigma^2}(\rho_N-m)\right)\left(1+\frac{\alpha^*}{(1+r)^N}(\rho_N-r)\right)=1,$$

and we arrive again at the solved equation (3.70′).

As a result, *the investment problem* (3.52)–(3.52′) *for the Cox–Ross–Rubinstein model with logarithmic utility function admits a solution with price function* (3.67), *terminal value* (3.68), *and optimal proportion* (3.71).

References for Chapter 3

[**1**], [**13**], [**15**], [**16**], [**21**], [**22**], [**27**], [**33**], [**34**], [**44**], [**46**], [**51**], [**72**], [**73**], [**79**], [**83**], [**86**], [**87**], [**90**], [**93**], [**95**], [**100**], [**106**], [**108**], [**113**], [**115**]–[**120**], [**126**], [**141**]–[**143**], [**152**], [**158**].

CHAPTER 4

Hedging and Incomplete Markets

A general description is given of incomplete markets, when calculations must be carried out working in a whole class of martingale measures. Here the theory of minimal hedging "overflows" into the theory of minimal superhedging, when a minimal hedge turns out in general to be a strategy with consumption. The Black–Scholes model with stochastic volatility is studied at length as a very representative model of an incomplete market. It is shown that controlled diffusion processes and Bellman equations make up a suitable and efficient technique for financial calculations in such models. Methods for estimating volatility are presented.

§ 4.1. A methodology for superhedging

While keeping to the description of a (B, S)-market with the class of strategies and contingent claims given in § 3.1, we assume that the set $\mathcal{M}(X, \mathbf{P})$ of martingale measures *consists of more than one measure*. Such markets are said to be *incomplete*. But how are the statements about the martingale characterization of self-financing strategies and strategies with consumption transformed in this more complicated situation, and also the theory of hedging contingent claims on the whole? The following two theorems are devoted to a reflection of exactly these changes.

THEOREM 4.1 (Martingale characterization of strategies). *Assume that a (B, S)-market is incomplete and Y is a positive random process. Then:*

1) *Y is the discounted value of a strategy $\pi \in SF \iff Y$ is a local martingale with respect to any measure $\widetilde{\mathbf{P}} \in \mathcal{M}(X, \mathbf{P})$.*
2) *Y is the discounted value of a strategy with consumption $\iff Y$ is a supermartingale with respect to any measure $\widetilde{\mathbf{P}} \in \mathcal{M}(X, \mathbf{P})$.*

To prove this we first observe that (3.3), (3.4), and (3.4′) do not depend at all on the *number* of martingale measures. Hence, using these relations in the first and second cases in Theorem 4.1, respectively, we must use the martingale and optional decompositions with respect to the family $\mathcal{M}(X, \mathbf{P})$, and this concludes the proof.

THEOREM 4.2 (Existence and structure of a minimal hedge). *Suppose that a (B, S)-market is incomplete, and let g be a discounted contingent claim (for f with exercise date T) such that*

$$\sup_{\widetilde{\mathbf{P}} \in \mathcal{M}(X, \mathbf{P})} \widetilde{\mathbf{E}} g_T < \infty. \tag{4.1}$$

Then in the class of strategies with consumption there is a minimal hedge π^* *with discounted value* Y^*, *risky component* γ^*, *and discounted consumption* D^* *determined by the formulas*

$$\frac{X_t^{\pi^*}}{B_t} = Y_t^* = \operatorname*{ess\,sup}_{\widetilde{\mathbf{P}} \in \mathcal{M}(X,\mathbf{P})} \widetilde{\mathbf{E}}(g_T \mid \mathcal{F}_t), \tag{4.2}$$

$$\frac{X_t^{\pi^*}}{B_t} = \frac{X_0^{\pi^*}}{B_0} + \int_0^t \gamma_u^* \, dX_u - D_t^*. \tag{4.3}$$

PROOF. By (2.25), the condition (4.1) ensures the existence and the supermartingale property of the process on the right-hand side of (4.2). The optional decomposition (2.23) makes it possible to write the equality (4.3) for this process, and Theorem 4.1 makes it possible to characterize this as the value of the strategy π^* with consumption D^*. It remains to prove its *minimality* in the class of strategies with consumption hedging g. For any such strategy (π, D) its discounted value is a nonnegative supermartingale with respect to each measure $\widetilde{\mathbf{P}} \in \mathcal{M}(X,\mathbf{P})$, and hence

$$\frac{X_t^{\pi}}{B_t} \geqslant \widetilde{\mathbf{E}}\Big(\frac{X_T^{\pi}}{B_T} \Bigm| \mathcal{F}_t\Big) \geqslant \widetilde{\mathbf{E}}(g \mid \mathcal{F}_t) \quad \text{(a.s.)}$$

for $t \leqslant T$ and any $\widetilde{\mathbf{P}} \in \mathcal{M}(X,\mathbf{P})$. Therefore,

$$\frac{X_t^{\pi}}{B_t} \geqslant \operatorname*{ess\,sup}_{\widetilde{\mathbf{P}} \in \mathcal{M}(X,\mathbf{P})} \widetilde{\mathbf{E}}(g \mid \mathcal{F}_t) = \frac{X_t^{\pi^*}}{B_t} \quad \text{(a.s.)},$$

which concludes the proof.

A number of conclusions can be drawn from Theorem 4.2

In incomplete markets it is natural to replace perfect hedging of a claim f by *superhedging*, which is based on strategies with consumption. Then the natural initial price of f is the "upper" price

$$\mathbb{C}^{\sup}(T) = B_t \operatorname*{ess\,sup}_{\widetilde{\mathbf{P}} \in \mathcal{M}(X,\mathbf{P})} \widetilde{\mathbf{E}}\Big(\frac{f}{B_T} \Bigm| \mathcal{F}_t\Big). \tag{4.4}$$

It should be noted that in the case of incomplete markets *certain* contingent claims (but not all) can have a replicating strategy that reproduces this claim at the terminal moment of time.

As an example we consider the $(n+1)$-dimensional Black–Scholes model consisting of

- a nonrisky asset growing in time according to an interest rate $r > 0$:

$$dS_t^0 = rS_t^0 \, dt,$$

- n risky assets whose evolution is described by the stochastic differential equations

$$dS_t^i = S_t^i \Big(\mu^i \, dt + \sum_{j=1}^d \sigma_{ij} \, dW_t^j\Big), \qquad 1 \leqslant i \leqslant n.$$

Here $\vec{W} = (W_t^j)_{t\geqslant 0,\,1\leqslant j\leqslant d}$ is a d-dimensional Wiener process on the probability space $(\Omega, \mathbb{F}, \mathcal{F}_t, \mathbf{P})$. It is assumed that $\det(\sigma_{ij}) \neq 0$.

In the case $n = d$ Girsanov's theorem gives us the existence of a unique martingale measure with respect to which the processes $(e^{-rt}S_t^i)_{t\geqslant 0,\,1\leqslant i\leqslant d}$ of discounted prices of risky assets are martingales, and hence the market is complete. In the case $d > n$ a martingale measure is not unique, and the market is correspondingly incomplete.

Let f be an $\mathcal{F}_T$-measurable random variable corresponding to a payment according to a contingent claim at a fixed time $T > 0$. We point out a certain general fact of separate interest.

Let $X = (X_t)_{t\geqslant 0}$ be a continuous semimartingale with respect to the filtration $\{\mathcal{F}_t\}_{t\geqslant 0}$, and let $\{\mathcal{F}_t^X\}_{t\geqslant 0}$ be the canonical filtration for X. Suppose that there is at least one measure $\widehat{\mathbf{P}} \equiv \mathbf{P}$ such that with respect to $\widehat{\mathbf{P}}$ we have:

1) X is a special semimartingale admitting the decomposition
$$X_t = X_0 + \widehat{M}_t + \widehat{A}_t,$$
where $\widehat{M}_t$ is a local martingale and $\widehat{A}_t$ a predictable process of bounded variation;

2) X is a Markov process with respect to $\{\mathcal{F}_t\}_{t\geqslant 0}$ (and hence also with respect to $\{\mathcal{F}_t^X\}_{t\geqslant 0}$).

Let us consider the $\widehat{\mathbf{P}}$-martingale $N_t = \widehat{\mathbf{E}}\big[f(X_T)\,|\,\mathcal{F}_t^X\big]$. Then by the Markov property 2),
$$N_t = \widehat{\mathbf{E}}\big[f(X_T)\,|\,\mathcal{F}_t^X\big] = \widehat{\mathbf{E}}\big[f(X_T)\,|\,X_t\big] = v(t,x), \qquad x = X_t.$$
Suppose that $v(t,x) \in C^{1,2}([0,\infty), \mathbb{R}_+)$. Then the process N admits the decomposition
$$N_t = \widehat{\mathbf{E}}\big[f(X_T)\,|\,X_0\big] + \int_0^t \frac{\partial v(s,X_s)}{\partial x}\,d\widehat{M}_s.$$
To prove this we must apply the Kolmogorov–Itô formula to $v(t,X_t)$ and use the properties 1)–2) of the process X.

In particular,
$$f(X_T) = \widehat{\mathbf{E}}\big[f(X_T)\,|\,X_0\big] + \int_0^T \frac{\partial v(s,X_s)}{\partial x}\,d\widehat{M}_s$$
for $t = T$.

We now consider the discounted contingent claim $\widetilde{f} = e^{-rT}f(\widetilde{S}_T)$, where $\widetilde{S}_T = e^{-rT}S_T$. Then with respect to the measure $\widehat{\mathbf{P}}$ we have the representation
$$\widetilde{f}(\widetilde{S}_T) = \widehat{\mathbf{E}}\big[\widetilde{f}(\widetilde{S}_T)\,|\,S_0\big] + \int_0^T e^{-ru}\frac{\partial v(u,\widetilde{S}_u)}{\partial \widetilde{S}}\,d\widetilde{S}_u,$$
where $v(u,\widetilde{S}_u) = e^{ru}\,\widehat{\mathbf{E}}\big[\widetilde{f}(\widetilde{S}_T)\,|\,\mathcal{F}_u\big]$ and $\dfrac{\partial v}{\partial \widetilde{S}} = \left(\dfrac{\partial v}{\partial \widetilde{S}^1}, \dots, \dfrac{\partial v}{\partial \widetilde{S}^n}\right)$.

Since $\dfrac{\partial v}{\partial \widetilde{S}} = e^{ru}\dfrac{\partial v}{\partial S}$, we have
$$\widehat{\mathbf{E}}\big[\widetilde{f}(\widetilde{S}_T)\,|\,\mathcal{F}_t\big] = \widehat{\mathbf{E}}\big[\widetilde{f}(\widetilde{S}_T)\,|\,S_0\big] + \int_0^t \frac{\partial v}{\partial S}(u, e^{-ru}S_u)\,d\widetilde{S}_u,$$

and hence

$$\begin{aligned} v(t,\widetilde{S}_t) &= e^{rt}\,\widehat{\mathbf{E}}\big[\widetilde{f}(\widetilde{S}_T)\,|\,\mathcal{F}_t\big] \\ &= \widehat{\mathbf{E}}\big[\widetilde{f}(\widetilde{S}_T)\,|\,S_0\big] + \int_0^t e^{ru}\frac{\partial v}{\partial S}(u,e^{-ru}S_u)\,d\widetilde{S}_u \\ &\quad + r\int_0^t v(u,e^{-ru}S_u)\,du \\ &= v(0,S_0) + \int_0^t \frac{\partial v}{\partial S}(u,e^{-ru}S_u)\,dS_u \\ &\quad + r\int_0^t \Big(v(u,e^{-ru}S_u) - \frac{\partial v}{\partial S}(u,e^{-ru}S_u)\Big)\,du. \end{aligned}$$

It follows from this representation that a strategy containing

- $\gamma_t = \dfrac{\partial v}{\partial S}(t,e^{-rt}S_t)$ shares of stock,
- $\beta_t = e^{-rt}\Big(v(t,e^{-rt}S_t) - \dfrac{\partial v}{\partial S}(t,e^{-rt}S_t)S_t\Big)$ in the nonrisky asset

at each moment of time t with $0 \leqslant t \leqslant T$ will be a hedging strategy for $v(t,X_t)$.

The cost of the contingent claim $\widetilde{f}(\widetilde{S}_T)$ is uniquely determined and is equal to $\widehat{\mathbf{E}}\big[\widetilde{f}(\widetilde{S}_T)\,|\,S_0\big]$, and the value of the hedging portfolio at a time t with $0 \leqslant t \leqslant T$ is $e^{rt}\widehat{\mathbf{E}}\big[\widetilde{f}(\widetilde{S}_T)\,|\,\mathcal{F}_t\big]$.

The Black–Scholes model with stochastic volatility, a comprehensive and meaningful model of an incomplete market, is considered in the next section.

§ 4.2. The Black–Scholes model with stochastic volatility

In the Black–Scholes model it was assumed that the stock price is subject to the stochastic differential equation

$$\frac{dS_t}{S_t} = \mu\,dt + \sigma\,dW_t,$$

that is, the relative change in the stock price is a Gaussian process with stationary independent increments. It follows from the form of the Black–Scholes formula that the basic (unknown) parameter for calculating the price of an option is the variability parameter σ: the *volatility* of the stock, which was assumed to be constant.

However, numerous investigations have shown that the volatility is not a constant quantity. Moreover, the assumption of constant volatility dissipates with the so-called "smile" effect of volatility, which amounts to the volatility determined from the prices of the options being traded having a tendency to grow as the exercise price of the option increases or decreases from a definite value (for which one says that the option is an "in-the-money option").

In this connection it is natural to consider generalizations of the Black–Scholes model when the volatility is random.

EXAMPLE 4.1: *Volatility depending on the price of a stock.* One such model for describing the evolution of a stock price is the following:

$$\frac{dS_t}{S_t} = \mu\,dt + \sigma(S_t)\,dW_t,$$

where the volatility is a function of the stock price, $\sigma = \sigma(S_t)$.

It is interesting to note that the model $\sigma(x) = \sigma x^{-(1-\alpha)}$ (constant elasticity of variance model, CEV) turns out to be complete, and the price of a contingent claim is uniquely determined in it.

EXAMPLE 4.2: *Volatility as a solution of a stochastic differential equation.* Let the price of the stock $S = (S_t)_{t\geqslant 0}$ and the volatility $\sigma = (\sigma_t)_{t\geqslant 0}$ be determined by the system of stochastic differential equations

$$\begin{aligned} dS_t &= S_t(\mu\, dt + \sigma_t\, dW_t^1), \\ dv_t &= v_t(\gamma\, dt + \delta\, dW_t^2), \end{aligned}$$

where $v_t = \sigma_t^2$, and $W^1 = (W_t^1)_{t\geqslant 0}$ and $W^2 = (W_t^2)_{t\geqslant 0}$ are two Wiener processes with covariance $dW_t^1 dW_t^2 = \rho\, dt$, $|\rho| < 1$.

If there is an asset on the market (for example, an option to buy) for which the only source of randomness is the volatility (one says that volatility is a tradable asset), then the model with two risky assets (that is, the stock and the option) is complete.

In the general case, because of the presence on the market of two sources of randomness (W^1 and W^2), the model is incomplete, and the price of a contingent claim is not uniquely determined.

EXAMPLE 4.3: *Volatility as an autoregression process of conditional inhomogeneity.* Autoregression models with conditional inhomogeneity (ARCH, GARCH, HARCH, ...) serve for constructing time series with nonconstant volatility in discrete time. In the case of the simplest GARCH(1, 1) model, the logarithm $\ln S_t$ of the stock price and the conditional variance σ_t satisfy the equations

$$\begin{aligned} \ln S_t &= \ln S_{t-1} + \mu_t + \sigma_t \varepsilon_t, \\ \sigma_t^2 &= \omega + \beta\sigma_{t-1}^2 + \alpha\varepsilon_{t-1}^2\sigma_{t-1}^2, \end{aligned}$$

where ε_t, $t \in \mathbb{Z}_+$, are independent identically distributed random variables with $\mathbf{E}\varepsilon_t = 0$ and $\mathbf{E}\varepsilon_t^2 = 1$. In the general case of GARCH models, σ_t is a certain function of the "past" values of the conditional variance and ε_t.

Models of GARCH type are more adequate than the lognormal Black–Scholes model in describing the distribution of stock prices observed in practice, and they enable one to "grasp" the so-called "heavy tail effect" of these distributions. Their basic deficiencies include the difficulty of constructing hedging portfolios for contingent claims.

We proceed to a consideration of the pricing of contingent claims on an incomplete market representing the following generalization of the Black–Scholes model. Suppose that the evolution of capital in a bank account B and the dynamics of a stock price S are described by the stochastic differential equations

$$\begin{aligned} dB_t &= B_t R_t\, dt, & B_0 > 0, \\ dS_t &= S_t(\mu\, dt + \Sigma_t\, dW_t), & S_0 > 0. \end{aligned} \tag{4.5}$$

Here, as before, $W = (W_t)_{0\leqslant t\leqslant T}$ is a Wiener process, but unlike the classical model the processes $R = (R_t)_{0\leqslant t\leqslant T}$ and $\Sigma = (\Sigma_t)_{0\leqslant t\leqslant T}$ describing the behavior of the interest rate and the volatility are assumed to be *random*. In the special case $R_t \equiv r = \text{const}$ and $\Sigma_t \equiv \sigma = \text{const}$ we arrive at the classical Black–Scholes model. We show below that the model (4.5) is not complete as a rule.

For a random process X defined on the stochastic base $(\Omega, \mathcal{F}_T, \mathbb{F}, \mathbf{P})$ we denote by $\mathcal{M}(X, \mathbf{P})$ the family of martingale measures of X. The method of constructing a minimal hedge for arbitrary incomplete models of financial markets is based on Theorems 4.1 and 4.2.

Let $\mathcal{M} = \mathcal{M}(S/B, \mathbf{P})$ denote the family of martingale measures of the process S/B. We write

$$W_t^* = W_t + \int_0^t \frac{\mu - R_u}{\Sigma_u}\, du, \qquad 0 \leqslant t \leqslant T. \tag{4.6}$$

By (4.5), the equation for the price of the stock can be written in the form

$$dS_t = S_t R_t\, dt + S_t \Sigma_t\, dW_t^*, \tag{4.7}$$

and hence

$$d\frac{S_t}{B_t} = \frac{S_t}{B_t} \Sigma_t\, dW_t^*.$$

Thus, an equivalent measure $\mathbf{Q}$ belongs to $\mathcal{M}$ if and only if W^* is a local martingale with respect to $\mathbf{Q}$ (moreover, W^* is a Wiener process with respect to $\mathbf{Q}$).

In view of Girsanov's theorem the density process of an arbitrary $\mathbf{Q} \in \mathcal{M}$ with respect to $\mathbf{P}$ has the form

$$Z_t^{\mathbf{Q}} = \exp\biggl\{\int_0^t \biggl(\frac{R_u - \mu}{\Sigma_u}\, dW_u - \frac{(R_u - \mu)^2}{2\Sigma_u^2}\, du\biggr)\biggr\} N_t^{\mathbf{Q}}, \quad 0 \leqslant t \leqslant T, \tag{4.8}$$

where $N^{\mathbf{Q}}$ is a local martingale orthogonal to W with respect to $\mathbf{P}$, that is, the quadratic covariance $\langle N^{\mathbf{Q}}, W\rangle$ is zero.

Suppose that the family of martingale measures of the process S/B is not empty, which is equivalent to the absence of arbitrage opportunities in the model. This assumption holds if, for example, the local martingale

$$Z_t^* = \exp\biggl\{\int_0^t \biggl(\frac{R_u - \mu}{\Sigma_u}\, dW_u - \frac{(R_u - \mu)^2}{2\Sigma_u^2}\, du\biggr)\biggr\}, \quad 0 \leqslant t \leqslant T, \tag{4.9}$$

is a uniformly integrable martingale, and hence is the density process for some martingale measure. A sufficient condition for the uniform integrability of Z^* is the Novikov condition

$$\mathbf{E} \exp\biggl\{\frac{1}{2}\int_0^T \biggl(\frac{R_u - \mu}{\Sigma_u}\biggr)^2 du\biggr\} < +\infty. \tag{4.10}$$

We consider a European option with exercise time T and payment function $f_T \geqslant 0$ that is an $\mathcal{F}_T$-measurable random variable. The structure of a minimal hedge for a model with random volatility and random interest rate is given by the theorem below, which is convenient for the subsequent consideration of concretizations of Theorem 4.2.

THEOREM 4.3. *Assume that $\mathcal{M}(S/B, \mathbf{P}) \neq \varnothing$. The set of hedging strategies is nonempty if and only if*

$$\sup_{\mathbf{Q} \in \mathcal{M}} \mathbf{E}^{\mathbf{Q}} f_T e^{-\int_0^T R_u du} < +\infty.$$

In this case a minimal hedging strategy exists, and its value at the time t is

$$\widehat{V}_t = \operatorname*{ess\,sup}_{\mathbf{Q}\in\mathcal{M}} \mathbf{E}^{\mathbf{Q}}\big[f_T e^{-\int_t^T R_u\,du} \,|\, \mathcal{F}_t\big]. \tag{4.11}$$

Here the structure of a minimal hedge (the number γ of shares and the consumption C) is determined from the optional decomposition

$$\widehat{V}_t e^{-\int_0^t R_u\,du} = \widehat{V}_0 + \int_0^t \gamma_u\, d\big(S_u e^{-\int_0^u R_u\,du}\big) - \int_0^t e^{-\int_0^u R_u\,du}\, dC_u. \tag{4.12}$$

Let us consider a contingent claim of the classical form

$$f_T = g(S_T),$$

where $g \geqslant 0$ is a Borel function of polynomial growth "at infinity". We study the structure of an optimal hedge described by the formulas (4.11) and (4.12) by using the example of a model in which the interest rate is constant,

$$R_t = r = \text{const},$$

while the volatility satisfies an equation of "telegraph signal" type:

$$\Sigma_t^2 = \sigma^2 + (-1)^{\Pi_t}\Delta\sigma^2, \tag{4.13}$$

where $0 \leqslant \Delta\sigma^2 < \sigma^2$ is a positive constant describing the upper and lower limits of the volatility, and Π is a Poisson process with intensity λ.

Setting

$$\sigma_{\min} = \sqrt{\sigma^2 - \Delta\sigma^2}, \qquad \sigma_{\max} = \sqrt{\sigma^2 + \Delta\sigma^2},$$

we can write the equation (4.13) in the form

$$d\Sigma_t = (\sigma_{\max} - \sigma_{\min})\big(I_{\{\Sigma_{t-}=\sigma_{\min}\}} - I_{\{\Sigma_{t-}=\sigma_{\max}\}}\big)\, d\Pi_t, \quad \Sigma_0 = \sigma_{\max}.$$

From Theorem 4.3 it is clear that the value of a minimal hedge should be sought in the form

$$\widehat{V}_t = \widehat{v}(S_t, t) \tag{4.14}$$

with the number of shares in an optimal hedge equal to

$$\widehat{\gamma}_t = \frac{\partial \widehat{v}}{\partial x}(S_t, t), \tag{4.15}$$

where the function $\widehat{v}$ is the price in the optimal control problem

$$\widehat{v}(x,t) = e^{-r(T-t)} \sup_{\alpha} \mathbf{E} g(S_{T-t}^{(\alpha)}). \tag{4.16}$$

Here $S^{(\alpha)}$ is a *controlled* diffusion process described by the stochastic differential equation

$$dS_u^{(\alpha)} = rS_u^{(\alpha)}\, du + S_u^{(\alpha)}\alpha_u\, dW_u, \qquad S_0^{(\alpha)} = x,$$

where the *controlling* process $\alpha = (\alpha_t)_{0\leqslant t\leqslant T}$ takes its values in the two-point set $\{\sigma_{\min}, \sigma_{\max}\}$.

As follows from the general theory of controlled diffusion processes, the price function $\widehat{v}(x,t)$ belongs to $C^{2,1}(\mathbb{R}_+, [0,T))$, and among the functions $v = v(x,t) \in$

$C^{2,1}(\mathbb{R}_+, [0,T))$ satisfying a polynomial growth condition there is a unique solution of the Bellman equation

$$\begin{cases} \dfrac{\partial \widehat{v}}{\partial t} + \mathfrak{L}^0 \widehat{v} - r\widehat{v} + \dfrac{1}{2} x^2 \left| \dfrac{\partial^2 \widehat{v}}{\partial x^2} \right| \Delta\sigma^2 = 0, \\ \widehat{v}(x,T) = g(x), \end{cases} \tag{4.17}$$

where the differential operator $\mathfrak{L}^0$ is

$$\mathfrak{L}^0 = \frac{1}{2}\sigma^2 x^2 \frac{\partial^2}{\partial x^2} + rx\frac{\partial}{\partial x}.$$

We note that for $\Delta r = \Delta\sigma^2 = 0$ the equation (4.17) reduces to the classical Black–Scholes partial differential equation for finding a fair price of a contingent claim of European type.

Let us clarify the origin of the equation (4.17). In the case when the volatility of the stock price is *not constant*, the fundamental Black–Scholes equation for the value of a hedging strategy passes into the nonlinear Bellman equation

$$\begin{cases} \left(\dfrac{\partial}{\partial t} + \mathfrak{L}^0 - r \right) \widehat{v} + \left| \dfrac{\partial \mathfrak{L}^0}{\partial \Delta\sigma} \widehat{v} \right| \Delta\sigma = 0, \\ \widehat{v}(x,T) = g(x). \end{cases}$$

For the value of a hedging portfolio we can write the approximation

$$\widehat{v} \approx v + \left. \frac{\partial \widehat{v}}{\partial \Delta\sigma} \right|_{\Delta\sigma = 0} \Delta\sigma,$$

where

$$\frac{\partial \widehat{v}}{\partial \Delta\sigma} = \mathbf{E}_{x,t} \int_0^{T-t} \left| \frac{\partial \mathfrak{L}^0}{\partial \Delta\sigma} v(S_u, t+u) \right| e^{-ru}\, du.$$

As $\Delta\sigma \to 0$,

$$\widehat{v},\ \widehat{v}_x,\ \widehat{v}_{xx} \ \to\ v,\ v_x,\ v_{xx}.$$

Therefore,

$$\begin{aligned} \frac{\partial \widehat{v}}{\partial \Delta\sigma} &\approx \mathbf{E}_x \int_0^{T-t} \left| \frac{\partial \mathfrak{L}^0}{\partial \sigma} v(S_u, t+u) \right| e^{-ru}\, du \\ &= \mathbf{E}_x \int_0^{T-t} \left| \sigma x^2 \widehat{v}_{xx}(S_u, t+u) \right| e^{-ru}\, du, \end{aligned}$$

which leads to (4.17).

In connection with problems of hedging contingent claims in markets with stochastic volatility, the concept of the *tracking error* of hedging turns out to be useful.

We denote by $v(t, S_t) = \mathbb{C}(t, S_t, \sigma_{\max}, K, T)$ the value of a self-financing strategy in the classical Black–Scholes model with constant volatility $\sigma_{\max}$, and by V_t the cost of a self-financing portfolio with initial value $V_0 = v(0, S_0)$ containing $(\partial/\partial S)v(t, S_t)$ shares at any time $0 \leqslant t \leqslant T$. Thus, the evolution of the value of the portfolio is described by the stochastic differential equation

$$dV_t = rV_t\, dt + \frac{\partial v}{\partial S}(t, S_t)[dS_t - rS_t\, dt].$$

The *tracking error* e_t at the time $0 \leqslant t \leqslant T$ of a hedging strategy is defined to be the difference between the actual and theoretical values of the hedging strategies:

$$e_t \overset{\text{def}}{=} V_t - v(t, S_t).$$

Since $v(T, S_T) = (S_T - K)^+$, it follows that e_T indicates the *deviation* between the actual terminal value and the payment according to the option. In particular, if the tracking error is positive, $e_T > 0$, then the terminal value for the investor completely covers the contingent claim.

Assume that the stock price evolves according to (4.5), and

$$\Sigma_t \leqslant \sigma_{\max}, \qquad \forall t,\ 0 \leqslant t \leqslant T. \tag{4.18}$$

Then the tracking error is subject to the stochastic differential equation

$$de_t = re_t\,dt + \frac{1}{2}(\sigma_{\max}^2 - \Sigma_u^2)S_u^2\frac{\partial^2 v}{\partial S^2}(u, S_u)\,du, \qquad e_0 = 0.$$

To prove this we use the Kolmogorov–Itô formula to get that

$$\begin{aligned} V_t = v(0, S_0) &+ \int_0^t \frac{\partial v(u, S_u)}{\partial S}\,dS_u \\ &+ \int_0^t \left(\frac{\partial v(u, S_u)}{\partial t} + \frac{1}{2}\Sigma_u^2 S_u^2 \frac{\partial^2 v(u, S_u)}{\partial^2 S} + rS_u\frac{\partial v(u, S_u)}{\partial S}\right)du. \end{aligned} \tag{4.19}$$

It follows from the fundamental Black–Scholes partial differential equation that

$$\frac{\partial v(u, S_u)}{\partial t} = -\frac{1}{2}\sigma_{\max}^2 S_u^2\frac{\partial^2 v(u, S_u)}{\partial S^2} - rS_u\frac{\partial v(u, S_u)}{\partial S}. \tag{4.20}$$

Substituting (4.20) in (4.19), we get the above equation for the tracking error.

It is easy to see that the nonnegativeness of the second derivative $\dfrac{\partial^2 v(u, S_u)}{\partial S^2}$ (for example, in the case of a standard option to buy) leads to the coincidence of the sign of the tracking error with the sign of $\sigma_{\max}^2 - \Sigma_t^2$. The superhedging strategy found is universal in the sense that it works for all models with stochastic volatility that satisfy the condition (4.18).

For an approximate solution of the equation (4.17) we use the *small parameter method*. Let $V_t^*(\sigma^2, \Delta\sigma^2)$ be the value of a minimal hedge in the model with random volatility, and let $V_t^*(\sigma^2, 0)$ be the value of a minimal hedge in the classical Black–Scholes model with constant volatility σ. Then up to small quantities of order greater than $\Delta\sigma^2$ we have

$$V_t^* = V_t^*(\sigma^2, \Delta\sigma^2) \approx V_t^*(\sigma^2, 0)\left(\frac{\partial V^*}{\partial \Delta\sigma^2}\right)_t\bigg|_{(\sigma^2, 0)}\Delta\sigma^2 = V^{(0)} + V^{(1)}\Delta\sigma^2.$$

It is natural to look for hedging portfolios in the class of strategies with value of the form $V_t = v(S_t, t)$, where v is a sufficiently smooth function. Then $\widehat{v} = \widehat{v}^{(0)} + \widehat{v}^{(1)}\Delta\sigma^2$ in our notation. After substituting this expression in (4.17), we get that

$$L\widehat{v}^{(0)} + L\widehat{v}^{(1)}\Delta\sigma^2 + \frac{1}{2}\left|\frac{\partial^2 \widehat{v}^{(0)}}{\partial S^2} + \frac{\partial^2 \widehat{v}^{(1)}}{\partial S^2}\Delta\sigma^2\right|\Delta\sigma^2 = 0.$$

Under the assumption that $\Delta\sigma^2$ is sufficiently small, the second derivatives of the functions $\widehat{v}^{(0)}$ and $\widehat{v}^{(1)}$ with respect to S will have the same sign, and therefore we can write

$$L\widehat{v}^{(0)} + L\widehat{v}^{(1)}\Delta\sigma^2 + \frac{1}{2}\left|\frac{\partial^2\widehat{v}^{(0)}}{\partial S^2}\right|\Delta\sigma^2 + \frac{1}{2}\left|\frac{\partial^2\widehat{v}^{(1)}}{\partial S^2}\right|(\Delta\sigma^2)^2 = 0.$$

Taking into account that $\widehat{v}^{(0)}$ is the value of a minimal hedge in the classical Black–Scholes model with constant volatility, and considering only terms of order at most $\Delta\sigma^2$, we arrive at the following equation for $\widehat{v}^{(1)}$:

$$L\widehat{v}^{(1)}\Delta\sigma^2 + \frac{1}{2}\left|\frac{\partial^2\widehat{v}^{(0)}}{\partial S^2}\right|\Delta\sigma^2 = 0$$

or

$$L\widehat{v}^{(1)} = -\frac{1}{2}\left|\frac{\partial^2\widehat{v}^{(0)}}{\partial S^2}\right|.$$

Thus, the problem (4.17) reduces to the solution of the boundary-value problem

$$\text{(4.21)}\qquad \begin{cases} L\widehat{v}^{(1)}(x,u) = h(x,u), \\ \widehat{v}^{(1)}(x,T) = 0, \end{cases}$$

where $h(x,u) = -\frac{1}{2}\left|\frac{\partial^2\widehat{v}^{(0)}}{\partial x^2}\right|$.

Using the Kolmogorov–Itô formula, we get that

$$\begin{aligned} e^{-rt}\widehat{v}^{(1)}(S_t,t) &- \int_0^t h(S_u,u)\,du \\ &= \widehat{v}^{(1)}(S_0,0) + \int_0^t \frac{\partial\widehat{v}^{(1)}}{\partial S}(S_u,u)\,d(e^{-ru}S_u) \\ &\qquad + \int_0^t \big(L\widehat{v}^{(1)}(u,S_u) - h(S_u,u)\big)\,du. \end{aligned}$$

From this, using (4.21), we find that

$$\begin{aligned} \widehat{v}^{(1)}(x,0) &= \mathbf{E}\bigg[e^{-rT}\widehat{v}^{(1)}(S_T,T) - \int_0^T h(S_u,u)\,du \\ &\qquad - \int_0^T \frac{\partial\widehat{v}^{(1)}}{\partial S}(S_u,u)\,d(e^{-ru}S_u)\bigg] \\ &= -\mathbf{E}\int_0^T h(S_u,u)\,du. \end{aligned}$$

Thus, we get expressions for the ask and bid prices of the option:

$$\overline{\mathbb{C}} = J_0(\sigma) + \left\{\sup_{\mathbf{Q}\in\mathcal{M}} \mathbf{E}_{\mathbf{Q}} \int_0^T \frac{1}{2}\left|\frac{\partial^2\widehat{v}^{(0)}}{\partial S^2}\right| du\right\}\Delta\sigma^2,$$

$$\underline{\mathbb{C}} = J_0(\sigma) + \left\{\inf_{\mathbf{Q}\in\mathcal{M}} \mathbf{E}_{\mathbf{Q}} \int_0^T \frac{1}{2}\left|\frac{\partial^2\widehat{v}^{(0)}}{\partial S^2}\right| du\right\}\Delta\sigma^2,$$

where $J_0(\sigma)$ is the fair price in the standard Black–Scholes model with constant volatility σ.

As noted above, introduction in a model of a *stochastic* volatility can lead to the model of a financial market becoming *incomplete*. However, *there are models of markets with stochastic volatility for which the market is complete, and hence the cost of a contingent claim is uniquely determined.* We give an example of such a market.

Denote the process of the logarithm of the discounted stock price by X, $X = (X_t)_{t\geqslant 0}$:

$$X_t = \ln(e^{-rt}S_t).$$

We define the function

$$Z_t^{(m)} = \int_0^\infty \lambda e^{-\lambda u}(X_t - X_{t-u})^m\,du, \qquad m \in \mathbb{Z}_+, \tag{4.22}$$

where the parameter λ corresponds to the rate of discounting of the "latest" information about the increments of the process X.

LEMMA 4.1. *The process $Z_t^{(m)}$ defined in* (4.22) *satisfies the stochastic differential equation*

$$dZ_t^{(m)} = mZ_t^{(m-1)}\,dX_t + \frac{m(m-1)}{2}Z_t^{(m-2)}\,d\langle X\rangle_t - \lambda Z_t^{(m)}\,dt. \tag{4.23}$$

For the proof we note that in view of (4.22)

$$\begin{aligned} e^{\lambda t}Z_t^{(m)} &= \int_{-\infty}^t \lambda e^{\lambda u}(X_t - X_u)^m\,du \\ &= \sum_{k=0}^m \binom{m}{k}(X_t)^k \int_{-\infty}^t \lambda e^{\lambda u}(-X_u)^{m-k}\,du, \end{aligned}$$

and hence

$$\begin{aligned} &\lambda e^{\lambda t}Z_t^{(m)}\,dt + e^{\lambda t}\,dZ_t^{(m)} = d(e^{\lambda t}Z_t^{(m)}) \\ &= \sum_{k=0}^m \binom{m}{k}\biggl\{\int_{-\infty}^t \lambda e^{\lambda u}(-X_u)^{m-k}\,du\biggl[kX_t^{k-1}\,dX_t + \frac{k(k-1)}{2}X_t^{k-2}\,d\langle X\rangle_t\biggr] \\ &\qquad\qquad + (X_t)^k\lambda e^{\lambda t}(-X_t)^{m-k}\,dt\biggr\} \\ &= m\sum_{k=1}^m \binom{m-1}{k-1}\biggl\{\int_{-\infty}^t \lambda e^{\lambda u}(-X_u)^{(m-1)-(k-1)}\,du\biggr\}X_t^{k-1}\,dX_t \\ &\quad + \frac{m(m-1)}{2}\sum_{k=2}^m \binom{m-2}{k-2}\biggl\{\int_{-\infty}^t \lambda e^{\lambda u}(-X_u)^{(m-2)-(k-2)}\,du\biggr\}X_t^{k-2}\,d\langle X\rangle_t \\ &= e^{\lambda t}\biggl\{mZ_t^{(m-1)}\,dX_t + \frac{m(m-1)}{2}Z_t^{(m-2)}\,d\langle X\rangle_t\biggr\}. \end{aligned}$$

We assume that the process $X = (X_t)_{t\geqslant 0}$ is controlled by the stochastic differential equation

$$dX_t = \mu(Z_t^{(1)},\dots,Z_t^{(n)})\,dt + \sigma(Z_t^{(1)},\dots,Z_t^{(n)})\,dW_t, \tag{4.24}$$

where the functions $\mu(\,\cdot\,)$ and $\sigma(\,\cdot\,)$ satisfy a Lipschitz condition, and $\sigma(\,\cdot\,) > 0$, $n \in \mathbb{Z}_+$.

The only source of randomness in this model is the Wiener process, and hence the model is complete. Nevertheless, as is clear from (4.24), the volatility in the model changes after a change in the stock price.

We show how to calculate the cost of a contingent claim of European type in this model. For simplicity we assume that $n = 1$ and denote $Z^{(1)}$ by Z. Then the equation (4.23) is transformed to

$$dZ_t = dX_t - \lambda Z_t\,dt.$$

By (4.24),

$$dZ_t = \big(\mu(Z_t) - \lambda Z_t\big)\,dt + \sigma(Z_t)\,dW_t. \tag{4.25}$$

Note that in this case the logarithm of the discounted stock price process is

$$X_t = \ln(e^{-rt}S_t) = X_0 + (Z_t - Z_0) + \lambda\int_0^t Z_u\,du.$$

From (4.25) it follows that Z_t is $\mathcal{F}_t$-measurable, where $\mathcal{F}_t = \sigma(u \leqslant t; W_u)$.

On the original probability space we define a new measure $\widehat{\mathbf{P}}$ as follows:

$$\frac{d\widehat{\mathbf{P}}}{d\mathbf{P}}\bigg|_{\mathcal{F}_t} = \exp\bigg\{-\int_0^t\bigg[\frac{\mu(Z_u)}{\sigma(Z_u)} + \frac{1}{2}\sigma(Z_u)\bigg]\,dW_u - \frac{1}{2}\int_0^t\bigg[\frac{\mu(Z_u)}{\sigma(Z_u)} + \frac{1}{2}\sigma(Z_u)\bigg]^2\,du\bigg\}.$$

We remark that with respect to $\widehat{\mathbf{P}}$ the process

$$\widehat{W}_t = W_t + \int_0^t\bigg[\frac{\mu(Z_u)}{\sigma(Z_u)} + \frac{1}{2}\sigma(Z_u)\bigg]\,du$$

is a Wiener process. Hence, (4.25) can be rewritten as

$$dZ_t = \sigma(Z_t)\,d\widehat{W}_t - \bigg(\frac{1}{2}\sigma^2(Z_t) + \lambda Z_t\bigg)\,dt.$$

Thus,

$$dX_t = dZ_t + \lambda Z_t\,dt = \sigma(Z_t)\,d\widehat{W}_t - \frac{1}{2}\sigma^2(Z_t)\,dt$$

and the process $\big(\ln(e^{-rt}S_t)\big)_{t\geqslant 0}$ of the logarithm of the discounted stock price is a martingale with respect to the measure $\widehat{\mathbf{P}}$.

According to Theorem 3.2, the cost of a contingent claim of European type with payment function $f(S_T)$ in this model is

$$V_t = e^{-r(T-t)}\widehat{\mathbf{E}}\big[f(S_T)\,|\,\mathcal{F}_t\big].$$

Further, as follows from the Feynman–Kac formula, the price of the contingent claim is a solution of the partial differential equation

$$\bigg(rS\frac{\partial v}{\partial S} - rv - \lambda z\frac{\partial v}{\partial z} - \frac{\partial v}{\partial t}\bigg) + \bigg(-\frac{1}{2}\frac{\partial v}{\partial z} + \frac{1}{2}S^2\frac{\partial^2 v}{\partial S^2} + \frac{1}{2}\frac{\partial^2 v}{\partial z^2} + S\frac{\partial^2 v}{\partial z\partial S}\bigg)\sigma^2(z) = 0$$

with the boundary condition

$$v_T = f(S_T).$$

We note that in the case of constant volatility $\dfrac{d\langle X\rangle_t}{dt} = \sigma^2(Z_t) = \sigma^2$ the price function for the contingent claim satisfies the fundamental Black–Scholes equation

$$rS\frac{\partial v}{\partial S} - rv - \frac{\partial v}{\partial t} + \frac{1}{2}\sigma^2 S^2 \frac{\partial^2 v}{\partial S^2} = 0.$$

§ 4.3. Estimation of volatility

We present some methods for estimating the unknown parameters used in financial mathematics.

The method of moments.

The method of moments is based on the intuitive notion that the sample moments calculated on the basis of the postulated model are approximately equal to the empirical moments calculated on the basis of observations of the random variable. Let $(x_1, \dots, x_n)$ be a sample of observations of a random variable X whose distribution function depends on k unknown parameters θ_i, $i = 1, \dots, k$, and let $\mathbf{E}X^r = h_r(\theta_1, \dots, \theta_k)$, $r = 1, 2, \dots$, be the moments of X, where $h_r(\,\cdot\,)$ are known functions of the unknown parameters.

The corresponding empirical sample moments are

$$m_r = \sum_{j=1}^{n} \frac{x_j^r}{n}, \qquad r = 1, 2, \dots .$$

Then the moment estimators $\widehat{\theta}_r$, $r = 1, 2, \dots, k$, are found by solving the system of equations

$$h_r(\widehat{\theta}_1, \dots, \widehat{\theta}_k) = m_r, \qquad r = 1, 2, \dots, k.$$

1. *Estimation of the parameters of a normal distribution.*

In the case of a normal distribution $N(\mu, \sigma)$,

$$\mathbf{E}X = \mu,$$
$$\mathbf{E}X^2 = \sigma^2 + \mu^2.$$

Consequently, the equations of the moments take the form

$$\begin{cases} \widehat{\mu} = \sum \dfrac{x_j}{n}, \\ (\widehat{\mu})^2 + (\widehat{\sigma})^2 = \sum \dfrac{x_j^2}{n}. \end{cases}$$

From this we find the moment estimators:

$$\widehat{\mu} = \overline{x}, \qquad \widehat{\sigma}^2 = \frac{1}{n}\Big(\sum x_j^2 - \overline{x}\sum x_j\Big).$$

2. *Estimation of the parameters of a gamma-distribution.*

The density of the gamma-distribution depends on a shape parameter α and a scale parameter β: $f(x; \alpha, \beta) = x^{\alpha-1}e^{-x/\beta}/\beta^\alpha\Gamma(\alpha)$, $x > 0$.

The first two moments are equal to

$$\mathbf{E}X = \int_0^\infty x f(x; \alpha, \beta)\, dx = \alpha\beta,$$
$$\mathbf{E}X^2 = \int_0^\infty x^2 f(x; \alpha, \beta)\, dx = \alpha(\alpha+1)\beta^2.$$

For the moment estimators we get the system of equations

$$\begin{cases} \widehat{\alpha}\widehat{\beta} = \overline{x} = \sum \dfrac{x_j}{n}, \\ \widehat{\alpha}(\widehat{\alpha}+1)\widehat{\beta}^2 = \overline{x^2} = \sum \dfrac{x_j^2}{n}, \end{cases}$$

whose solution is

$$\widehat{\alpha} = \frac{(\overline{x})^2}{\overline{x^2} - (\overline{x})^2}, \qquad \widehat{\beta} = \frac{\overline{x^2} - (\overline{x})^2}{\overline{x}}.$$

Estimation of the parameters in some models of markets.

1. In the framework of the classical Black–Scholes model the stock price is described by a geometric Brownian motion

$$S_t = S_0 \exp\left\{\left(\mu - \frac{\sigma^2}{2}\right)t + \sigma W_t\right\}, \qquad S_0 = s_0 > 0.$$

The quantities $x_j = \ln S_j/S_{j-1}$ are thus independent $N(\mu-\sigma^2/2, \sigma^2)$-distributed random variables. A moment estimator of the volatility is given by

$$\widehat{\sigma}_t^2 = \frac{1}{t-1}\sum_{j=1}^{t}(x_j - \overline{x})^2, \qquad \overline{x} = \frac{1}{t}\sum_{j=1}^{t} x_j.$$

2. Let us consider the following discrete model of a market:

$$\begin{cases} \ln S_t - \ln S_{t-1} = \mu + \sigma_{t-1}U_t, \\ \ln \sigma_t = \alpha + \phi[\ln \sigma_{t-1} - \alpha] + \theta\eta_t, \end{cases}$$

where (U_t, η_t) are independent identically distributed random variables with correlation δ, and the parameter t takes integer values $t \in \mathbb{Z}_+$. The given autoregression model of a market with random volatility is stationary if and only if $|\phi| < 1$. Under this condition the random variable $\ln \sigma_t$ has mean α and variance $\beta^2 = \theta^2/(1-\phi^2)$. Estimators of the unknown parameters are found by solving the system of equations

$$\begin{cases} \sqrt{\dfrac{2}{\pi}}\, e^{\alpha+\beta^2/2} = \sum \dfrac{|x_j - \overline{x}|}{n}, \\ e^{2\alpha+2\beta^2} = \sum \dfrac{(x_j - \overline{x})^2}{n}, \\ 3e^{4\beta^2} = \sum \dfrac{(x_j - \overline{x})^3}{n}. \end{cases}$$

Method of estimation from the price statistics of derivative instruments, and the "smile" effect.

The volatility, which enters as a parameter in various models, is not an observable random variable. Stock prices and prices of derivative instruments "written" on these stocks are observable variables. The following method of estimation is based on the observation that if the volatility enters explicitly (for example, through the Black–Scholes formula) in the formation of the prices of derivative instruments, then it is necessary to "extract" an estimator of the volatility from the prices of these derivative instruments.

For example, $\mathbb{C}^{\text{market}}(t, S_t, K, T)$ is the price of some tradable option at the time t, $0 \leqslant t \leqslant T$. Here S_t is the price of the stock at this time, K is the exercise price, and T is the exercise time for the option. Let $\mathbb{C}^{\text{theor}}(t, S_t, K, T, \sigma)$ be the price of an option with the same characteristics and volatility σ, calculated theoretically by starting from the assumptions of the model.[1] Then an estimator of the volatility, also called the *intrinsic volatility*, can be determined from the relation

$$\mathbb{C}^{\text{theor}}(t, S_t, K, T, \widehat{\sigma}) = \mathbb{C}^{\text{market}}(t, S_t, K, T).$$

It should be noted that this estimator of the volatility is a function $\widehat{\sigma} = \widehat{\sigma}(t, S_t, K, T)$ of the parameters of the option contract. Thus, for different K and T the estimators of volatility do not coincide in general. For example, in the case of a standard option to buy, the estimators of volatility, as functions $\widehat{\sigma}(K)$ of the exercise price, form a curve resembling a smile for fixed t, S_t, and T. Furthermore, the less the time before the exercise time of the option, the sharper the smile. Thus, the volatility has, so to speak, a "time structure", and this should be taken into account in constructing models of volatility.

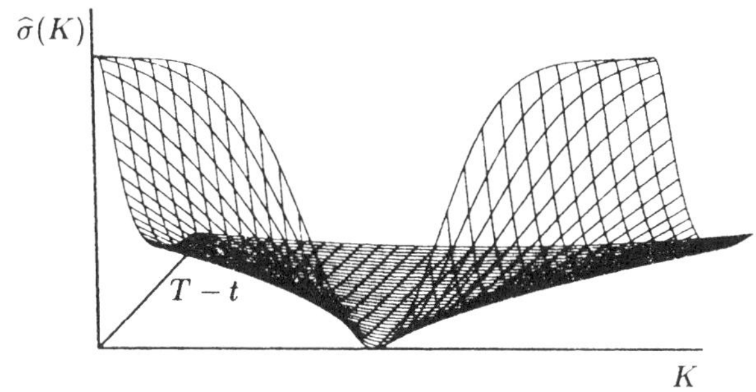

FIGURE 4.1. The "volatility smile".

Confidence estimation of options in the Black–Scholes model.

Let us now look at the Black–Scholes formula (3.23) from a statistical point of view, considering that the *randomness* arises from errors in estimating the *unknown* volatility. We shall study the statistical properties of the Black–Scholes price $\mathbb{C}(t, S_t, K, T, \sigma) = \mathbb{C}(\sigma)$.

Let $S_0, \dots, S_n$, $n = 1, 2, \dots$, be a sequence of stock prices (for example, the sequence of closing prices of the stock over $n + 1$ days). As mentioned earlier, because the variables $x_j = \ln S_j/S_{j-1}$ are independent $N(\mu - \sigma^2/2, \sigma^2)$-distributed random variables, an unbiased moment estimator of σ^2 is given by

$$\widehat{\sigma}_n^2 = \frac{1}{n-1} \sum_{j=1}^{n} (x_j - \overline{x})^2,$$

where $\overline{x} = \frac{1}{n} \sum_{j=0}^{n} x_j$.

Further, the variable $\widehat{\sigma}_n^2$ is distributed like $\sigma^2 \cdot \chi_{n-1}/(n-1)$, where χ_m is the chi-square distribution with m degrees of freedom and with density function

$$f(x) = \frac{1}{2^{m/2}} \frac{1}{\Gamma(m/2)} x^{m/2-1} e^{x/2}, \qquad x > 0.$$

[1]If it does not seem possible to get an analytic expression for the cost of the option, then $\mathbb{C}^{\text{theor}}(t, S_t, K, T)$ can be taken to be an estimator of the cost obtained as a result of some simulation.

Thus,

$$\mathbf{P}(\widehat{\sigma}_n^2 < \sigma_0^2) = \mathbf{P}\left(\chi_{n-1}^2 < \frac{\sigma_0^2}{\sigma^2}(n-1)\right).$$

Since the Black–Scholes price is an increasing function of the parameter σ, it follows that

$$\mathbf{P}(\mathbb{C}(\widehat{\sigma}_n) < \mathbb{C}(\sigma_0)) = \mathbf{P}\left(\chi_{n-1}^2 < \frac{\sigma_0^2}{\sigma^2}(n-1)\right). \tag{4.26}$$

Thus, (4.26) gives the distribution function of the Black–Scholes price and can be used for constructing a confidence interval for the price $\mathbb{C}(\sigma)$ of the option. Namely, we select a number $1-\alpha$ (the *confidence level* $(0<\alpha<1)$) and choose constants χ_* and χ^* such that

$$\mathbf{P}(\chi_* \leqslant \chi_{n-1}^2 \leqslant \chi^*) = 1-\alpha.$$

Then for these constants

$$\mathbf{P}\left(\chi_* \leqslant \frac{\widehat{\sigma}_n^2}{\sigma^2}(n-1) \leqslant \chi^*\right) = 1-\alpha,$$

and hence

$$\mathbf{P}\left(\frac{\widehat{\sigma}_n^2}{\chi^*}(n-1) \leqslant \sigma^2 \leqslant \frac{\widehat{\sigma}_n^2}{\chi_*}(n-1)\right) = 1-\alpha.$$

Again using the monotonicity of the Black–Scholes price, we get that

$$\mathbf{P}\left(\mathbb{C}\left(\sqrt{\frac{\widehat{\sigma}_n^2}{\chi^*}(n-1)}\right) \leqslant \mathbb{C}(\sigma) \leqslant \mathbb{C}\left(\sqrt{\frac{\widehat{\sigma}_n^2}{\chi_*}(n-1)}\right)\right) = 1-\alpha. \tag{4.27}$$

Consequently, the $(1-\alpha)\cdot 100\%$ confidence interval for $\mathbb{C}(\sigma)$ has the form

$$\left(\mathbb{C}\left(\sqrt{\frac{\widehat{\sigma}_n^2}{\chi^*}(n-1)}\right), \mathbb{C}\left(\sqrt{\frac{\widehat{\sigma}_n^2}{\chi_*}(n-1)}\right)\right).$$

Letting $\sigma_0 \to 0$ and $\sigma_0 \to \infty$ in (4.26), we get as a consequence the following bounds for the price of an option defined on the interval $[n, n+T]$ and having exercise price K:

$$S_n - Ke^{-rT} < \mathbb{C}(\sigma) < S_n.$$

REMARK. Since the price of the option depends in a nonlinear way on the volatility, an unbiased estimator of the volatility does not lead to an unbiased estimator of the option price. This may be one of the possible explanations for the "volatility smile" effect observed in practice.

References for Chapter 4

[5], **[8]**–**[10]**, **[15]**, **[20]**, **[38]**, **[40]**, **[42]**, **[43]**, **[45]**, **[49]**, **[50]**, **[51]**, **[59]**, **[60]**, **[62]**, **[76]**, **[77]**, **[80]**, **[81]**, **[90]**, **[91]**, **[94]**, **[95]**, **[98]**, **[99]**, **[107]**, **[127]**, **[152]**, **[153]**, **[158]**.

CHAPTER 5

Markets with Structural Constraints and Transaction Costs

In this chapter we study the general theory of hedging (and investment) for financial markets subjected to various constraints (for example, different credit and deposit interest rates, prohibition on "short" selling, and so on). Applications of general theoretical statements in concrete models are given. A study is made of hedging and investment with transaction costs taken into account in the formation of hedging and investment strategies (Leland's approach and maximization of mean yield).

§ 5.1. Calculations in models of markets with structural constraints: A general methodology and its concrete realization

We consider contingent claims in markets with initially imposed constraints on the class of strategies that can be used, constraints having a *structural* character, for example, prohibition on "short" selling, distinction between deposit accounts and credit accounts, and so on.

Optional decompositions in markets with constraints.

We present a model of a securities market consisting of d assets. Let $X = (X^i)_{1\leqslant i\leqslant d}$ denote the process of prices of these assets. We determine a portfolio (strategy) $\pi = (H, C)$, where $H = (H^i)_{1\leqslant i\leqslant d}$ is a predictable X-integrable process specifying the quantity of each asset contained in the portfolio, and $C = (C_t)_{t\geqslant 0}$ is an increasing process of total consumption in the portfolio up to and including the time t. The change in the value $V = (V_t)_{t\geqslant 0}$ of this portfolio π is given by the equation

$$V_t = v + \int_0^t H_s\, dX_s - C_t, \qquad t \geqslant 0, \tag{5.1}$$

where v is the initial value of π.

As earlier, such a portfolio will be called a *strategy with consumption.* The equality $C \equiv 0$ means that π is a *self-financing* portfolio.

All processes are defined on a standard stochastic base $(\Omega, \mathbb{F} = (\mathcal{F}_t)_{t\geqslant 0}, \mathbf{P})$ and, with the exception of integrands of stochastic integrals, are real-valued and right-continuous, have limits from the left, and are measurable with respect to the given filtration $\mathbb{F}$. For processes X and Y the relation $X \prec Y$ means that $Y - X$ is an increasing process.

To present a unified methodology for hedging in markets with constraints we need some results in the theory of stochastic integration in addition to what was given in Chapter 2.

The stochastic integral of a predictable process H with respect to a semimartingale X will be denoted by $H * X = \int H\,dX$, and $L(X)$ will denote the space of predictable processes integrable with respect to X.

A process $H \in L(X)$ is said to be (locally) *admissible* if $H * X$ is (locally) bounded below, that is, there exists a sequence of stopping times $\tau_n \uparrow \infty$ such that $(H * X)_{\tau_n}$ are bounded below.

The corresponding classes of admissible (locally admissible) integrands are denoted by $L^a(X)$ ($L^a_{\text{loc}}(X)$).

The *Emery distance* between semimartingales X and Y is defined to be

$$D(X,Y) = \sup_{|H|\leqslant 1} \left(\sum_{n\geqslant 1} \mathbf{E}\big[\min(|H * (X-Y)_n|, 1)\big] \right),$$

where the sup is taken over all *predictable* processes H bounded in modulus by 1.

The quantity $D(X,Y)$ defines a metric in the space of semimartingales. The corresponding topology is called the semimartingale topology or the *Emery topology.* If X is a semimartingale, then the space $L(X)$ is complete with respect to the metric

$$d_X(H,G) = D(H * X, G * X). \tag{5.2}$$

Suppose that $\aleph \subseteq L^a_{\text{loc}}(X)$ is a family of locally admissible integrands for X. We assume that $\aleph \supseteq H \equiv 0$, $\aleph$ is closed in $L^a_{\text{loc}}(X)$ with respect to the metric d_X defined by (5.2), and $\aleph$ is convex in the sense that *the process $h * H + (1-h) * G$ is in $\aleph$ for any $H, G \in \aleph$ and every predictable process $0 \leqslant h \leqslant 1$.*

If $H \in \aleph$, then a portfolio $\pi = (H, C)$ satisfying (5.1) is called a *strategy with $\aleph$-constraints.*

Such a securities market in which strategies with constraints satisfying (5.1) operate is called a *market with constraints.*

Let us look at some examples of markets with constraints, differing from each other in the choice of the class $\aleph$. We consider the following variants of $\aleph$.

1) $\aleph = L^a_{\text{loc}}(X)$: no constraints.
2) $\aleph = \{H \in L^a_{\text{loc}}(X) : H^i \geqslant 0,\ 1 \leqslant i \leqslant d\}$, where d is the number of risky assets: no short selling for the first d assets.
3) $\aleph = \{H \in L^a_{\text{loc}}(X) : \underline{G}^i \leqslant H^i \leqslant \overline{G}^i,\ \underline{G}^i \leqslant 0,\ \overline{G}^i \geqslant 0,\ 1 \leqslant i \leqslant d\}$, where $\underline{G}^i$, $\overline{G}^i \in L^a_{\text{loc}}(X)$: the number of assets in the portfolio is bounded above and below.
4) A market with a *deposit* account, a *credit* account, and stocks in which the interest rates for borrowing capital and allocating capital are different is also a market with constraints. In this case the set $\aleph$ has the following form: $\aleph = \{H \in L^a_{\text{loc}}(X) : H^1 \geqslant 0,\ H^2 \leqslant 0,\ H^3 \in \mathbb{R}\}$, where H^1 and H^2 are the numbers of units in the deposit account and the credit account, and H^3 is the number of stocks in the portfolio.

A portfolio π with $\aleph$-constraints is said to be *admissible* if $V_t \geqslant 0$ for all $t \geqslant 0$.

Assume that $\Im$ is a family of special semimartingales Y that are locally bounded below with initial value $Y_0 = 0$, and let $Y \equiv 0 \in \Im$.

We introduce the class of measures $P(\Im) = \{\mathbf{Q} : \mathbf{Q} \sim \mathbf{P} : \exists$ an increasing predictable process A depending only on $\mathbf{Q}$ and $\Im$ such that $Y - A$ is a local supermartingale with respect to $\mathbf{Q}$ for any $Y \in \Im$, that is,

$$A^Y(\mathbf{Q}) \prec A \text{ for any } Y \in \Im, \tag{5.3}$$

where $A^Y(\mathbf{Q})$ is the compensator of Y with respect to $\mathbf{Q}\}$.

An increasing predictable process $A^{\Im}(\mathbf{Q})$ is called an *upper variation* of the family $\Im$ with respect to the measure $\mathbf{Q}$ if $A^Y(\mathbf{Q}) \prec A^{\Im}(\mathbf{Q})$ for any $Y \in \Im$, and the process $A^{\Im}(\mathbf{Q})$ is minimal in the sense that $A^{\Im}(\mathbf{Q}) \prec A$ for any increasing predictable process A satisfying the condition (5.3).

The family $\Im$ is said to be *predictably convex* if

$$h * Y^1 + (1-h) * Y^2 \in \Im$$

for any $Y^1, Y^2 \in \Im$ and any predictable process $0 \leqslant h \leqslant 1$.

In what follows we are considering a predictably convex family $\Im$.

LEMMA 5.1. *A measure $\mathbf{Q}$ equivalent to $\mathbf{P}$ is in $P(\Im) \iff$ any $Y \in \Im$ is a special semimartingale with respect to $\mathbf{Q}$ and*

$$\operatorname*{ess\,sup}_{Y \in \Im} A^Y(\mathbf{Q})_t < +\infty \quad \textit{for all } t \geqslant 0. \tag{5.4}$$

In this case the upper variation of the family $\Im$ with respect to the measure $\mathbf{Q}$ exists, is unique, and is determined from the equations

$$A^{\Im}(\mathbf{Q})_\tau = \operatorname*{ess\,sup}_{Y \in \Im} A^Y(\mathbf{Q})_\tau,$$
$$\mathbf{E}[A^{\Im}(\mathbf{Q})_\tau] = \sup_{Y \in \Im} \mathbf{E}[A^Y(\mathbf{Q})_\tau]$$

for any stopping time τ.

Moreover, there exists a sequence $Y^n \in \Im$ such that the compensators $A^n = A^{Y^n}(\mathbf{Q})$ satisfy the conditions $A^n \prec A^{n+1}$ and $\lim_{n\to\infty} \sup_{0 \leqslant t \leqslant \tau}(A^{\Im}(\mathbf{Q})_t - A^n_t) = 0$ (a.s.) for any stopping time τ such that $A^{\Im}(\mathbf{Q})_\tau < +\infty$ (a.s.).

EXAMPLE 5.1. Suppose that X is a semimartingale, $\overline{G} \geqslant 0$, and $\underline{G} \leqslant 0$, where $\overline{G}$ and $\underline{G}$ are locally admissible integrands for X, that is, $\overline{G} * X$ and $\underline{G} * X$ are defined and locally bounded below. Let $\aleph = \{H \text{ a predictable process} \mid \underline{G} \leqslant H \leqslant \overline{G}\}$.

All the stochastic integrals $H * X$ for $H \in \aleph$ are locally bounded below, and $\Im = \{H * X : H \in \aleph\}$ is predictably convex.

Let a measure $\mathbf{Q} \sim \mathbf{P}$ be such that X is a special semimartingale with respect to $\mathbf{Q}$. Then X has a canonical decomposition $X = M + A$, where $M \in \mathcal{M}_{\text{loc}}(\mathbf{Q})$ and A is a predictable process of bounded variation. The compensator of an arbitrary process $Y = H * X \in \Im$ has the form $H \circ A$ with $H \in \aleph$. Further, $H \circ A = H \circ A^+ - H \circ A^- \leqslant \overline{G} \circ A^+ - \underline{G} \circ A^-$, with equality for $H = h\overline{G} + (1-h)\underline{G}$, where $h = dA^+/dA$. By Lemma 5.1,

$$A^{\Im}(\mathbf{Q})_t = \int_0^t \overline{G}_s \, dA_s^+ - \int_0^t \underline{G}_s \, dA_s^- \quad \text{for} \quad t \geqslant 0.$$

In particular, $P(\Im)$ contains all probability measures $\mathbf{Q} \sim \mathbf{P}$ with respect to which X is a special semimartingale.

We can now formulate an equivalent definition of the family $P(\Im)$ of measures:

$$P(\Im) = \Big\{\mathbf{Q} : \mathbf{Q} \sim \mathbf{P} \text{ for which } A^{\Im}(\mathbf{Q})_t = \operatorname*{ess\,sup}_{Y \in \Im} A^Y(\mathbf{Q})_t < \infty\Big\}.$$

The next theorem gives a *(super)martingale*, or *dual*, characterization of strategies with $\aleph$-constraints and represents a criterion for the existence of an optional decomposition in such markets.

We consider the family

$$\Im = \{H * X : H \in \aleph\} \tag{5.5}$$

of semimartingales.

THEOREM 5.1. *Suppose that $P(\Im) \neq \varnothing$ and the process V is locally bounded below. Then the following assertions are equivalent:*

1) *V is the value of an admissible portfolio with $\aleph$-constraints, and it satisfies the relation*

$$V = V_0 + H * X - C, \tag{5.6}$$

where $H \in \aleph$ and C is an increasing process;

2) *for all measures $\mathbf{Q} \in P(\Im)$ the process $V - A^{\Im}(\mathbf{Q})$ is a local supermartingale with respect to $\mathbf{Q}$.*

Constraints are often imposed on the portions of a portfolio's value invested in various assets. Assume that the components X^i, $i \leqslant d$, of the process X are strictly positive. Let

$$R_t^i = \int_0^t dX_s^i / X_{s-}^i, \qquad t \geqslant 0, \tag{5.7}$$

be the relative growth process of the price of the ith asset, let

$$K_t^i = H_t^i X_{t-}^i / V_{t-} I_{\{V_{t-} > 0\}} \tag{5.8}$$

be the fraction of the portfolio's value in the ith asset at time t, and let

$$D_t = \int_0^t dC_s / V_{s-} I_{\{V_{s-} > 0\}}, \qquad t \geqslant 0, \tag{5.9}$$

be the fraction of the total consumption up to and including the time t.

Then by (5.7)–(5.9), the additive representation (5.6) can be written in the *multiplicative* form

$$V = V_0 \, \mathcal{E}(K * R - D). \tag{5.10}$$

Let $\Re$ be a family of integrands for R such that $K \Delta R \geqslant -1$ for all $K \in \Re$. We assume that $\Re \supseteq K \equiv 0$, $\Re$ is closed in $L^a_{\text{loc}}(R)$ with respect to the metric d_R defined by (5.2), and $\Re$ is convex in the sense that $h * K + (1 - h) * L \in \Re$ *for any $K, L \in \Re$ and every predictable process $0 \leqslant h \leqslant 1$.*

The next theorem gives a dual characterization of an admissible portfolio whose fractions of the value belong to the class $\Re$. It establishes a multiplicative form of the optional decomposition for markets with $\Re$-constraints.

We consider the family

$$\Im = \{K * R : K \in \Re\} \tag{5.11}$$

of semimartingales.

THEOREM 5.2. *Suppose that $P(\mathfrak{I}) \neq \varnothing$ and let the process V be nonnegative. Then the following assertions are equivalent*:

1) *V is the value of an admissible portfolio with proportions in the class $\mathfrak{R}$, and it satisfies the relation*

$$V = V_0\, \mathcal{E}(K * R - D), \tag{5.12}$$

where $K \in \mathfrak{R}$ and D is an increasing process;

2) *for all measures $\mathbf{Q} \in P(\mathfrak{I})$ the process $V/\mathcal{E}(A^{\mathfrak{I}}(\mathbf{Q}))$ is a supermartingale with respect to $\mathbf{Q}$.*

Thus, to get the optional decomposition of the value V_t of a portfolio of securities in a market with constraints we must calculate the upper variation $A^{\mathfrak{I}}(\mathbf{Q})$ for the family $\mathfrak{I}$ with respect to $\mathbf{Q} \in P(\mathfrak{I})$ as the essential supremum of the family of compensators with respect to $\mathbf{Q}$ and then use Theorem 5.1 or Theorem 5.2.

The structure of an optimal hedge in markets with constraints. As an application of the above theory we consider the problem of hedging a contingent claim in markets with constraints.

Let f_T, $T \geqslant 0$, a nonnegative random variable on $(\Omega, \mathcal{F}_T, \mathbf{P})$, be a contingent claim (a European option with such a payment function).

A strategy π with $\aleph$-constraints and value V is a *hedging* strategy for the contingent claim f_T if π is admissible and $V_T \geqslant f_T$.

A strategy $\widetilde{\pi}$ with $\aleph$-constraints and value $\widetilde{V}$ is called a *minimal hedging strategy with $\aleph$-constraints* if $V_t \geqslant \widetilde{V}_t \geqslant f_T$, $t \leqslant T$, for any hedging strategy π with $\aleph$-constraints whose value is given by the process V.

THEOREM 5.3. *Suppose that $\mathfrak{I}$ satisfies* (5.5). *The set of hedging strategies is nonempty if and only if*

$$\sup_{\mathbf{Q} \in P(\mathfrak{I})} \mathbf{E}_{\mathbf{Q}}(f_T - A^{\mathfrak{I}}(\mathbf{Q})_T) < +\infty.$$

In this case a minimal hedging strategy $\widetilde{\pi} = (\widetilde{H}, \widetilde{C})$ with $\aleph$-constraints exists, and its value at a time $t \leqslant T$ is

$$\widetilde{V}_t = \widetilde{v} + (\widetilde{H} * X)_t - \widetilde{C}_t = \operatorname*{ess\,sup}_{\mathbf{Q} \in P(\mathfrak{I})} \bigl(\mathbf{E}_{\mathbf{Q}}[f_T - A^{\mathfrak{I}}(\mathbf{Q})_T \mid \mathcal{F}_t] + A^{\mathfrak{I}}(\mathbf{Q})_t\bigr)^+, \tag{5.13}$$

where $a^+ = \max(a, 0)$.

Theorem 5.3 follows from Theorem 5.4 below, which characterizes the value process of a minimal hedging strategy with $\aleph$-constraints for a dynamic contingent claim.

A nonnegative process $f = (f_t)_{t \geqslant 0}$ will be called a *dynamic* contingent claim.[1] Since $f_t = f_T I_{\{t \geqslant T\}}$, we get that the contingent claim f_T introduced earlier is a special case of dynamic contingent claims.

A strategy π with $\aleph$-constraints and value $V = (V_t)_{t \geqslant 0}$ is called a *hedging strategy with $\aleph$-constraints* for f if $V_t \geqslant f_t$ for all $t \geqslant 0$.

A strategy $\widetilde{\pi}$ with $\aleph$-constraints and value $\widetilde{V} = (\widetilde{V}_t)_{t \geqslant 0}$ is called a *minimal hedging strategy with $\aleph$-constraints* for f if $V_t \geqslant \widetilde{V}_t \geqslant f_t$, $t \geqslant 0$, for any hedging strategy π with $\aleph$-constraints whose value is given by the process V.

[1] See also Chapter 7, which is devoted to a study of such contingent claims.

For $\mathbf{Q} \in P(\Im)$ and $t \geqslant 0$ we introduce the set

$$M_t(\mathbf{Q}) = \big\{\tau \text{ a stopping time} : \tau \in [t, +\infty) \text{ and the process } \big(A^{\Im}_{u\wedge t}(\mathbf{Q}) - A^{\Im}_t(\mathbf{Q})\big)_{u\geqslant 0} \text{ is bounded on } [0,\tau]\big\}. \tag{5.14}$$

LEMMA 5.2. *Let $(f_t)_{t\geqslant 0}$ be a nonnegative process such that*

$$\sup_{\mathbf{Q}\in P(\Im)} \sup_{\tau\in M_0(\mathbf{Q})} \mathbf{E}_{\mathbf{Q}}\big(f_\tau - A^{\Im}(\mathbf{Q})_\tau\big) < +\infty.$$

Then there exists a process $(U_t)_{t\geqslant 0}$ such that

$$U_t = \operatorname*{ess\,sup}_{\mathbf{Q}\in P(\Im),\, \tau\in M_t(\mathbf{Q})} \big(\mathbf{E}_{\mathbf{Q}}\big[f_\tau - A^{\Im}(\mathbf{Q})_\tau \,|\, \mathcal{F}_t\big] + A^{\Im}(\mathbf{Q})_t\big) \text{ for all } t \geqslant 0 \text{ (a.s.)}.$$

Moreover, for any measure $\mathbf{Q} \in P(\Im)$ the process $U - A^{\Im}(\mathbf{Q})$ is a local supermartingale with respect to $\mathbf{Q}$.

THEOREM 5.4. *Suppose that $\Im$ satisfies* (5.5). *The set of hedging strategies is nonempty if and only if*

$$\sup_{\mathbf{Q}\in P(\Im)} \sup_{\tau\in M_0(\mathbf{Q})} \mathbf{E}_{\mathbf{Q}}\big(f_\tau - A^{\Im}(\mathbf{Q})_\tau\big) < +\infty.$$

In this case a minimal hedging strategy $\widetilde{\pi} = (\widetilde{H}, \widetilde{C})$ with $\aleph$-constraints exists, and its value at a time $t \geqslant 0$ is

$$\begin{aligned} \widetilde{V}_t &= \widetilde{v} + (\widetilde{H} * X)_t - \widetilde{C}_t \\ &= \operatorname*{ess\,sup}_{\mathbf{Q}\in P(\Im),\, \tau\in M_t(\mathbf{Q})} \big(\mathbf{E}_{\mathbf{Q}}\big[f_\tau - A^{\Im}(\mathbf{Q})_\tau \,|\, \mathcal{F}_t\big] + A^{\Im}(\mathbf{Q})_t\big). \end{aligned} \tag{5.15}$$

PROOF. Let

$$\widetilde{V}_t = \operatorname*{ess\,sup}_{\mathbf{Q}\in P(\Im),\, \tau\in M_t(\mathbf{Q})} \big(\mathbf{E}_{\mathbf{Q}}\big[f_\tau - A^{\Im}(\mathbf{Q})_\tau \,|\, \mathcal{F}_t\big] + A^{\Im}(\mathbf{Q})_t\big).$$

Let $\mathbf{Q} \in P(\Im)$ and let $V = (V_t)_{t\geqslant 0}$ be the value process of a hedging strategy with $\aleph$-constraints. Let $(\tau_n)_{n\geqslant 1}$ be a sequence of stopping times such that $\mathbf{E}_{\mathbf{Q}} A^{\Im}(\mathbf{Q})_{\tau_n} \leqslant n$. Since $V \geqslant f \geqslant 0$ and since V has the representation (5.1), it follows from Theorem 5.1 that the process $V - A^{\Im}(\mathbf{Q})$ is a local supermartingale with respect to $\mathbf{Q}$ on $[0, \tau_n]$. Therefore,

$$\begin{aligned} V_{t\wedge\tau_n} &\geqslant \mathbf{E}_{\mathbf{Q}}\big[V_{\tau\wedge\tau_n} - A^{\Im}(\mathbf{Q})_{\tau\wedge\tau_n} \,|\, \mathcal{F}_{t\wedge\tau_n}\big] + A^{\Im}(\mathbf{Q})_{t\wedge\tau_n} \\ &\geqslant f_{\tau_n} I_{\{t>\tau_n\}} + \mathbf{E}_{\mathbf{Q}}\big[(f_{\tau\wedge\tau_n} - A^{\Im}(\mathbf{Q})_{\tau\wedge\tau_n} + A^{\Im}(\mathbf{Q})_t) I_{\{t\leqslant\tau_n\}} \,|\, \mathcal{F}_t\big] \end{aligned}$$

for all $t \geqslant 0$ and for every stopping time $\tau \in M_t(\mathbf{Q})$. It follows from the definition of $M_t(\mathbf{Q})$ that the sequence

$$\big[(f_{\tau\wedge\tau_n} - A^{\Im}(\mathbf{Q})_{\tau\wedge\tau_n} + A^{\Im}(\mathbf{Q})_t) I_{\{t\leqslant\tau_n\}}\big]_{n\geqslant 1}$$

is bounded below for each n. Then we conclude from Fatou's Lemma that $V_t \geqslant \mathbf{E}_{\mathbf{Q}}\big[f_\tau - A^{\Im}(\mathbf{Q})_\tau \,|\, \mathcal{F}_t\big] + A^{\Im}(\mathbf{Q})_t$, and hence $V_t \geqslant \widetilde{V}_t$ for $t \geqslant 0$.

We remark that $\widetilde{V}_t \geqslant f_t$ for $t \geqslant 0$. Therefore, we need only show that $\widetilde{V}$ is the value process corresponding to a portfolio with $\aleph$-constraints. This follows from Theorem 5.1 and Lemma 5.2. The proof is complete.

An analogous description applies to the value process of a minimal hedging strategy in the case of the set $\Re$ of constraints.

THEOREM 5.5. *Let $f_T \geqslant 0$ be a contingent claim, and suppose that $\Im$ satisfies* (5.11). *The set of hedging strategies is nonempty if and only if*

$$\sup_{\mathbf{Q}\in P(\Im)} \mathbf{E}_{\mathbf{Q}}\big[f_T/\mathcal{E}(A^{\Im}(\mathbf{Q}))_T\big] < +\infty.$$

In this case there exists a minimal hedging strategy $\widetilde{\pi}$ with fractions of the assets in the class $\Re$ and such that the value of the portfolio at time t is

$$\widetilde{V}_t = \operatorname*{ess\,sup}_{\mathbf{Q}\in P(\Im)} \big(\mathcal{E}(A^{\Im}(\mathbf{Q}))_t \mathbf{E}_{\mathbf{Q}}\big[f_T/\mathcal{E}(A^{\Im}(\mathbf{Q}))_T \mid \mathcal{F}_t\big]\big). \tag{5.16}$$

We consider a dynamic contingent claim. Let M_t^∞ denote the set of stopping times τ with values in the interval $[t, +\infty)$.

THEOREM 5.6. *Suppose that $(f_t)_{t\geqslant 0}$ is a nonnegative process and $\Im$ satisfies* (5.11). *Then the set of hedging strategies is nonempty if and only if*

$$\sup_{\mathbf{Q}\in P(\Im)} \sup_{\tau\in M_0^\infty} \mathbf{E}_{\mathbf{Q}}\big(f_\tau/\mathcal{E}(A^{\Im}(\mathbf{Q}))_\tau\big) < +\infty.$$

In this case there exists a minimal hedging strategy $\widetilde{\pi}$ with fractions of assets in the class $\Re$ and such that the value of the portfolio at time t is

$$\widetilde{V}_t = \operatorname*{ess\,sup}_{\mathbf{Q}\in P(\Im),\, \tau\in M_t^\infty} \big(\mathcal{E}(A^{\Im}(\mathbf{Q}))_t \mathbf{E}_{\mathbf{Q}}\big[f_\tau/\mathcal{E}(A^{\Im}(\mathbf{Q}))_\tau \mid \mathcal{F}_t\big]\big). \tag{5.17}$$

The proofs of Theorems 5.5 and 5.6 are analogous to the proof of Theorem 5.4 above.

The binomial model with different interest rates.

Description of the model and formulation of the hedging problem for this model. Let us consider a model of a discrete financial market with three assets: a deposit account, a credit account, and a stock. Suppose that the evolution of the deposit account value B^1 and the credit account value B^2 and the dynamics of the stock price S correspond to the recursion equations

$$\begin{aligned} \Delta B_n^1 &= r^1 B_{n-1}^1, & B_0^1 &> 0,\\ \Delta B_n^2 &= r^2 B_{n-1}^2, & B_0^2 &> 0,\\ \Delta S_n &= \rho_n(\omega) S_{n-1}, & S_0 &> 0. \end{aligned} \tag{5.18}$$

Here r^1 and r^2 are the constant interest rates for the deposit and credit accounts, respectively, $r^1 \leqslant r^2$, and $\rho = (\rho_n(\omega))_{0\leqslant n\leqslant N}$ is a sequence of independent identically distributed random variables with respect to any measure $\mathbf{P}$ in a family $\{\mathbf{P}\}$ of probability measures defined on some measurable space $(\Omega, \mathcal{F})$, where $\Omega = \{+1, -1\}^N$. Furthermore, the random variables $\rho_n(\omega)$ can take two values $a < b$:

$$\begin{aligned} \mathbf{P}(\rho_n = b) &= p > 0,\\ \mathbf{P}(\rho_n = a) &= 1 - p > 0. \end{aligned}$$

We introduce on the space $(\Omega, \mathcal{F})$ the discrete filtration $(\mathcal{F}_n)_{0\leqslant n\leqslant N}$ adapted to the stock prices:

$$\mathcal{F}_0 = \{\varnothing, \mathcal{F}\}, \quad \mathcal{F}_n = \sigma(S_0, S_1, \dots, S_n), \quad n > 0.$$

Thus, we have a discrete model of a securities market with three assets, and the sequence of prices of these assets has the form

$$X = (X_n)_{0\leqslant n\leqslant N} = (X_n^1, X_n^2, X_n^3)_{0\leqslant n\leqslant N} = (B_n^1, B_n^2, S_n)_{0\leqslant n\leqslant N}.$$

In this model an *investor strategy* is a stochastic sequence

$$H = (H_n^1, H_n^2, H_n^3) = (\beta_n^1, \beta_n^2, \gamma_n),$$

where β_n^1, β_n^2, and γ_n are the numbers of units in the deposit and credit accounts and the number of shares in the investor's portfolio at the time n.

Since the portfolio is put together solely on the basis of information about previous prices, the stochastic sequence determining the strategy is *predictable*. The investor's value at time n corresponding to the strategy H has the form

$$V_n^\pi = \beta_n^1 B_n^1 + \beta_n^2 B_n^2 + \gamma_n S_n. \tag{5.19}$$

A portfolio $\pi = (H, C)$ is a *strategy with consumption* if its value at a time n satisfies the equation

$$V_n = V_0 + \sum_{k=1}^{n} (\beta_k^1 \Delta B_k^1 + \beta_k^2 \Delta B_k^2 + \gamma_k \Delta S_k) - C_n, \tag{5.20}$$

where the additional parameter $C = (C_n)_{0\leqslant n\leqslant N}$ of the portfolio is an increasing stochastic sequence that is interpreted as the total consumption up to the time n.

Auxiliary results. Dual characterization of the model. We consider stochastic sequences V which have a representation

$$V_n = V_0 + \sum_{k=1}^{n} H_k^i \Delta X_k^i - C_n, \tag{5.21}$$

where $H_k = (H_k^1, H_k^2, H_k^3) = (\beta_k^1, \beta_k^2, \gamma_k)$, $X_k = (X_k^1, X_k^2, X_k^3) = (B_k^1, B_k^2, S_k)$, and C_n is an increasing stochastic sequence. The problem is to find a criterion for the existence of an optional decomposition for the sequence $V = (V_n)_{0\leqslant n\leqslant N}$ given by (5.21) in the model (5.18) of a discrete financial market. For this it is necessary to determine the family $\Im$, find the upper variation $A^{\Im}(\mathbf{Q})$ for all measures $\mathbf{Q} \in P(\Im)$, and use Theorem 5.1 or Theorem 5.2.

The representation (5.21) is a discrete analogue of the representation (5.1). In view of (5.5), the family $\Im$ has the form

$$\Im = \Im(H, X) = \Big\{ Y : Y = \sum H\Delta X,\ H \in \aleph \Big\}, \tag{5.22}$$

where the constraints $\aleph$ are determined by the inequalities $\beta^1 \geqslant 0$ and $\beta^2 \leqslant 0$.

LEMMA 5.3. *Any $Y \in \Im$ has a Doob decomposition*

$$Y = M^Y + A^Y, \tag{5.23}$$

where M^Y is a martingale and A^Y a predictable sequence.

Furthermore, the decomposition is unique, and A^Y is called the compensator of the stochastic sequence Y.

PROOF. Let $M_0^Y = Y_0$, $A_0^Y = 0$, and

$$M_n^Y = M_0^Y + \sum_{k=1}^{n} \bigl(Y_k - \mathbf{E}(Y_k \mid \mathcal{F}_{k-1})\bigr),$$

$$A_n^Y = \sum_{k=1}^{n} \mathbf{E}(\Delta Y_k \mid \mathcal{F}_{k-1}).$$

It is clear that the so-defined M^Y and A^Y do have the required properties. Suppose now that $Y = M'^Y + A'^Y$, where M'^Y is a martingale and A'^Y a predictable sequence. Then

$$A'^Y_{n+1} - A'^Y_n = (A^Y_{n+1} - A^Y_n) + (M^Y_{n+1} - M^Y_n) - (M'^Y_{n+1} - M'^Y_n).$$

Taking the conditional expectation of both sides, we get that

$$A'^Y_{n+1} - A'^Y_n = A^Y_{n+1} - A^Y_n,$$

from which, since $A'^Y_0 = A^Y_0 = 0$, it follows that $A^Y_n = A'^Y_n$ and $M^Y_n = M'^Y_n$ for all $n \geqslant 0$, that is, the decomposition is unique.

Let $U = (U_n)_{n \geqslant 0}$ be a stochastic sequence with $U_0 = 0$. Then the discrete stochastic exponential $\mathcal{E}_n(U)$ is the solution of the difference equation

$$\Delta X_n = X_{n-1} \Delta U_n, \qquad X_0 = 1. \tag{5.24}$$

By (5.24),

$$\mathcal{E}_n(U) = \prod_{k=1}^{n} (1 + \Delta U_k), \qquad \mathcal{E}_0(U) = 1. \tag{5.25}$$

Corresponding to (5.7)–(5.9), we define the sequences (R_n^i), (K_n^i), and D_n for $0 \leqslant n \leqslant N$ and $1 \leqslant i \leqslant 3$ as follows in the given model:

$$R_n^1 = r^1 n, \quad R_n^2 = r^2 n, \quad R_n^3 = \sum_{k=1}^{n} \rho_k, \tag{5.26}$$

$$\begin{gathered} K_n^1 = \frac{\beta_n^1 B_{n-1}^1}{V_{n-1}} I_{\{V_{n-1} > 0\}}, \quad K_n^2 = \frac{\beta_n^2 B_{n-1}^2}{V_{n-1}} I_{\{V_{n-1} > 0\}}, \\ K_n^3 = \frac{\gamma_n S_{n-1}}{V_{n-1}} I_{\{V_{n-1} > 0\}}, \end{gathered} \tag{5.27}$$

$$D_n = \sum_{k=1}^{n} \frac{\Delta C_k}{V_{k-1}} I_{\{V_{k-1} > 0\}}. \tag{5.28}$$

Let $\mathfrak{K}$ be the family of all predictable sequences $K = (K_n^1, K_n^2, K_n^3)_{0 \leqslant n \leqslant N}$ taking values in the set

$$\mathcal{A} = \Bigl\{ a \in \mathbb{R}^3 : \sum_{i=1}^{3} a_i \leqslant 1,\ a_1 \geqslant 0,\ a_2 \leqslant 0,\ a_3 \in \mathbb{R} \Bigr\}. \tag{5.29}$$

By analogy with (5.11) we consider the family

$$\mathfrak{I} = \mathfrak{I}(\mathfrak{K}, R) = \Bigl\{ Y : Y = \sum K \Delta R,\ K \in \mathfrak{K} \Bigr\}. \tag{5.30}$$

Then the family $P(\Im)$ of probability measures has the form

(5.31)
$$P(\Im) = \Big\{ \mathbf{Q} : \mathbf{Q} \sim \mathbf{P}, A^{\Im}(\mathbf{Q})_n = \operatorname*{ess\,sup}_{Y \in \Im} A_n^Y(\mathbf{Q}) < +\infty,$$
$$0 \leqslant n \leqslant N,\ A^Y(\mathbf{Q}) \text{ the compensator of } Y \text{ with respect to } \mathbf{Q} \Big\}.$$

We can now state the following result for the model (5.18).

PROPOSITION 5.1. *The positive stochastic sequence $V = (V_n)_{0 \leqslant n \leqslant N}$ is the value of a strategy with consumption if and only if for every $\mathbf{Q} \in P(\Im)$*

$$r^1 \leqslant \mathbf{E}_{\mathbf{Q}}(\rho_k \mid \mathfrak{F}_{k-1}) \leqslant r^2, \quad \forall k, 1 \leqslant k \leqslant N, \tag{5.32}$$

and the sequence $V_n \big/ \prod_{k=1}^n \big(1 + \mathbf{E}_{\mathbf{Q}}(\rho_k \mid \mathfrak{F}_{k-1})\big)$, $0 \leqslant n \leqslant N$, is a supermartingale with respect to $\mathbf{Q}$.

PROOF. We reduce the proof to Theorem 5.2. For this it is necessary to find the class of measures $\mathbf{Q} \in P(\Im)$ and compute the value of the upper variation $A^{\Im}(\mathbf{Q})$ for such measures.

Suppose that $\Re$ and $\Im$ satisfy (5.29) and (5.30), respectively. Then by (5.26)–(5.28) we can assert that V is the value of a strategy with consumption, that is, it satisfies the relation (5.19), if and only if

$$V_n = V_0 + \sum_{k=1}^n V_{k-1}(K_k \Delta R_k - \Delta D_k). \tag{5.33}$$

In view of the definition of the stochastic exponential, (5.33) is equivalent to

$$V_n = V_0 \prod_{k=1}^n (1 + K_k \Delta R_k - D_k). \tag{5.34}$$

Further, according to Lemma 5.3,

$$\begin{aligned} A_n^Y(\mathbf{Q}) &= \sum_{k=1}^n \mathbf{E}_{\mathbf{Q}}\big(a_1 r^1 + a_2 r^2 + a_3 \rho_k \mid \mathfrak{F}_{k-1}\big) \\ &= \sum_{k=1}^n \big(a_1 r^1 + a_2 r^2 + a_3 \mathbf{E}_{\mathbf{Q}}(\rho_k \mid \mathfrak{F}_{k-1})\big). \end{aligned}$$

Let $x_k = \mathbf{E}_{\mathbf{Q}}(\rho_k \mid \mathfrak{F}_{k-1})$. To find the class $P(\Im)$ of measures defined in (5.31) we need to investigate the boundedness of the function

$$G_k(a) = a_1 r^1 + a_2 r^2 + a_3 x_k, \qquad 1 \leqslant k \leqslant N.$$

Suppose that the measure $\mathbf{Q}$ is such that there is a k with $x_k > r^2$, that is, $x_k = r^2 + \alpha$, $\alpha > 0$. Then

$$G_k(a) = a_1 r^1 + a_2 r^2 + a_3(r^2 + \alpha) = a_1 r^1 + (a_2 + a_3) r^2 + a_3 \alpha.$$

For $a_1 = 1$ and $a_3 = -a_2$ the function $G_k(a)$ is $= r^1 + a_3\alpha$, and as $a_3 \to +\infty$ we get that $G_k(a) \to +\infty$, that is, $\mathbf{Q} \notin P(\Im)$.

Suppose now that $\mathbf{Q}$ is such that there is a k with $x_k < r^1$, that is, $x_k = r^1 - \alpha$, $\alpha > 0$. Then $G_k(a) = (a_1 + a_3) r^1 + a_2 r^2 - \alpha a_3$, and $G_k(a) \to +\infty$ as $a_3 \to -\infty$. Hence, $\mathbf{Q} \notin P(\Im)$.

It remains to examine the case

$$r^1 \leqslant x_k \leqslant r^2 \quad \text{for all} \quad k = 1, \dots, N,$$

so that $r^1 + \alpha^1 = x_k = r^2 - \alpha_2$, where $\alpha_1, \alpha_2 \geqslant 0$. Consequently,

$$\begin{aligned} G_k(a) &= a_1(x_k - \alpha_1) + a_2(x_k + \alpha_2) + a_3 x_k \\ &= x_k\Bigl(\sum_{i=1}^{3} a_i\Bigr) - a_1\alpha_1 + a_2\alpha_2 \\ &\leqslant x_k\Bigl(\sum_{i=1}^{3} a_i\Bigr) \leqslant x_k = \mathbf{E}_{\mathbf{Q}}(\rho_k \,|\, \mathcal{F}_{k-1}). \end{aligned}$$

Hence, a measure $\mathbf{Q} \sim \mathbf{P}$ belongs to $P(\Im) \iff$ the condition (5.32) holds for it. The upper variation sequence for the family $\Im$ with respect to a $\mathbf{Q} \in P(\Im)$ is

$$\mathcal{A}_n^{\Im}(\mathbf{Q}) = \sum_{k=1}^{n} \mathbf{E}_{\mathbf{Q}}(\rho_k \,|\, \mathcal{F}_{k-1}).$$

In view of (5.25) the proof follows immediately from Theorem 5.2.

The structure of an optimal hedge in the binomial model with different interest rates.

We consider a European option with exercise time N and payment function $f_N \geqslant 0$. Let $\Im$ and $P(\Im)$ satisfy the relations (5.30) and (5.31), respectively, and assume the condition (5.32). Then Theorem 5.5 takes the following form for the model of a financial market under discussion.

PROPOSITION 5.2. *The set of hedging strategies is nonempty if and only if*

$$\sup_{\mathbf{Q}\in P(\Im)} \mathbf{E}_{\mathbf{Q}}\Bigl[f_N \Big/ \prod_{k=1}^{N}\bigl(1 + \mathbf{E}_{\mathbf{Q}}(\rho_k \,|\, \mathcal{F}_{k-1})\bigr)\Bigr] < \infty.$$

In this case a minimal hedging strategy exists, and its value at the time n is

$$V_n = \operatorname*{ess\,sup}_{\mathbf{Q}\in P(\Im)} \mathbf{E}_{\mathbf{Q}}\Bigl[f_N \Big/ \prod_{k=n+1}^{N}\bigl(1 + \mathbf{E}_{\mathbf{Q}}(\rho_k \,|\, \mathcal{F}_{k-1})\bigr) \Bigm| \mathcal{F}_n\Bigr], \quad 0 \leqslant n \leqslant N. \tag{5.35}$$

The proof is analogous to that of Theorem 5.5.

If the price of a European option is now defined as the initial value of a minimal hedging strategy, then for the binomial model of a financial market with different interest rates the price $\mathbb{C}$ of a European option is

$$\mathbb{C} = V_0 = \sup_{\mathbf{Q}\in P(\Im)} \mathbf{E}_{\mathbf{Q}}\Bigl[f_N \Big/ \prod_{k=1}^{N}\bigl(1 + \mathbf{E}_{\mathbf{Q}}(\rho_k \,|\, \mathcal{F}_{k-1})\bigr)\Bigr]. \tag{5.36}$$

To conclude our investigation of the model (5.18) we present difference equations for the value of a minimal hedging strategy for a contingent claim of the form

$$f_N = g(S_N). \tag{5.37}$$

PROPOSITION 5.3. *If a contingent claim satisfies (5.37), then the value of a minimal hedging strategy can be expressed in the form*

$$\widehat{V}_n = \widehat{v}(S_n, n), \tag{5.38}$$

where the function $\widehat{v}(S_n, n)$ is a solution of the difference equation

$$\begin{aligned}\widehat{v}(S_{n-1}, n-1) = \max\Big\{ &\widehat{v}(S_{n-1}(1+b), n)\frac{r^2-a}{b-a} + \widehat{v}(S_{n-1}(1+a), n)\frac{b-r^2}{b-a},\\ &\widehat{v}(S_{n-1}(1+b), n)\frac{r^1-a}{b-a} + \widehat{v}(S_{n-1}(1+a), n)\frac{b-r^1}{b-a}\Big\}\end{aligned}$$

with the boundary condition

$$\widehat{v}(S_N, N) = g(S_N).$$

To prove this we use the method of backward induction. Indeed, the validity of the equality

$$\widehat{V}_k = \widehat{v}(S_k, k) \tag{5.39}$$

for $k = N$ follows from the definition of the function $\widehat{v}(S_n, n)$.

Suppose now that (5.39) is true for all $k \geqslant n$. Then it suffices to show that this equality remains valid for $k = n-1$. To this end we use the fact that

$$\operatorname*{ess\,sup}_{\mathbf{Q}\in P(\Im)} \mathbf{E}_{\mathbf{Q}}\big(\widehat{V}_n - \widehat{V}_{n-1} \,|\, \mathcal{F}_{n-1}\big) = 0. \tag{5.40}$$

Let $\mathbf{Q} \in P(\Im)$ be arbitrary. Since

$$\mathbf{E}_{\mathbf{Q}}\rho_n = p_{\mathbf{Q}} b + (1 - p_{\mathbf{Q}})a,$$

where $p_{\mathbf{Q}} = \mathbf{P}^{\mathbf{Q}}(\rho_n = b)$ and $1 - p_{\mathbf{Q}} = \mathbf{P}^{\mathbf{Q}}(\rho_n = a)$, it follows that $p_{\mathbf{Q}} = \dfrac{\mathbf{E}_{\mathbf{Q}}\rho_n - a}{b-a}$. From the condition (5.32) we get that

$$\frac{r^1 - a}{b-a} \leqslant p_{\mathbf{Q}} \leqslant \frac{r^2 - a}{b-a}.$$

Using the induction hypothesis, we now calculate the conditional expectation:

$$\begin{aligned}&\mathbf{E}_{\mathbf{Q}}\big(\widehat{V}_n - \widehat{V}_{n-1} \,|\, \mathcal{F}_{n-1}\big)\\ &\quad = \mathbf{E}_{\mathbf{Q}}\big(\widehat{v}(S_n, n) - \widehat{V}_{n-1} \,|\, \mathcal{F}_{n-1}\big)\\ &\quad = \mathbf{E}_{\mathbf{Q}}\big(\widehat{v}(S_{n-1}(1+\rho_n), n) - \widehat{V}_{n-1} \,|\, \mathcal{F}_{n-1}\big)\\ &\quad = \widehat{v}\big(S_{n-1}(1+a), n\big)(1-p_{\mathbf{Q}}) + \widehat{v}\big(S_{n-1}(1+b), n\big)p_{\mathbf{Q}} - \widehat{V}_{n-1}\\ &\quad = h(p_{\mathbf{Q}}).\end{aligned}$$

Consequently, (5.40) is equivalent to the equality

$$\sup_{\frac{r^1-a}{b-a} \leqslant p_{\mathbf{Q}} \leqslant \frac{r^2-a}{b-a}} h(p_{\mathbf{Q}}) = 0.$$

Since the function $h(x)$ is linear, its maximum is attained at an endpoint of the interval $\left[\dfrac{r^1-a}{b-a}, \dfrac{r^2-a}{b-a}\right]$, and hence

$$\widehat{V}_{n-1} = \max\Big\{\widehat{v}(S_{n-1}(1+b),n)\frac{r^2-a}{b-a} + \widehat{v}(S_{n-1}(1+a),n)\frac{b-r^2}{b-a},$$
$$\widehat{v}(S_{n-1}(1+b),n)\frac{r^1-a}{b-a} + \widehat{v}(S_{n-1}(1+a),n)\frac{b-r^1}{b-a}\Big\},$$

from which

$$\widehat{V}_{n-1} = \widehat{v}(S_{n-1}, n-1).$$

We now treat the case of a payment function of the form

$$f_N = \sup_{n\leqslant N} S_n - S_N. \tag{5.41}$$

PROPOSITION 5.4. *If a contingent claim satisfies* (5.41), *then the value of a minimal hedging strategy can be expressed in the form*

$$\widehat{V}_n = S_n\widehat{v}(X_n,n), \tag{5.42}$$

where

$$X_n = \frac{\sup_{k\leqslant n} S_k}{S_n}, \tag{5.43}$$

and the function $\widehat{v}(S_n,n)$ *is a solution of the difference equation*

$$\widehat{v}(X_{n-1},n-1) = \max\Big\{\widehat{v}\Big(\max\Big\{\frac{X_{n-1}}{1+a},1\Big\},n\Big)\frac{(1+a)(b-r^2)}{b-a}$$
$$+\widehat{v}\Big(\max\Big\{\frac{X_{n-1}}{1+b},1\Big\},n\Big)\frac{(1+b)(r^2-a)}{b-a},$$
$$\widehat{v}\Big(\max\Big\{\frac{X_{n-1}}{1+a},1\Big\},n\Big)\frac{(1+a)(b-r^1)}{b-a}$$
$$+\widehat{v}\Big(\max\Big\{\frac{X_{n-1}}{1+b},1\Big\},n\Big)\frac{(1+b)(r^1-a)}{b-a}\Big\}$$

with the boundary condition

$$\widehat{v}(X_N,N) = X_N - 1.$$

PROOF. We proceed by analogy with the proof of Proposition 5.3. The relation

$$\widehat{V}_k = S_k\widehat{v}(X_k,k) \tag{5.44}$$

for $k=N$ follows from the definition of the function $\widehat{v}(X_n,n)$.

Suppose now that (5.44) is true for all $k\geqslant n$. We prove that it remains true for $k=n-1$.

Since

$$X_n = \frac{\sup_{k\leqslant n} S_k}{S_k} = \frac{\max\{S_{n-1}X_{n-1}, S_n\}}{S_{n-1}(1+\rho_n)}$$
$$= \frac{S_{n-1}\max\{X_{n-1},1+\rho_n\}}{S_{n-1}(1+\rho_n)} = \max\Big\{\frac{X_{n-1}}{1+\rho_n},1\Big\},$$

the conditional expectation is

$$\begin{aligned}
&\mathbf{E}_{\mathbf{Q}}(\widehat{V}_n - \widehat{V}_{n-1} \,|\, \mathcal{F}_{n-1}) \\
&\quad = \mathbf{E}_{\mathbf{Q}}(S_n \widehat{v}(X_n, n) - \widehat{V}_{n-1} \,|\, \mathcal{F}_{n-1}) \\
&\quad = \mathbf{E}_{\mathbf{Q}}\biggl(S_{n-1}(1+\rho_n)\,\widehat{v}\biggl(\max\Bigl\{\frac{X_{n-1}}{1+\rho_n}, 1\Bigr\}, n\biggr) - \widehat{V}_{n-1} \biggm| \mathcal{F}_{n-1}\biggr) \\
&\quad = S_{n-1}\biggl((1+a)\,\widehat{v}\biggl(\max\Bigl\{\frac{X_{n-1}}{1+a}, 1\Bigr\}, n\biggr)(1-p_{\mathbf{Q}}) \\
&\qquad\qquad + (1+b)\,\widehat{v}\biggl(\max\Bigl\{\frac{X_{n-1}}{1+b}, 1\Bigr\}, n\biggr) p_{\mathbf{Q}}\biggr) - \widehat{V}_{n-1}.
\end{aligned}$$

Using arguments analogous to those in the proof of Proposition 5.3, we get that

$$\begin{aligned}
\widehat{V}(X_{n-1}, n-1) = S_{n-1} \max\biggl\{ &\widehat{v}\biggl(\max\Bigl\{\frac{X_{n-1}}{1+a}, 1\Bigr\}, n\biggr) \frac{(1+a)(b-r^2)}{b-a} \\
&+ \widehat{v}\biggl(\max\Bigl\{\frac{X_{n-1}}{1+b}, 1\Bigr\}, n\biggr) \frac{(1+b)(r^2-a)}{b-a}, \\
&\widehat{v}\biggl(\max\Bigl\{\frac{X_{n-1}}{1+a}, 1\Bigr\}, n\biggr) \frac{(1+a)(b-r^1)}{b-a} \\
&+ \widehat{v}\biggl(\max\Bigl\{\frac{X_{n-1}}{1+b}, 1\Bigr\}, n\biggr) \frac{(1+b)(r^1-a)}{b-a} \biggr\},
\end{aligned}$$

from which

$$\widehat{V}_{n-1} = S_{n-1}\widehat{v}(X_{n-1}, n-1).$$

The Black–Scholes model with different interest rates.

Description of the model. We consider a financial market with three assets: a deposit account, a credit account, and a stock. The changes in the value B^1 of the deposit account, the value B^2 of the credit account, and the dynamics of the stock price S have the forms

$$\begin{aligned}
dB_t^1 &= B_t^1 r^1\,dt, \quad B_0^1 = 1, \\
dB_t^2 &= B_t^2 r^2\,dt, \quad B_0^2 = 1, \\
dS_t &= S_{t-}(\mu\,dt + \sigma\,dW_t), \quad S_0 > 0.
\end{aligned} \tag{5.45}$$

Here r^1 and r^2 are the interest rates received and paid for the deposit and credit accounts, respectively, $r^1 \leqslant r^2$; μ is the mean rate of return on the stock, $\mu > r^1$; σ is the volatility, $\sigma > 0$; $W = (W_t)_{0 \leqslant t \leqslant T}$ is a Wiener process.

We assume that all processes are defined on a standard stochastic base $(\Omega, \mathcal{F}_T, \mathbb{F} = (\mathcal{F}_t)_{0\leqslant t\leqslant T}, \mathbf{P})$, where $T < +\infty$ is a time in the future interpreted as the exercise time for the contingent claim. We assume that $\mathbb{F} = \mathbb{F}^S = \mathbb{F}^W$.

An *investor strategy* in the market model (5.45) is defined to be an adapted process $H = (H_t)_{0\leqslant t\leqslant T} = (\beta_t^1, \beta_t^2, \gamma_t)_{0\leqslant t\leqslant T}$, where β_t^1, β_t^2, and γ_t are the numbers of units in the deposit and credit accounts and the number of shares in the portfolio at the time t. Here it is assumed that $\beta^1 \geqslant 0$ and $\beta^2 \leqslant 0$. Then the value of the strategy at time t is

$$V_t = \beta_t^1 B_t^1 + \beta_t^2 B_t^2 + \gamma_t S_t. \tag{5.46}$$

A portfolio $\pi = (H, C) = (\beta^1, \beta^2, \gamma, C)$ is called a *strategy with consumption* if its value at time t satisfies the equation

$$V_t = V_0 + \int_0^t (\beta_u^1\, dB_u^1 + \beta_u^2\, dB_u^2 + \gamma_u\, dS_u) - C_t, \tag{5.47}$$

where $C = (C_t)_{0\leqslant t\leqslant T}$ is an increasing adapted process interpreted as the total consumption process in the portfolio up to and including the time t.

The equality $C_t \equiv 0$ corresponds to the definition of a *self-financing strategy.*

Auxiliary results. Dual characterization of the model. We consider processes V satisfying (5.1) with $H_s = (\beta_s^1, \beta_s^2, \gamma_s)$ and $X_s = (B_s^1, B_s^2, S_s)$. Suppose that the value (5.47) of the strategy with consumption is positive at each moment of time. Then the following representation is valid:

$$V = V_0\, \mathcal{E}\Big(\int K\, dR - D\Big), \tag{5.48}$$

where, by (5.7)–(5.9),

$$R = (R_t^1, R_t^2, R_t^3) = (r^1 t, r^2 t, \mu t + \sigma W_t) \tag{5.49}$$

is the process of relative growth of the stock prices,

$$\begin{aligned} K &= (K_t^1, K_t^2, K_t^3) \\ &= \left(\frac{\beta_t^1 B_{t-}^1}{V_{t-}} I_{\{V_{t-}>0\}}, \frac{\beta_t^2 B_{t-}^2}{V_{t-}} I_{\{V_{t-}>0\}}, \frac{\gamma_t S_{t-}}{V_{t-}} I_{\{V_{t-}>0\}}\right) \end{aligned} \tag{5.50}$$

is the process of the fractions of the portfolio's value invested at the time t in the deposit account, the credit account, and the stock, and

$$D = (D_t) = \int_0^t \frac{dC_u}{V_{u-}} I_{\{V_{u-}>0\}} \tag{5.51}$$

is the process of relative consumption in the portfolio up to the time t.

The process $K = (K_t^i)_{1\leqslant i\leqslant 3,\, 0\leqslant t\leqslant T}$ defined in (5.50) belongs to a set $\mathfrak{K}$ which satisfies the conditions of convexity and closedness and contains the process $K \equiv 0$.

Let

$$\mathfrak{I} = \{Y : Y = K * R,\ K \in \mathfrak{K}\}. \tag{5.52}$$

The problem is to find the class $\mathbf{P}(\mathfrak{I})$ of measures and calculate the value of the upper variation $A^{\mathfrak{I}}(\mathbf{Q})$ for this class. We present some theoretical facts to be used in arguments below.

A variant of Girsanov's theorem for semimartingales was mentioned in Chapter 2. Here we need a concretization of it.

Girsanov's theorem for local martingales.

THEOREM 5.7. *Assume that* $\mathbf{Q} \overset{\text{loc}}{\ll} \mathbf{P}$ *and* Z *is the density process. Let* M *be a* $\mathbf{P}$*-local martingale with* $M_0 = 0$ *such that the* $\mathbf{P}$*-quadratic covariance* $[M, Z]$ *has* $\mathbf{P}$*-locally integrable variation. Let* $\langle M, Z\rangle$ *be its* $\mathbf{P}$*-compensator. Then the process*

$$M' = M - \frac{1}{Z_-} \circ \langle M, Z\rangle \tag{5.53}$$

is defined $\mathbf{Q}$*-a.s. and is a* $\mathbf{Q}$*-local martingale. Moreover, the* $\mathbf{P}$*-quadratic characteristic* $\langle M^c, M^c\rangle$ *of the continuous component* M^c *(with respect to* $\mathbf{P}$*) of the local martingale* M *is also a version of the* $\mathbf{Q}$*-quadratic characteristic of the continuous part (with respect to* $\mathbf{Q}$*) of the local martingale* M'.

We consider a d-dimensional semimartingale $X = (X^1, \dots, X^d)$ defined on the stochastic base $(\Omega, \mathcal{F}_T, \mathbb{F} = (\mathcal{F}_t)_{0\leqslant t\leqslant T}, \mathbf{P})$.

Denote by C_t^d the class of all bounded functions $h\colon \mathbb{R}^d \to \mathbb{R}^d$ with compact support such that $h(x) = x$ in a neighborhood of zero.

Let $h \in C_t^d$. Then $\Delta X_s - h(\Delta X_s) \neq 0$ only if $|\Delta X_s| > b$ for some $b > 0$. Therefore, the formulas

$$\begin{cases} \widetilde{X}(h)_t = \sum\limits_{s\leqslant t} (\Delta X_s - h(\Delta X_s)), \\ X(h) = X - \widetilde{X}(h) \end{cases}$$

define a process $\widetilde{X}(h) \in \mathcal{V}$ and a d-dimensional semimartingale $X(h)$. Since the process $\Delta X(h) = h(\Delta X)$ is bounded, it follows that $X(h)$ is a special semimartingale. We consider its canonical decomposition:

$$X(h) = X_0 + M(h) + B(h), \quad M(h) \in \mathcal{M}_{\mathrm{loc}}, \ B(h) \in \mathcal{V} \text{ and is predictable.}$$

Let h be a fixed function in C_t^d. The *characteristics* of the semimartingale X associated with the cutoff function h are defined to be the triplet (B, C, ν), where:

(i) $B = (B^i)_{i\leqslant d}$ is a predictable process in $\mathcal{V}^d$, namely, the $B(h)$ appearing in the canonical decomposition for $X(h)$;

(ii) $C = (C^{ij})_{i,j\leqslant d}$ is a continuous process in $\mathcal{V}^{d\times d}$, namely, $C^{ij} = \langle X^{i,c}, X^{j,c}\rangle$;

(iii) ν is a predictable random measure on $\mathbb{R}_+ \times E$, namely, the compensator of the jump measure μ^X of the process X, where $E = \mathbb{R}^d \setminus \{0\}$.

Girsanov's theorem for semimartingales. Let $X = (X^i)_{i\leqslant d}$ be a semimartingale with characteristics (B, C, ν) with respect to a given cutoff function h, let X^c be the continuous component of X with respect to $\mathbf{P}$, and let A be an increasing predictable process such that $C^{ij} = c^{ij} A$.

THEOREM 5.8. *Assume that* $\mathbf{Q} \overset{\mathrm{loc}}{\ll} \mathbf{P}$ *with local density* Z_t*, and let* X *be the semimartingale introduced above. Then there exist a* $\mathbf{P}\otimes E$*-measurable nonnegative function* Y *and a predictable process* $\beta = (\beta^i)_{i\leqslant d}$ *satisfying the conditions*

$$|h(x)(Y-1)| * \nu_t < \infty \quad (\mathbf{Q}\text{-}a.s.) \quad \textit{for any } t \in \mathbb{R}_+,$$

$$\Big|\sum_{j\leqslant d} c^{ij}\beta^i\Big| \circ A_t < \infty \quad (\mathbf{Q}\text{-}a.s.) \quad \textit{for any } t \in \mathbb{R}_+,$$

$$\Big(\sum_{j,k\leqslant d} \beta^j c^{jk} \beta^k\Big) \circ A_t < \infty \quad (\mathbf{Q}\text{-}a.s.) \quad \textit{for any } t \in \mathbb{R}_+,$$

and such that the predictable characteristics of X *with respect to* $\mathbf{Q}$ *are*

$$\begin{aligned} B'^i &= B^i + \Big(\sum_{j\leqslant d} c^{ij}\beta^j\Big) \circ A + h^i(x)(Y-1) * \nu, \\ C' &= C, \\ \nu' &= Y * \nu. \end{aligned} \tag{5.54}$$

Moreover,

$$\langle Z^c, X^{i,c}\rangle = \Big(\sum_{j\leqslant d} c^{ij}\beta^j Z_-\Big)\circ A. \tag{5.55}$$

To get a representation of the Wiener process W_t with respect to a measure $\mathbf{Q}\in P(\Im)$, we can use both Theorem 5.1 and Theorem 5.2. It follows from Girsanov's theorem for local martingales that with respect to $\mathbf{Q}\sim\mathbf{P}$ the process W_t can be represented in the form $W = M + \dfrac{1}{Z_-}\circ\langle W, Z\rangle$, where M is a local martingale with respect to $\mathbf{Q}$.

We now consider Girsanov's theorem for semimartingales when the process X is a standard Wiener process W_t. This W_t has characteristics $(B, C, \nu) = (0, t, 0)$ with respect to $\mathbf{P}$. By Theorem 5.8, W_t is a semimartingale with characteristics $(B', C', \nu') = \left(\int_0^t b_s\,ds, t, 0\right)$ with respect to $\mathbf{Q}\sim\mathbf{P}$, where $b_s(\omega)$ is a predictable process satisfying the conditions

$$\begin{gathered}\int_0^t |b_s|\,ds<\infty, \qquad \int_0^t b_s^2<\infty,\\ \langle Z^c, W\rangle = \int_0^t b_s Z_{s-}\,ds \quad (\mathbf{Q}\text{-a.s. for all } t\in\mathbb{R}_+).\end{gathered} \tag{5.56}$$

Then with respect to $\mathbf{Q}\sim\mathbf{P}$ the process W_t can be represented in the form $W_t = M_t + \int_0^t b_s\,ds$, where M_t is a local martingale with respect to $\mathbf{Q}$ and $b_s(\omega)$ satisfies (5.56).

There is a process, unique up to a negligible set and called the *compensator* $A^p = \widetilde{A}$ of A, which is a predictable process in the class $\mathcal{A}_{\mathrm{loc}}$ and such that $A - A^p$ is a local martingale.

By Theorem 5.7, $[W, Z]$ is in $\mathcal{A}_{\mathrm{loc}}$ with respect to $\mathbf{P}$ and is continuous. Its compensator satisfies the condition $[W, Z]^p = \langle W, Z\rangle = [W, Z]\in\mathcal{A}_{\mathrm{loc}}$ and is predictable by virtue of continuity.

In this case

$$\frac{1}{Z_-}\circ\langle W, Z\rangle = \frac{1}{Z_-}\circ[W, Z] \tag{5.57}$$

with respect to $\mathbf{P}$. Let us prove that there is a process $b_s(\omega)$ satisfying (5.56) such that

$$\frac{1}{Z_-}\circ[W, Z] = \int_0^t b_s\,ds \quad (\mathbf{Q}\text{-a.s. for } t\in\mathbb{R}_+).$$

Indeed, by Theorem 5.8 and the condition (5.57),

$$\frac{1}{Z_-}\circ[W, Z]_t = \int_0^t \frac{1}{Z_{s-}} b_s Z_{s-}\,ds = \int_0^t b_s\,ds \quad (\mathbf{Q}\text{-a.s. for all } t\in\mathbb{R}_+).$$

Thus, the two Girsanov's theorems above give the same representation for the Wiener process W_t with respect to a measure $\mathbf{Q}\sim\mathbf{P}$ with $\mathbf{Q}\in P(\Im)$:

$$W_t = M_t + \int_0^t b_s\,ds, \tag{5.58}$$

where M_t is a local martingale with respect to $\mathbf{Q}$ and $b_s(\omega)$ satisfies (5.56).

We consider the family $\Im$ of semimartingales defined in (5.52). Then by (5.58) a process $Y \in \Im$ has the following form with respect to $\mathbf{Q}$:

$$Y = \int_0^t \left(K_u^1 r^1 + K_u^2 r^2 + K_u^3(\mu + \sigma b_u)\right) du + \int_0^t \sigma K_u^3 \, dM_u. \tag{5.59}$$

By properties of stochastic integrals, the second term in (5.59) is a local martingale with respect to $\mathbf{Q}$. We then conclude from the definition of the canonical decomposition of a special semimartingale that the compensator $A^Y(\mathbf{Q})_t$ has the form

$$A^Y(\mathbf{Q})_t = \int_0^t \left(K_u^1 r^1 + K_u^2 r^2 + K_u^3(\mu + \sigma b_u)\right) du, \quad 0 \leqslant t \leqslant T. \tag{5.60}$$

We remark that the predictable process K_t takes values in the set

$$\mathcal{A} = \left\{ a \in \mathbb{R}^3 : \sum_{i=1}^3 a_i \leqslant 1,\ a_1 \geqslant 0,\ a_2 \leqslant 0,\ a_3 \in \mathbb{R} \right\}.$$

Then to find the upper variation of the family $\Im$ with respect to $\mathbf{Q}$ we must investigate the boundedness of the function $G_u(a) = r^1 a_1 + r^2 a_2 + x_u a_3$, where $x_u = \mu + \sigma b_u$. By arguments analogous to the preceding model, we conclude that a measure $\mathbf{Q} \sim \mathbf{P}$ belongs to the set $P(\Im)$ if and only if

$$r^1 \leqslant \mu + \sigma b_u \leqslant r^2, \qquad 0 \leqslant u \leqslant T. \tag{5.61}$$

Further, the upper variation process of the family $\Im$ with respect to a measure $\mathbf{Q} \in P(\Im)$ is

$$A^{\Im}(\mathbf{Q})_t = \int_0^t (\mu + \sigma b_u) \, du. \tag{5.62}$$

Then by the definition of the stochastic exponential,

$$\mathcal{E}\left(A^{\Im}(\mathbf{Q})\right)_t = \exp\left\{ \int_0^t (\mu + \sigma b_u) \, du \right\}. \tag{5.63}$$

As a result, we arrive at the following criterion for the existence of an optional decomposition in the Black–Scholes model with different interest rates.

THEOREM 5.9. *The positive process* $V = (V_t)_{0 \leqslant t \leqslant T}$ *satisfies the relation*

$$V = V_0 \, \mathcal{E}(K * R - D),$$

where the processes K, R, *and* D *are defined according to* (5.49)–(5.51), *if and only if for all measures* $\mathbf{Q} \in P(\Im)$ *the condition* (5.61) *holds and the process* $V_t \exp\left\{ -\int_0^t (\mu + \sigma b_s) \, ds \right\}$ *is a supermartingale with respect to* $\mathbf{Q}$.

Structure of an optimal hedge in the Black–Scholes model with different interest rates. Let us consider a European option with exercise time $T < +\infty$ and payment function $f_T \geqslant 0$.

THEOREM 5.10. *The set of hedging strategies is nonempty if and only if*

$$\sup_{\mathbf{Q}\in P(\Im)} \mathbf{E}_{\mathbf{Q}}\big(f_T e^{-\int_0^T(\mu+\sigma b_s)\,ds}\big) < +\infty. \tag{5.64}$$

In this case there is a minimal hedging strategy, and its value at time t is

$$\widehat{V}_t = \operatorname*{ess\,sup}_{\mathbf{Q}\in P(\Im)} \mathbf{E}_{\mathbf{Q}}\big[f_T e^{-\int_t^T(\mu+\sigma b_s)\,ds} \mid \mathcal{F}_t\big]. \tag{5.65}$$

Hence, the price of a European option in the market (5.45) as the initial value of a minimal hedge is equal to

$$\mathbb{C} = \widehat{V}_0 = \sup_{\mathbf{Q}\in P(\Im)} \mathbf{E}_{\mathbf{Q}}\bigg[f_T \exp\bigg\{-\int_0^T (\mu+\sigma b_s)\,ds\bigg\}\bigg], \tag{5.66}$$

where $r^1 \leqslant \mu + \sigma b_s \leqslant r^2$ for all $0 \leqslant s \leqslant T$.

Merton's model with different interest rates.

Description of the model. We consider a financial market with three assets: a deposit account, a credit account, and a stock. The changes in the value B^1 of the deposit account and the value B^2 of the credit account and the dynamics of the stock price S have the form

$$\begin{aligned} dB_t^1 &= B_{t-}^1 r^1\,dt, & B_0^1 &= 1,\\ dB_t^2 &= B_{t-}^2 r^2\,dt, & B_0^2 &= 1,\\ dS_t &= S_{t-}(\mu\,dt - \nu\,d\Pi_t), & S_0 &> 0. \end{aligned} \tag{5.67}$$

Here r^1 and r^2 are the interest rates of the deposit account and the credit account, $r^1 \leqslant r^2$; μ is the average rate of return from the stock, $\mu > r^1$; $\nu < 1$ is a constant; $\Pi = (\Pi_t)_{0\leqslant t\leqslant T}$ is a standard Poisson process with jump intensity $\lambda > 0$ and with the representation

$$\Pi_t = \widetilde{\Pi}_t + \lambda t, \tag{5.68}$$

where $\mathbf{E}\Pi_t = \lambda t$ and $\widetilde{\Pi}_t$ is a martingale with respect to $\mathbf{P}$.

All the processes are bounded on the standard stochastic base $(\Omega, \mathcal{F}_T, \mathbb{F} = (\mathcal{F}_t)_{0\leqslant t\leqslant T}, \mathbf{P})$, where $T < +\infty$ is a moment of time in the future that is interpreted as the exercise time for a contingent claim. It is assumed that $\mathbb{F} = \mathbb{F}^S = \mathbb{F}^\Pi$.

The earlier definitions of a strategy and of a strategy with consumption will be preserved also in this model.

Auxiliary results. Dual characterization of the model. As in the above Black–Scholes model, we characterize the processes V satisfying the equation (5.1), in which $H_s = (\beta_s^1, \beta_s^2, \gamma_s)$ and $X_s = (B_s^1, B_s^2, S_s)$. Assume that the value (5.47) of the strategy with consumption is positive at each moment of time. Then the following representation is valid:

$$V_t = V_0\,\mathcal{E}\Big(\int K\,dR - D\Big), \tag{5.69}$$

where, by (5.7)–(5.9),

$$R = (R_t^1, R_t^2, R_t^3) = (r^1 t, r^2 t, \mu t - \nu\Pi_t) \tag{5.70}$$

is the process of relative growth of the asset prices,

$$
\begin{aligned}
K &= (K_t^1, K_t^2, K_t^3) \\
&= \left(\frac{\beta_t^1 B_{t-}^1}{V_{t-}} I_{\{V_{t-}>0\}}, \frac{\beta_t^2 B_{t-}^2}{V_{t-}} I_{\{V_{t-}>0\}}, \frac{\gamma_t S_{t-}}{V_{t-}} I_{\{V_{t-}>0\}} \right)
\end{aligned}
\tag{5.71}
$$

is the process of fractions of the portfolio value invested at the time t in the deposit account, the credit account, and the stock, and

$$
D = (D_t) = \int_0^t \frac{dC_u}{V_{u-}} I_{\{V_{u-}>0\}}
\tag{5.72}
$$

is the process of relative consumption in the portfolio up to the time t.

The process $K = (K_t^i)_{1 \leqslant i \leqslant 3,\, 0 \leqslant t \leqslant T}$ defined in (5.71) belongs to a set $\mathfrak{R}$ which is convex and closed and contains $K \equiv 0$. Suppose that the family $\mathfrak{I}$ of semimartingales has the form

$$
\mathfrak{I} = \{Y : Y = K * R,\ K \in \mathfrak{R}\}.
\tag{5.73}
$$

The problem is to find a criterion for the existence of an optional decomposition in Merton's model with different interest rates. For this we must find the upper variation of the family $\mathfrak{I}$ with respect to a measure $\mathbf{Q} \in P(\mathfrak{I})$ and then use Theorem 5.2.

Let us consider a cutoff function $h(x) = xI_{\{|x| \leqslant 1\}}$, with $h \in C_t^1$. Since a standard Poisson process Π_t with respect to $\mathbf{P}$ has a representation in the form (5.2), Π_t is a one-dimensional semimartingale with characteristics

$$
\begin{aligned}
B_t(h) &= h(1)\lambda t, \\
C_t &= 0, \\
\nu(\omega; dt, dx) &= \lambda\, dt \times \varepsilon_1(dx).
\end{aligned}
\tag{5.74}
$$

Taking a measure $\mathbf{Q} \sim \mathbf{P}$ with $\mathbf{Q} \in P(\mathfrak{I})$, we see from Girsanov's theorem for semimartingales that with respect to $\mathbf{Q}$ the process Π_t is a semimartingale with characteristics

$$
\begin{aligned}
B_t'(h) &= B_t(h) + \int_0^t \int_{\mathbb{R}} h(x)(Y(\omega; s, x) - 1)\nu(\omega; ds, dx) \\
&= B_t(h) + \int_0^t h(1)(Y(\omega; s, 1) - 1)\lambda\, ds \\
&= h(1)\lambda \int_0^t Y(\omega; s, 1)\, ds = \int_0^t \lambda_s^{\mathbf{Q}}(\omega)\, ds, \\
C_t' &= 0, \\
\nu' &= \int_0^t \int_{\mathbb{R}} Y(\omega; s, x)\varepsilon_1(dx) \times \lambda\, ds.
\end{aligned}
\tag{5.75}
$$

Since Y is a nonnegative function and $\lambda > 0$, it follows that $\lambda_s^{\mathbf{Q}}(\omega) \geqslant 0$.

Thus, with respect to a measure $\mathbf{Q} \sim \mathbf{P}$ with $\mathbf{Q} \in P(\mathfrak{I})$ the Poisson process Π_t admits a representation of the form

$$
\Pi_t = M_t + \int_0^t \lambda_s^{\mathbf{Q}}(\omega)\, ds,
\tag{5.76}
$$

where M_t is a local martingale with respect to $\mathbf{Q}$.

Since $\Im$ satisfies (5.73) and the predictable process K_t takes values in the set

$$\mathcal{A} = \left\{ a \in \mathbb{R}^3 : \sum_{i=1}^{3} a_i \leqslant 1,\ a_1 \geqslant 0,\ a_2 \leqslant 0,\ a_3 \in \mathbb{R} \right\},$$

a process $Y \in \Im$ has the following form with respect to $\mathbf{Q}$:

$$\begin{aligned} Y &= \int_0^t (K_u^1 r^1 + K_u^2 r^2 + K_u^3 \mu)\, du - \int_0^t \nu K_u^3\, d\Pi_u \\ &= \int_0^t \big(K_u^1 r^1 + K_u^2 r^2 + K_u^3(\mu - \nu\lambda_u^{\mathbf{Q}})\big)\, du - \int_0^t \nu K_u^3\, dM_u. \end{aligned} \tag{5.77}$$

By properties of stochastic integrals, the last term is a martingale with respect to $\mathbf{Q} \sim \mathbf{P}$. Using the definition of the canonical decomposition of a special semimartingale, we conclude from (5.77) that the compensator of a process $Y \in \Im$ with respect to $\mathbf{Q} \in P(\Im)$ has the form

$$A^Y(\mathbf{Q})_t = \int_0^t \big(K_u^1 r^1 + K_u^2 r^2 + K_u^3(\mu - \nu\lambda_u^{\mathbf{Q}})\big)\, du \quad \text{for all } t,\ 0 \leqslant t \leqslant T. \tag{5.78}$$

To find the upper variation of the family $\Im$ with respect to $\mathbf{Q}$ we need to investigate the boundedness of the function $G_u(a) = r^1 a_1 + r^2 a_2 + x_u a_3$, where $x_u = \mu - \nu\lambda_u^{\mathbf{Q}}$ is an arbitrary number. By arguments analogous to the preceding, we conclude that a measure $\mathbf{Q} \sim \mathbf{P}$ belongs to $P(\Im)$ if and only if

$$r^1 \leqslant \mu - \nu\lambda_u^{\mathbf{Q}} \leqslant r^2, \qquad 0 \leqslant u \leqslant T. \tag{5.79}$$

Further, the process of the upper variation of the family $\Im$ with respect to $\mathbf{Q} \in P(\Im)$ is

$$A^{\Im}(\mathbf{Q})_t = \int_0^t (\mu - \nu\lambda_u^{\mathbf{Q}})\, du, \qquad 0 \leqslant t \leqslant T. \tag{5.80}$$

According to the definition of the stochastic exponential,

$$\mathcal{E}(A^{\Im}(\mathbf{Q}))_t = \exp\left\{ \int_0^t (\mu - \nu\lambda_u^{\mathbf{Q}})\, du \right\}. \tag{5.81}$$

The next theorem gives a criterion for the existence of an optional decomposition in Merton's model with different interest rates.

THEOREM 5.11. *A positive process $V = (V_t)_{0 \leqslant t \leqslant T}$ satisfies the relation*

$$V = V_0\, \mathcal{E}(K * R - D),$$

where the processes K, R, and D are defined in (5.70)–(5.72), *if and only if the condition* (5.79) *holds and the process $V_t \exp\left\{ -\int_0^t (\mu - \nu\lambda_s^{\mathbf{Q}})\, ds \right\}$ is a supermartingale with respect to $\mathbf{Q}$ for every measure $\mathbf{Q} \in P(\Im)$.*

Structure of an optimal hedge in Merton's model with different interest rates. Let us consider a European option with exercise time $T < +\infty$ and payment function $f_T \geqslant 0$.

THEOREM 5.12. *The set of hedging strategies is nonempty if and only if*

$$\sup_{\mathbf{Q}\in P(\Im)} \mathbf{E_Q}\big[f_T e^{-\int_0^T (\mu-\nu\lambda_s^{\mathbf{Q}})ds}\big] < +\infty. \tag{5.82}$$

In this case a minimal hedging strategy exists, and its value at the time t is

$$\widehat{V}_t = \operatorname*{ess\,sup}_{\mathbf{Q}\in P(\Im)} \mathbf{E_Q}\big[f_T e^{-\int_t^T (\mu-\nu\lambda_s^{\mathbf{Q}})ds} \mid \mathcal{F}_t\big]. \tag{5.83}$$

The proof follows from the proofs of Theorems 5.4–5.6.

As the initial value of a minimal hedge, the price of a European option in the market (5.67) is equal to

$$\mathbb{C} = \widehat{V}_0 = \sup_{\mathbf{Q}\in P(\Im)} \mathbf{E_Q}\left[f_T \exp\left\{-\int_0^T (\mu-\nu\lambda_s^{\mathbf{Q}})\,ds\right\}\right], \tag{5.84}$$

where $r^1 \leqslant \mu-\nu\lambda_s^{\mathbf{Q}} \leqslant r^2$ for all $0\leqslant s\leqslant T$.

In conclusion we present brief calculations in the models (binomial, Black–Scholes, Merton) under an additional constraint known as "prohibition of short selling".

In the presence of this constraint the set $\mathcal{A}$ of values of the process $K = (K_t)_{0\leqslant t\leqslant T}$ has the form

$$\mathcal{A} = \left\{a\in\mathbb{R}^3 : \sum_{i=1}^3 a_i \leqslant 1,\ a_1\geqslant 0,\ a_2\leqslant 0,\ a_3\geqslant 0\right\}.$$

To find the upper variation of a process $Y\in\Im$ with respect to a measure $\mathbf{Q}\in P(\Im)$, we must investigate the boundedness of the function $G_k(a) = a_1r^1 + a_2r^2 + a_3x_k$ for all k, $1\leqslant k\leqslant N$. Here x_k is an arbitrary number.

Suppose that $\mathbf{Q}$ is such that there exists a k with $x_k > r^2$, that is, $x_k = r^2+\alpha$, $\alpha>0$. In this case

$$G_k(a) = a_1r^1 + a_2r^2 + a_3(r^2+\alpha) = a_1r^1 + (a_2+a_3)r^2 + a_3\alpha.$$

From this with $a_1 = 1$ and $a_3 = -a_2$ it is clear that $G_k(a) = r^1 + a_3\alpha \to +\infty$ as $a_3\to+\infty$, that is, $\mathbf{Q}\notin P(\Im)$.

We now consider the case when there exists a k with $x_k < r^1$, that is, $x_k = r^1-\alpha$, $\alpha>0$. Then, since $r^1\leqslant r^2$, we have

$$G_k(a) = (a_1+a_3)r^1 + a_2r^2 - a_3\alpha \leqslant r^2\left(\sum_{i=1}^3 a_i\right) - a_3\alpha \leqslant r^2,$$

that is, $\mathbf{Q}\in P(\Im)$.

It remains to check the case when $r^1\leqslant x_k\leqslant r^2$ for all k, $1\leqslant k\leqslant N$. We have $r^1+\alpha_1 = x_k = r^2-\alpha_2$, where $\alpha_1,\alpha_2\geqslant 0$. By the definition of the set $\mathcal{A}$, this gives us that

$$\begin{aligned}G_k(a) &= a_1(x_k-\alpha_1) + a_2(x_k+\alpha_2) + a_3x_k\\ &= x_k\left(\sum_{i=1}^3 a_i\right) - a_1\alpha_1 + a_2\alpha_2 \leqslant x_k\left(\sum_{i=1}^3 a_i\right) \leqslant x_k,\end{aligned}$$

that is, $\mathbf{Q}\in P(\Im)$.

Thus,

$$\sup_{a\in\mathcal{A}} G_k(a) = \begin{cases} r^2 & \text{if } x_k < r^1, \\ x_k & \text{if } r^1 \leqslant x_k \leqslant r^2. \end{cases}$$

Hence, $\mathbf{Q} \in P(\Im)$ if and only if $x_k \leqslant r^2$. Further, $\sup_{a\in\mathcal{A}} G_k(a) = r^2$.

1. Let us consider the binomial model with different interest rates when there is a prohibition on short selling. In this case, $x_k = \mathbf{E}(\rho_k \mid \mathcal{F}_{k-1}) = \mathbf{E}\rho_k$. Then $\mathbf{Q} \in P(\Im)$ if and only if $\mathbf{E}\rho_k \leqslant r^2$. Furthermore,

$$A^{\Im}(\mathbf{Q}) = r^2 n, \qquad \mathcal{E}(A^{\Im}(\mathbf{Q}))_n = \prod_{k=1}^{n}(1+r^2) = (1+r^2)^n,$$

and the price of a European option is equal to

$$\mathbb{C} = \widehat{V}_0 = \sup_{\mathbf{Q}\in P(\Im)} \mathbf{E}_{\mathbf{Q}}\big[f_N(1+r^2)^{-N}\big].$$

2. In the case of the Black–Scholes model we have $k = t \in [0,T]$, $x_t = \mu + \sigma b_t$, and $\mathbf{Q} \in P(\Im)$ if and only if $\mu+\sigma b_t \leqslant r^2$, where $b = (b_t)_{0\leqslant t\leqslant T}$ is a process satisfying (5.61). Furthermore,

$$A^{\Im}(\mathbf{Q})_t = \int_0^t r^2\,ds = r^2 t, \qquad \mathcal{E}(A^{\Im}(\mathbf{Q}))_t = e^{r^2 t},$$

and the price of a European option is equal to

$$\mathbb{C} = \widehat{V}_0 = \sup_{\mathbf{Q}\in P(\Im)} \mathbf{E}_{\mathbf{Q}}[f_T e^{-r^2 T}].$$

3. In the case of the Merton model $x_t = \mu - \nu\lambda_t^{\mathbf{Q}}$, and $\mathbf{Q} \in P(\Im)$ if and only if $\mu - \nu\lambda_t^{\mathbf{Q}} \leqslant r^2$, where $\lambda^{\mathbf{Q}} = (\lambda_t^{\mathbf{Q}})_{0\leqslant t\leqslant T}$ is a process satisfying (5.79). Furthermore,

$$A^{\Im}(\mathbf{Q})_t = \int_0^t r^2\,ds = r^2 t, \qquad \mathcal{E}(A^{\Im}(\mathbf{Q}))_t = e^{r^2 t},$$

and the price of a European option is equal to

$$\mathbb{C} = \widehat{V}_0 = \sup_{\mathbf{Q}\in P(\Im)} \mathbf{E}_{\mathbf{Q}}[f_T e^{-r^2 T}].$$

§5.2. Hedging and investment with transaction costs

In this section we consider the Black–Scholes model (3.19) in which the formation of dynamic strategies is implemented with *transaction costs* taken into account. Under these constraints we solve two fundamental financial problems:

- *hedging* contingent claims $f = f(S_T)$ on a time interval $[0,T]$;
- *investment* when the search for an optimal strategy is implemented by the criterion of maximizing the mean yield.

In the solution of the first problem the possibilities of re-structuring the portfolio are limited to discrete times $t_i = \frac{i}{N}T$, $\Delta = t_{i+1} - t_i$, and to the taking of a certain "payment" for re-balancing the portfolio. It is clear that this assumption brings us closer to a more adequate description of financial realities. We present Leland's approach (on a physical level of rigor) to the pricing of options in this situation, assuming for simplicity that $B_t \equiv B_0 = 1$.

As before, suppose that the value X_t^π of a strategy $\pi_t = (\beta_t, \gamma_t)$ is determined by the balance relation

$$X_t^\pi = \beta_t + \gamma_t S_t.$$

Here its change over a time $\Delta t = t_{i+1} - t_i$ is realized according to the equation

$$\Delta X_t^\pi = \gamma_t \Delta S_t - \frac{k}{2}\, S_t |\Delta \gamma_t|, \tag{5.85}$$

where k is a nonnegative quantity that can depend in general on N.

The last term $\frac{k}{2}\, S_t|\Delta\gamma_t|$ on the right-hand side of (5.85) is called the *transaction costs* (*operational costs, overhead costs, commission fees*) borne by the investor managing the portfolio π. The form of the size of these costs, called *proportional* costs in this case, can be completely explained and constitutes $\frac{k}{2}\, 100\%$ of the size of the change in the risky part of the portfolio.

One additional assumption about re-balancing the portfolio π is that for each time t_i

$$X_{t_i}^\pi = X^\pi(t_i, S_{t_i}) = \mathbb{C}^{BS}(t_i, S_{t_i}), \tag{5.86}$$

that is, the value of the selected portfolio at the *re-hedging* time t_i is equal to the price of the contingent claim f, calculated in the framework of the ideal Black–Scholes model (3.19).

According to the Kolmogorov–Itô formula, we get the following stochastic equation for $\mathbb{C}^{BS}$ (see, for example, (3.25)):

$$d\mathbb{C}^{BS}(t, S_t) = \frac{\partial \mathbb{C}^{BS}}{\partial S}(t, S_t)\, dS_t + \left[\frac{\partial \mathbb{C}^{BS}}{\partial t}(t, S_t) + \frac{\sigma^2}{2} S_t^2 \frac{\partial^2 \mathbb{C}^{BS}}{\partial S^2}(t, S_t)\right] dt. \tag{5.87}$$

Comparing (5.85) and (5.87) and using (5.86), we arrive at the following equalities, valid up to small quantities of order $\Delta = \Delta t \to 0$:

$$\begin{aligned} \gamma_t &= \frac{\partial \mathbb{C}^{BS}}{\partial S}(t, S_t), \\ -\frac{k}{2} S_t |\Delta\gamma_t| &= \left[\frac{\partial \mathbb{C}^{BS}}{\partial t}(t, S_t) + \frac{\sigma^2}{2} S_t^2 \frac{\partial^2 \mathbb{C}^{BS}}{\partial S^2}\right] \Delta t. \end{aligned} \tag{5.88}$$

Further, by the properties $\mathbf{E}|\Delta W_t| = \sqrt{\frac{2}{\pi}}\sqrt{\Delta t}$ and $\mathbf{E}|\Delta W_t|^2 = \Delta t$ of a Wiener process, we have with the same order of smallness $\Delta t \to 0$ that

$$\begin{aligned} \Delta\gamma_t &= \frac{\partial^2 \mathbb{C}^{BS}}{\partial S^2}(t, S_t)\Delta S_t, \\ \frac{1}{2} k S_t |\Delta\gamma_t| &= \frac{1}{2} k S_t \left|\frac{\partial^2 \mathbb{C}^{BS}}{\partial S^2}(t, S_t)\right| |\Delta S_t| \\ &= \frac{1}{2} k S_t \sigma^2 \left|\frac{\partial^2 \mathbb{C}^{BS}}{\partial S^2}(t, S_t)\right| |\Delta W_t| \\ &= \frac{\sigma^2}{2}\sqrt{\frac{2}{\pi}}\left(\frac{k}{\sigma\sqrt{\Delta t}}\right) S_t^2 \left|\frac{\partial^2 \mathbb{C}^{BS}}{\partial S^2}\right| \Delta t. \end{aligned}$$

On the other hand,

$$\beta_t = X_t^\pi - \gamma_t S_t = X_t^\pi - \frac{\partial \mathbb{C}^{BS}}{\partial S}(t, S_t) S_t,$$

and hence the value $X^\pi(t, S_t)$ of the selected strategy π with transaction costs taken into account must satisfy the equation

$$\frac{\partial X^\pi(t,S_t)}{\partial t} + \frac{\widetilde{\sigma}^2(\Gamma)}{2} S_t^2 \frac{\partial^2 X^\pi(t,S_t)}{\partial S^2} = 0 \tag{5.89}$$

with the boundary condition $X^\pi(T, S_T) = f(S_T)$, where

$$\widetilde{\sigma}^2(\Gamma) = \sigma^2\left(1 + \sqrt{\frac{2}{\pi}}\left(\frac{k}{\sigma\sqrt{\Delta t}}\right)\right)\operatorname{sgn}(\Gamma), \qquad \Gamma = \frac{\partial^2 \mathbb{C}^{BS}(t,S_t)}{\partial S^2}.$$

In the case of an option to buy with payment function $f = (S_T - K)^+$ we have $\Gamma > 0$, and the value of the hedging strategy satisfies the Black–Scholes equation (3.27) with the increased volatility

$$\widetilde{\sigma}^2 = \sigma^2\left(1 + \sqrt{\frac{2}{\pi}}\left(\frac{k}{\sigma\sqrt{\Delta t}}\right)\right) > \sigma^2.$$

Thus, our attempt to replicate the contingent claim f in this situation is accompanied by an increase in volatility, which is a kind of "payment" for the presence of the transaction costs.

However, Kabanov and Safaryan have shown that the asymptotic behavior as $\Delta \to 0$ of the convergence of the contingent claim price, determined with the help of the above discrete strategy (*Leland strategy*), is not unique. For example, for $k = k_0 N^{-\alpha}$, $\alpha \in [0, 1/2]$, and $T = 1$ the indicated strategy ensures perfect hedging, while for $k \equiv k_0$ a *hedging error* $J_1 - J_2 < 0$ arises in the "limit", where

$$\begin{aligned}
J_1 &= \min\{K, S_1\},\\
J_2 &= \frac{k_0\sigma}{\sqrt{2\pi}}\int_0^\infty \frac{S_1}{\sqrt{v}} G(S_1, v)\exp\left\{-\frac{v}{2}\left(\frac{\ln S_1/K}{v} + \frac{1}{2}\right)^2\right\} dv,\\
G(S, v) &= \frac{1}{\sqrt{2\pi}}\int_{-\infty}^\infty \left| x - \frac{2k_0 \ln S_1/K}{\sqrt{2\pi}\, v} + \frac{k_0}{\sqrt{2\pi}}\right| e^{-x^2/2}\, dx.
\end{aligned}$$

In solving the second investment problem we start out from the assumption that the strategies $\pi_t = (\beta_t, \gamma_t)$ satisfy the "balance relation"

$$d\beta_t + S_t\, d\gamma_t = -\delta S_t\, |d\gamma_t|, \qquad \delta > 0, \tag{5.90}$$

which can be interpreted in a natural way as the taking of $\delta \cdot 100\%$ in commission fees, equal to $\delta S_t |\Delta\gamma_t|$, when stocks are bought (sold). Of course, it is assumed here that the corresponding differentials $d\beta_t$ and $d\gamma_t$ exist.

We define the (random) yield on $[0, T]$ of a portfolio π satisfying (5.90) by the relation

$$R^\pi(\beta, \delta) = \frac{1}{\beta T}\left(\left(\frac{X_T^\pi}{X_0}\right)^\beta - 1\right), \qquad \beta \in [0, 1). \tag{5.91}$$

The parameter β in (5.91) "is responsible for" compound interest, and in particular, we get for $\beta = 0$ that

$$R^\pi(0, \delta) = \frac{1}{T}\ln\frac{X_T^\pi}{X_0},$$

which corresponds to a rate of growth with continuous compounding of interest.

The selection of an *optimal portfolio* π^* can be realized by solving the extremal problem

$$R^*(\beta,\delta) = \mathbf{E}R^{\pi^*}(\beta,\delta) = \sup_{\pi}\mathbf{E}R^{\pi}(\beta,\delta), \tag{5.92}$$

where the "sup" is taken over the class of strategies π satisfying (5.90).

Getting the exact theoretical result in the problem (5.92) is obviously impossible. However, the assumption that the operational expenses are "small" ($\delta \to 0$) leads to the following important result in this direction.

THEOREM 5.13. *R^* has an asymptotic representation*

$$R^*(\beta,\delta) = R^*(\beta,0) - \exp\Bigl(\beta\,\frac{\alpha^*\mu}{2}\,T\Bigr)v^*\delta^{2/3} + o(\delta^{2/3}),$$

where

$$R^*(\beta,0) = \frac{1}{\beta T}\Bigl(\exp\Bigl(\beta\,\frac{\alpha^*\mu}{2}\,T\Bigr) - 1\Bigr), \tag{5.93}$$

$$\alpha^* = \frac{1}{1-\beta}\,\frac{\mu}{\sigma^2}, \tag{5.94}$$

$$v^* = \sigma^2\Bigl(\frac{9(1-\beta)}{32}\Bigr)^{1/3}\bigl(\alpha^*(1-\alpha^*)\bigr)^{4/3}. \tag{5.95}$$

Further, a strategy π^* for which a change in the structure of the portfolio is implemented at the times the process $\alpha = (\alpha_t)_{t\geqslant 0}$, $\alpha_t = \gamma_t S_t/X_t^{\pi}$, first leaves the interval

$$[\alpha^* - x^*\delta^{1/3}, \alpha^* + x^*\delta^{1/3}], \qquad x^* = \Bigl(\frac{3}{2(1-\beta)}\Bigr)^{1/3}\bigl(\alpha^*(1-\alpha^*)\bigr)^{2/3},$$

in an amount minimally necessary for returning to the boundary of this interval, is asymptotically optimal.

PROOF. From (5.90) we get the equation

$$dV_t = V_t\bigl(\alpha_t\mu\,dt + \alpha_t\sigma\,dW_t - \delta|\alpha_t|\,|dK_t|\bigr)$$

for the evolution of the value of the portfolio, where $dK = d\gamma/\gamma$ is the process of relative change of the stocks in the portfolio. Using the Kolmogorov–Itô formula, we arrive at the relation

$$\Bigl(\frac{V_T}{V_0}\Bigr)^{\beta} = \mathcal{E}\Bigl(\int_0^T \beta\alpha_t\sigma\,dW_t\Bigr)\mathcal{E}\Bigl(\beta\int_0^T \alpha_t\mu\,dt - \frac{1-\beta}{2}\alpha_t^2\sigma^2\,dt - \delta|\alpha_t|\,|dK_t|\Bigr).$$

Let α^* be the constant defined in (5.94). We denote by $\xi = \alpha - \alpha^*$ the deviation from this quantity of the process that is the proportion of the risky part of the value of the portfolio. It is clear that $\alpha \approx \alpha^*$ for "small" operational expenses, and therefore in the first approximation with respect to δ

$$\begin{aligned}\left(\frac{V_T}{V_0}\right)^{\beta} &\approx \mathcal{E}\left(\int_0^T \beta\alpha^*\sigma\,dW_t\right)\\ &\quad\times \mathcal{E}\left(\beta\int_0^T \frac{\alpha^*\mu}{2}\,dt - \frac{1-\beta}{2}\xi_t^2\sigma^2\,dt - \delta|\alpha^*|\,|dK_t|\right)\\ &\approx \mathcal{E}\left(\int_0^T \beta\alpha^*\sigma\,dW_t\right)\\ &\quad\times \exp\left(\beta\frac{\alpha^*\mu}{2}T\right)\left(1-\beta\int_0^T \frac{1-\beta}{2}\xi_t^2\sigma^2\,dt + \delta|\alpha^*|\,|dK_t|\right).\end{aligned}$$

We define a new probability measure $\widehat{\mathbf{P}}$ as follows:

$$\frac{d\widehat{\mathbf{P}}}{d\mathbf{P}} = \mathcal{E}\left(\int_0^T \beta\alpha^*\sigma\,dW_t\right).$$

Note that the process $\widehat{W}_t = W_t - \beta\alpha^*\sigma\,dt$ is a Wiener process with respect to $\widehat{\mathbf{P}}$. Then the extremal problem (5.92) is equivalent to the problem

$$\widehat{\mathbf{E}}\int_0^T \left(\frac{1-\beta}{2}\xi_t^2\sigma^2\,dt + \delta|\alpha^*|\,|dK_t|\right) \to \min_K.$$

Making the change of variables

$$\begin{gathered} t = s\delta^{2/3}, \quad T = S\delta^{2/3}, \quad \xi_t = \zeta_s\delta^{1/3},\\ \widehat{W}_t = B_s\delta^{1/3}, \quad |\alpha^*|K_t = L_s\delta^{1/3}, \end{gathered}$$

we arrive at the problem

$$\lim_{S\to\infty}\frac{1}{S}\int_0^S \left(\frac{1-\beta}{2}\zeta^2\sigma^2\,ds + |dL_s|\right) \to \min_L\ (= v^*). \tag{5.96}$$

According to the Tauberian theorem, under very general conditions the asymptotic behavior of a function at infinity is uniquely determined by the behavior of the Laplace transform of the function near zero. Hence, the problem (5.96) is equivalent to the problem

$$v^* = \lim_{\lambda\to 0}\lambda v(\lambda, 0),$$

where

$$v(\lambda, x) = \inf \mathbf{E}\int_0^\infty e^{-\lambda t}\left[\frac{1-\beta}{2}\sigma^2\zeta^2\,dt + |dL_t|\right], \quad \zeta_0 = x.$$

Thus, we come to the following Bellman equation for the function $v(\lambda, x)$:

$$\min\left(\frac{\sigma^2}{2}\left(\alpha^*(1-\alpha^*)\right)^2\frac{\partial^2 v(\lambda,x)}{\partial x^2} - \lambda v(\lambda,x) + \frac{1-\beta}{2}\sigma^2x^2, \left|\frac{\partial v(\lambda,x)}{\partial x}\right| - 1\right) = 0. \tag{5.97}$$

Let us consider formally the following decomposition of $v(\lambda, x)$ with respect to powers of λ:

$$v(\lambda, x) = \frac{1}{\lambda}v^* + w(x) + o(\lambda).$$

Passing to the limit as $\lambda \to 0$ in the Bellman equation (5.97), we get that

$$\min\left(\frac{\sigma^2}{2}\big(\alpha^*(1-\alpha^*)\big)^2\frac{\partial^2 w(x)}{\partial x^2} - v^* + \frac{1-\beta}{2}\sigma^2 x^2, \left|\frac{\partial w(x)}{\partial x}\right| - 1\right) = 0.$$

According to the general theory of optimal stopping rules, the solution of the last equation must have the structure

$$\begin{cases} \dfrac{\sigma^2}{2}\big(\alpha^*(1-\alpha^*)\big)^2\dfrac{\partial^2 w(x)}{\partial x^2} - v^* + \dfrac{1-\beta}{2}\sigma^2 x^2 = 0, & |x| \leqslant x^*, \\ \left|\dfrac{\partial w(x)}{\partial x}\right| = 1, & |x| > x^*. \end{cases}$$

Three additional conditions are needed to find the unknown parameters. Two are the so-called *conditions of "smooth pasting"* on the boundary separating the regions of "continuing and stopping the observations":

$$\frac{\partial w}{\partial x}(x^*) = 1,$$
$$\frac{\partial^2 w}{\partial x^2}(x^*) = 0.$$

The third condition

$$\frac{\partial w}{\partial x}(0) = 0$$

arises because by the sense of the problem the solution $w(x)$ is a symmetric function. Solving the system of equations with the indicated boundary conditions, we find the desired expressions for x^* and v^*, and this completes the proof of the theorem.

We remark that the quantity $R^*(\beta, 0)$ is the solution of the problem (5.92) in the absence of operational expenses ($\delta = 0$). Further, the value of α^* given in (5.94) determines the optimal proportion of stocks in the portfolio.

§ 5.3. Appendix: Examples of the simulation of hedging strategies

In this section we give an example of the hedging of a "life" option contract, taken from Russian financial practice. It is known that a characteristic feature of option contracts is the lack of symmetry between profits and losses for the parties in the contract. As a rule, the maximal profit of the buyer of an option is unlimited, while the amount of his maximal loss is limited to the premium paid for the option. On the other hand, the maximal profit of the seller of the option is equal to the premium received from emission of the option, while his losses are without limit.

As an illustration of the fact that buying an option can turn out to be an extremely profitable financial operation, we present an example taken from the history of a financial market. Table 5.1 contains prices for the sale of options to buy on domestic currency loan (DCL) bonds of the 2nd tranche with exercise times 30, 60, and 90 days from 04/19/95, and with exercise price equal to the weighted mean price over transactions with this bond on 04/19/95. Table 5.2 shows the yields from the operation of buying these options, computed with respect to the weighted mean prices of the DCL bonds at the exercise times of the options.

TABLE 5.1. Quotations of options to buy (4/19/95).

Exercise time	Exercise price	Premium (percent of face value)
05/19/95	79.60	1.80
06/18/95	79.60	2.60
07/18/95	79.60	3.10

TABLE 5.2. Return to buyers of options (5/19/95, 6/18/95, 7/18/95).

Exercise time	Price of purchase at exercise time	Yield of operation (annual %)
05/19/95	83.90	1667%
06/18/95	85.80	831%
07/18/95	88.40	735%

As is clear from Table 5.2, the buyers of the options garnered a fairly nice yield for themselves.[2] Correspondingly, a seller of the options would incur sizable losses with his "passive" strategy of waiting for the exercise time of the option.

In this connection the seller of the option must solve the following problems for himself:

- how to protect himself against possible losses connected with fulfilling obligations according to the option;
- how to minimize costs by constructing a strategy ensuring fulfillment of the obligation according to the option.

Of course, it is necessary to take into account factors connected with the imperfectness of a real financial market, namely, the presence of stochastic volatility, structural constraints on the hedging portfolio, transaction costs, and so on.

Below we give the results of a simulation (with operational expenses and the leasing of stocks taken into account) of hedging processes on the basis of real data.

Assume that on September 20, 1994 we wanted to purchase an option to buy on a DCL bond of the 2nd tranche with exercise price 77.00 (% of face value) and exercise time October 20, 1994. An approximate value in estimating the historical volatility over this time period is $\sigma \approx 16\%$, and the interest rate with respect to US dollars is $r \approx 14\%$. If we worked in the market of the ideal Black–Scholes model, then hedging with the given parameters σ and r would lead to a zero value of the minimal hedge at the terminal time.

Table 5.3 represents a process of hedging an option to buy in the Black–Scholes model "in real time mode", that is, with respect to calendar days on which there are records of the weighted mean prices of DCL bonds of the 2nd tranche. For each such date the table shows the structure of the hedging portfolio:

- the *value* of a minimal hedging portfolio;
- the *amount of DCL bonds* in a minimal hedging portfolio;

[2]DCL bonds are traded in US dollars.

TABLE 5.3. Simulation of a hedging procedure for a purchased call option in the classical Black–Scholes model.

Date	Closing price DCL bonds	Value of a min. hedge	Amount DCL bonds	Amount US dollars	Disbalance
09/20/94	77.13	1.60	−0.66	49.34	
09/21/94	76.94	1.44	−0.63	47.04	−0.01
09/22/94	77.82	2.01	−0.75	56.48	0.03
09/23/94	77.83	1.98	−0.75	56.59	0.01
09/26/94	77.85	1.89	−0.76	56.99	−0.03
09/27/94	77.99	1.96	−0.78	58.51	−0.05
09/28/94	78.50	2.33	−0.84	63.48	−0.05
09/29/94	78.88	2.62	−0.88	66.71	−0.05
09/30/94	79.31	2.98	−0.92	69.79	−0.05
10/03/94	80.24	3.76	−0.97	74.28	−0.04
10/04/94	80.87	4.34	−0.99	75.62	−0.04
10/05/94	80.61	4.05	−0.99	75.44	−0.04
10/06/94	80.97	4.39	−0.99	76.03	−0.04
10/07/94	81.11	4.50	−1.00	76.26	−0.04
10/10/94	81.57	4.86	−1.00	76.65	−0.04
10/11/94	81.96	5.22	−1.00	76.72	−0.04
10/12/94	81.47	4.71	−1.00	76.74	−0.04
10/13/94	80.28	3.49	−1.00	76.57	−0.04
10/14/94	79.17	2.36	−0.98	75.16	−0.04
10/17/94	78.42	1.51	−0.97	74.38	−0.06
10/18/94	78.03	1.11	−0.95	73.07	−0.06
10/19/94	79.23	2.26	−1.00	76.97	−0.02
10/20/94	79.51	2.51	−1.00	77.00	−0.02

- the *amount of US dollars* in a minimal hedging portfolio;
- the *disbalance* of the hedging portfolio, that is, the sum of the differences between the values of a real portfolio and a minimal hedging portfolio from the initial date to the current date.

The disparity between theory and practice is, of course, a consequence of the fact that the parameters of a minimal hedging portfolio are calculated in a continuous model, while hedging "in practice" is implemented at discrete moments of time. If the value of the portfolio is greater than the theoretical value of the minimal hedge, then we put a certain amount of US dollars into our portfolio, in order to bring it into correspondence with the theoretical value. And if the value of the portfolio is less than the theoretical value of the minimal hedge, then we withdraw a certain amount of US dollars from our portfolio (for example, by consumption).

In this example the size of the "disbalance" at the exercise time of the option turned out to be less than zero, and the buyer of the option (the hedger) "earned" 0.02% of the face value of the bond, or 200 dollars. This quantity is comparable with the usual broker's commission (constituting $\approx$ 100 dollars) on the purchase/sale of DCL bonds.

TABLE 5.4. Simulation of a classical Black–Scholes hedge for a purchased call option in a market with operational expenses.

Date	Closing price DCL bonds	Value of a min. hedge	Amount DCL bonds	Amount US dollars	Disbalance
09/20/94	77.13	1.39	−0.61	45.64	
09/21/94	76.94	1.25	−0.58	43.31	0.02
09/22/94	77.82	1.79	−0.71	53.33	0.01
09/23/94	77.83	1.77	−0.71	53.51	0.03
09/26/94	77.85	1.70	−0.72	54.11	0.08
09/27/94	77.99	1.77	−0.74	55.81	0.10
09/28/94	78.50	2.14	−0.81	61.30	0.11
09/29/94	78.88	2.42	−0.85	64.94	0.13
09/30/94	79.31	2.78	−0.90	68.46	0.14
10/03/94	80.24	3.58	−0.96	73.76	0.14
10/04/94	80.87	4.17	−0.98	75.37	0.14
10/05/94	80.61	3.89	−0.98	75.16	0.15
10/06/94	80.97	4.24	−0.99	75.88	0.15
10/07/94	81.11	4.35	−0.99	76.15	0.15
10/10/94	81.57	4.75	−1.00	76.63	0.15
10/11/94	81.96	5.12	−1.00	76.71	0.15
10/12/94	81.47	4.62	−1.00	76.73	0.15
10/13/94	80.28	3.41	−1.00	76.52	0.15
10/14/94	79.17	2.29	−0.98	74.92	0.15
10/17/94	78.42	1.48	−0.96	74.15	0.18
10/18/94	78.03	1.09	−0.95	72.80	0.19
10/19/94	79.23	2.25	−1.00	76.97	0.16
10/20/94	79.51	2.51	−1.00	77.00	0.16

Introduction into the hedging process of the spread between the purchase/sale prices of DCL bonds (it usually constitutes 0.50% of face value) leads to a depressing result (see Table 5.4.).

In this case the disbalance at the exercise time of the option is 0.16 or 1 600 dollars for a single DCL bond. Thus, a trader who used the classical Black–Scholes hedge in the real market would suffer fairly appreciable material losses.

We give an example of hedging a purchased call option with use of the technique developed in Chapter 4. There we noted that the presence of a spread between the purchase and sale price can be reduced to the case of stochastic volatility "squeezed" in the interval $[\sigma_{\min}, \sigma_{\max}]$. The results of simulation of the hedging process in this situation are shown in Table 5.5.

The disbalance of the hedging portfolio is now equal to zero. We mention also that in this case the (purchase) price of the option at the initial moment of time is 1.89% (or 18 900 dollars), while this quantity was 1.60% (or 16 000 dollars) in the Black–Scholes model.

TABLE 5.5. Simulation of a minimal hedge for a purchased call option with leasing fee and operational costs taken into account.

Date	Closing price DCL bonds	Value of a min. hedge	Amount DCL bonds	Amount US dollars	Disbalance
09/20/94	77.13	1.89	−0.63	46.52	
09/21/94	76.94	1.73	−0.60	46.55	−0.02
09/22/94	77.82	2.27	−0.70	46.57	−0.02
09/23/94	77.83	2.23	−0.70	46.61	−0.04
09/26/94	77.85	2.13	−0.71	52.77	−0.09
09/27/94	77.99	2.18	−0.72	52.81	−0.11
09/28/94	78.50	2.52	−0.78	52.83	−0.11
09/29/94	78.88	2.78	−0.82	61.61	−0.07
09/30/94	79.31	3.11	−0.86	61.64	−0.08
10/03/94	80.24	3.82	−0.93	70.79	−0.02
10/04/94	80.87	4.38	−0.96	70.81	−0.02
10/05/94	80.61	4.09	−0.95	70.86	−0.04
10/06/94	80.97	4.41	−0.97	74.18	−0.02
10/07/94	81.11	4.51	−0.98	74.21	−0.03
10/10/94	81.57	4.87	−0.99	74.30	−0.03
10/11/94	81.96	5.22	−1.00	76.51	−0.02
10/12/94	81.47	4.71	−1.00	76.54	−0.02
10/13/94	80.28	3.50	−0.98	75.44	−0.01
10/14/94	79.17	2.39	−0.94	72.07	−0.01
10/17/94	78.42	1.54	−0.92	70.73	−0.04
10/18/94	78.03	1.14	−0.90	70.79	−0.06
10/19/94	79.23	2.26	−1.00	76.96	0.00
10/20/94	79.51	2.51	−1.00	76.99	0.00

Bibliography for Chapter 5

[**7**], [**14**], [**37**], [**55**], [**69**], [**70**], [**82**], [**85**], [**90**], [**92**], [**102**], [**125**], [**144**], [**145**], [**157**], [**158**].

CHAPTER 6

Imperfect Forms of Hedging

In this chapter we consider two approaches for solving the problem of imperfect hedging: mean-variance hedging and quantile hedging. The first approach is characterized by the "quality" of the hedging strategy being measured by averaging the square of the difference between the terminal value and the contingent claim. In the second approach the probability of the "set of trajectories of asset prices admissible for hedging" is maximized. In both cases the possibility of loss is preserved, which makes these forms of hedging "imperfect".

§ 6.1. Mean-variance hedging

Suppose that an investor with some initial capital (value) x accepts a contingent claim f_T presentable at the exercise time T.

Perfect hedging ensures the possibility of choosing a strategy π with terminal value greater ($\mathbf{P}$-a.s.) than the contingent claim f_T.

At the same time, this hedging of contingent claims can be so expensive that the initial value x is insufficient for implementing the strategy π.

In this case one natural solution yields a strategy π^* with terminal value $X_T^{\pi^*}$ such that

$$\mathbf{E}\big[(f_T - X_T^{\pi^*})^2\big] \leqslant \mathbf{E}\big[(f_T - X_T^{\pi})^2\big] \tag{6.1}$$

for any strategy π attainable by the investor, with terminal value X_T^{π}.

Of course, the investor prefers to measure his returns in real and not nominal terms, and therefore the criterion for optimality should be taken to be the condition

$$\mathbf{E}\left[\left(\frac{f_T - X_T^{\pi^*}}{B_T}\right)^2\right] \leqslant \mathbf{E}\left[\left(\frac{f_T - X_T^{\pi}}{B_T}\right)^2\right] \tag{6.2}$$

and not (6.1), where B_T is the discounting factor. If the value of B_T is a constant, then the inequalities (6.1) and (6.2) are equivalent, but this is not true in the general case.

Such a strategy π^* is naturally said to be *mean-variance optimal*, since the contingent claim f_T is approximated as well as possible in the sense of the distance in the space L^2. We proceed to a precise exposition of this approach.

Let processes $B = \{B_t\}_{t\geqslant 0}$ and $S = \{S_t\}_{t\geqslant 0}$ representing respectively the price of a nonrisky asset (bank account) and the price of a risky asset (a stock) be given on a probability base $(\Omega, \mathcal{F}, \mathbb{F}, \mathbf{P})$ with filtration $\mathbb{F} = \{\mathcal{F}_t\}_{t\geqslant 0}$, where $t \in [0, T]$ or $t \in \{0, 1, 2, \dots\}$. It is assumed that they form a complete (B, S)-market, and that the density process Z_t^* of the unique martingale measure $\mathbf{P}^*$ with respect to $\mathbf{P}$ is *square integrable*.

The hedging of a contingent claim $f_T \in \mathcal{F}_T$ such that

$$\frac{f_T}{B_T} \in L^2(\mathcal{F}_T, \mathbf{P})$$

is implemented by means of self-financing strategies π with discounted terminal value $Y_T^\pi = X_T^\pi / B_T$ belonging to the space $L^2(\mathcal{F}_T, \mathbf{P})$. The set of such strategies is denoted by $SF_2(\mathbf{P})$.

We consider the terminal value X_T^π of an arbitrary strategy $\pi \in SF_2(\mathbf{P})$. In view of the martingale characterization of self-financing strategies (see Chapter 3) any $\pi \in SF_2(\mathbf{P})$ has a discounted value Y_t^π that is a square-integrable martingale. Consequently,

$$\mathbf{E}_{\mathbf{P}^*}[Y_T^\pi] = \mathbf{E}_{\mathbf{P}}[Z_T^* Y_T^\pi] = Y_0^\pi. \tag{6.3}$$

Setting $G_t^\pi = Y_t^\pi - Y_0^\pi$, we get from (6.3) that for any strategy $\pi \in SF_2(\mathbf{P})$

$$\mathbf{E}_{\mathbf{P}^*}[G_T^\pi] = \mathbf{E}_{\mathbf{P}}[Z_T^* G_T^\pi] = 0.$$

Note that for any strategy $\pi \in SF_2(\mathbf{P})$ the result G_T^π of using it (gain or loss) is represented as a random variable *orthogonal* (in the sense of the inner product in $L^2(\mathbf{P})$) to the variable Z_T^*.

Further, for any random variable $V \in L^2(\mathbf{P}, \mathcal{F}_T)$ orthogonal to Z_T^* we define $\widetilde{X}_T = B_T(x/B_0 + V)$. Since the market is complete, there is a strategy $\widetilde{\pi} \in SF$ such that $X_T^{\widetilde{\pi}} = \widetilde{X}_T$ ($\mathbf{P}$-a.s.). The discounted value $Y_t^{\widetilde{\pi}}$ of $\widetilde{\pi}$ is a square-integrable martingale with respect to $\mathbf{P}^*$. Then

$$\frac{X_0^{\widetilde{\pi}}}{B_0} = Y_0^{\widetilde{\pi}} = \mathbf{E}_{\mathbf{P}^*}[Y_T^{\widetilde{\pi}}] = \mathbf{E}_{\mathbf{P}}\left[\frac{\widetilde{X}_T}{B_T} Z_T^*\right] = \frac{x}{B_0},$$

and hence $X_0^{\widetilde{\pi}} = x$.

It follows from the definition of $\widetilde{X}_T$ that $X_T^{\widetilde{\pi}}/B_T \in L^2(\mathbf{P})$, that is, $\widetilde{\pi} \in SF_2(\mathbf{P})$.

Thus, *for any strategy $\pi \in SF_2(\mathbf{P})$ the result G_T^π of using it belongs to the subspace $L_0(Z_T^*)$ of vectors in $L^2(\mathbf{P})$ orthogonal to Z_T^*, and conversely, for any vector $V \in L_0(Z_T^*)$ there is a strategy $\widetilde{\pi} \in SF_2(\mathbf{P})$ with initial value x such that $G_T^{\widetilde{\pi}} = V$.*

In other words, the space $L_0(Z_T^*)$ represents the set of all results $G_T^\pi = Y_T^\pi - Y_0^\pi$ of using the strategies π in the family $SF_2(\mathbf{P})$.

We rewrite the condition (6.2) in the form

$$\mathbf{E}\left[\left(\frac{f_T}{B_T} - Y_0^{\pi^*} - G_T^{\pi^*}\right)^2\right] \leqslant \mathbf{E}\left[\left(\frac{f_T}{B_T} - Y_0^\pi - G_T^\pi\right)^2\right]. \tag{6.4}$$

Considering only the strategies with initial value x and writing

$$\widehat{h} = \frac{f_T}{B_T} - \frac{x}{B_0}, \tag{6.5}$$

we use (6.4) to reduce the problem (6.2) to the determination of a strategy $\pi^* \in SF_2(\mathbf{P})$ corresponding to the condition

$$\mathbf{E}\big[(\widehat{h} - G_T^{\pi^*})^2\big] \leqslant \mathbf{E}\big[(\widehat{h} - G_T^\pi)^2\big], \tag{6.6}$$

where $\widehat{h} \in L^2(\mathbf{P})$, and $G_T^\pi \in L_0(Z_T^*) \subset L^2(\mathbf{P})$.

In a Euclidean space the element of a subspace closest to a given vector is the projection of this vector on the given subspace, and therefore the strategy π^* should be chosen so that

$$G_T^{\pi^*} = \widehat{h} - Z_T^* \frac{\mathbf{E_P}[Z_T^* \widehat{h}]}{\mathbf{E_P}[(Z_T^*)^2]}. \tag{6.7}$$

Since $G_T^{\pi^*} \in L_0(Z_T^*)$, it follows from the foregoing that there is a strategy $\pi^* \in SF_2(\mathbf{P})$ such that

$$X_0^{\pi^*} = x \quad \text{and} \quad G_T^{\pi^*} - \widehat{h} - Z_T^* \frac{\mathbf{E_P}[Z_T^* \widehat{h}]}{\mathbf{E_P}[(Z_T^*)^2]} \in L^2(\mathbf{P})$$

If we call the quantity

$$R(x, \pi) = \mathbf{E_P}\left[\left(\frac{f_T - X_T^\pi}{B_T}\right)^2\right]$$

the *risk* of the strategy π, then it is natural for us to choose π *so that this risk is minimized.* It is clear that R depends also on the initial value x.

Therefore, it is natural to consider the problem of finding a pair (x^*, π^*) such that for any strategy $\pi \in SF_2(\mathbf{P})$ and for some initial value $x \in \mathbb{R}$

$$R(x^*, \pi^*) = \mathbf{E_P}\left[\left(\frac{f_T - X_T^{\pi^*(x^*)}}{B_T}\right)^2\right] \leqslant \mathbf{E_P}\left[\left(\frac{f_T - X_T^{\pi(x)}}{B_T}\right)^2\right] = R(x, \pi),$$
$$X_0^{\pi^*} = x^*, \quad X_0^\pi = x.$$

As was shown, for any initial value x there is a strategy $\pi^*(x) \in SF_2(\mathbf{P})$ such that for any other strategy $\pi \in SF_2(\mathbf{P})$ with the same initial value

$$R\big(x, \pi^*(x)\big) \leqslant R\big(x, \pi(x)\big).$$

Consequently, to find an optimal initial value it suffices to minimize the quantity $r(x) = R\big(x, \pi^*(x)\big)$.

We remark that

$$r(x) = \mathbf{E_P}\left[\left(\frac{f_T - X_T^{\pi^*(x)}}{B_T}\right)^2\right] = \mathbf{E_P}\left[(\widehat{h} - G_T^{\pi^*(x)})^2\right]$$
$$= \mathbf{E_P}\left[\left(-Z_T^* \frac{\mathbf{E_P}[Z_T^* \widehat{h}]}{\mathbf{E_P}[(Z_T^*)^2]}\right)^2\right] = \frac{(\mathbf{E_P}[Z_T^* \widehat{h}])^2}{\mathbf{E_P}[(Z_T^*)^2]}.$$

Using the definition of $\widehat{h}$ in (6.6) and (6.7), we get that

$$\begin{aligned} r(x) &= \frac{1}{\mathbf{E_P}[(Z_T^*)^2]}\left(\mathbf{E_P}\left[\frac{f_T}{B_T} Z_T^*\right] - \frac{x}{B_0}\right)^2 \\ &= \frac{1}{\mathbf{E_P}[(Z_T^*)^2]}\left(\mathbf{E_{P^*}}\left[\frac{f_T}{B_T}\right] - \frac{x}{B_0}\right)^2. \end{aligned} \tag{6.8}$$

It is clear that the minimum of (6.8) is equal to zero and is attained for

$$x^* = B_0\, \mathbf{E_{P^*}}\left[\frac{f_T}{B_T}\right], \tag{6.9}$$

which coincides with the price of the contingent claim for perfect hedging, and in particular, if $dB_t = B_t r\,dt$, then $x^* = e^{-rT}\mathbf{E}_{\mathbf{P}^*}[f_T]$.

The Black–Scholes model.

$$\begin{aligned} dB_t &= rB_t\,dt, & B_0 &= 1,\\ dS_t &= S_t(\mu\,dt + \sigma\,dw_t), & S_0 &> 0, \end{aligned}$$

where $r, \mu \in \mathbb{R}^1_+$, $0 < \sigma$ (the volatility), and $w = (w_t)_{t\geqslant 0}$ is a standard Wiener process generating the filtration $\mathbb{F}$ (see (3.19)).

As was shown, this (B, S)-market is a complete no-arbitrage market, and the density of the martingale measure $\mathbf{P}^*$ with respect to the initial measure $\mathbf{P}$ is represented by the process

$$Z_t^* = \frac{d\mathbf{P}_t^*}{d\mathbf{P}_t} = \mathcal{E}_t\Big(-\frac{\mu - r}{\sigma}\,w\Big) = \exp\Big\{-\frac{\mu - r}{\sigma}\,w_t - \frac{1}{2}\Big(\frac{\mu - r}{\sigma}\Big)^2 t\Big\}.$$

Suppose that an investor is faced with the problem of hedging a contingent claim at the time T with payment function $f_T \in L^2(\mathbf{P})$; since B_t is deterministic, this is equivalent to the requirement that $f_T/B_T \in L^2(\mathbf{P})$.

If we assume that f_T is a constant, then the solution of the problem posed follows immediately from (6.7) and the Kolmogorov–Itô formula. Indeed,

$$G_T^{\pi^*} = \widehat{h} - Z_T^* \frac{\mathbf{E}_{\mathbf{P}}[Z_T^*\widehat{h}]}{\mathbf{E}_{\mathbf{P}}[(Z_T^*)^2]} = \widehat{h}\Big(1 - \frac{Z_T^*}{\mathbf{E}_{\mathbf{P}}[(Z_T^*)^2]}\Big), \tag{6.10}$$

from which, using the Kolmogorov–Itô formula, we get that

$$\begin{aligned} dG_t^{\pi^*} &= \widehat{h}\,\frac{Z_t^*}{\mathbf{E}_{\mathbf{P}}[(Z_t^*)^2]}\Big(\frac{\mu - r}{\sigma}\,dw_t + \Big(\frac{\mu - r}{\sigma}\Big)^2 dt\Big)\\ &= \widehat{h}\,\frac{Z_t^*}{\mathbf{E}_{\mathbf{P}}[(Z_t^*)^2]} \cdot \frac{\mu - r}{\sigma^2} \cdot \frac{B_t}{S_t}\,d\frac{S_t}{B_t}. \end{aligned}$$

Since

$$G_t^{\pi^*} = Y_t^{\pi^*} - Y_0 = \int_0^t \gamma_t^{\pi^*}\,d\frac{S_t}{B_t},$$

we have

$$\gamma_t^{\pi^*} = \widehat{h}\,\frac{Z_t^*}{\mathbf{E}_{\mathbf{P}}[(Z_t^*)^2]}\,\frac{\mu - r}{\sigma^2}\,\frac{B_t}{S_t}. \tag{6.11}$$

Further, it follows from (6.11) and (6.10) that

$$\gamma_t^{\pi^*} = (\widehat{h} - G_t^{\pi^*})\,\frac{\mu - r}{\sigma^2}\,\frac{B_t}{S_t}.$$

Let $f_T = (S_T - K)^+$ be a standard European option to buy. Then we get from (6.7) that

$$G_T^{\pi^*} = \Big(\frac{(S_T - K)^+}{B_T} - x\Big) - \frac{Z_T^*}{\mathbf{E}_{\mathbf{P}}[(Z_T^*)^2]}\Big(\mathbf{E}_{\mathbf{P}^*}\Big[\frac{(S_T - K)^+}{B_T}\Big] - x\Big). \tag{6.12}$$

Since $\mathbb{C}_0 = \mathbf{E}_{\mathbf{P}^*}[B_T^{-1}(S_T - K)^+]$ is the cost of the option in the Black–Scholes model, we get by adding and subtracting $\mathbb{C}_0$ on the right-hand side of (6.12) that

$$G_T^{\pi^*} = \left(\frac{(S_T - K)^+}{B_T} - \mathbb{C}_0\right) + (\mathbb{C}_0 - x)\left(1 - \frac{Z_T^*}{\mathbf{E}_{\mathbf{P}}[(Z_T^*)^2]}\right). \tag{6.13}$$

Note that the first term of the sum in (6.13) is the stochastic integral

$$\int_0^T \widehat{\gamma}_t \, d\frac{S_t}{B_t} = \left(\frac{(S_T - K)^+}{B_T} - \mathbb{C}_0\right),$$

where $\widehat{\gamma}$ is the investment strategy in the Black–Scholes model.

As in the case of (6.10), the second term of the sum in (6.13) can be represented in the form

$$\int_0^T \widetilde{\gamma}_t \, d\frac{S_t}{B_t} = (\mathbb{C}_0 - x)\left(1 - \frac{Z_T^*}{\mathbf{E}_{\mathbf{P}}[(Z_T^*)^2]}\right),$$

where (analogous to (6.11))

$$\widetilde{\gamma}_t = (\mathbb{C}_0 - x)\frac{Z_t^*}{\mathbf{E}_{\mathbf{P}}[(Z_t^*)^2]}\,\frac{\mu - r}{\sigma^2}\,\frac{B_t}{S_t}\,.$$

Setting $\gamma_t^{\pi^*} = \widehat{\gamma}_t + \widetilde{\gamma}_t$, we get that

$$G_T^{\pi^*} = \int_0^T \widehat{\gamma}_t \, d\frac{S_t}{B_t} + \int_0^T \widetilde{\gamma}_t \, d\frac{S_t}{B_t} = \int_0^T \gamma_t^{\pi^*} \, d\frac{S_t}{B_t},$$

and hence the optimality of the strategy $\gamma_t^{\pi^*}$.

It is not hard to interpret the result obtained. The strategy $\gamma_t^{\pi^*}$ has two components: one operates according to the Black–Scholes strategy and replicates the contingent claim **P**-a.s.; the other minimizes the (possible) deficiency of the initial value x.

The formula (6.9) representing the optimal initial value shows that the value $x = \mathbb{C}_0$ is optimal, and in this case $\widetilde{\gamma}_t \equiv 0$, that is, the perfect hedging strategy is optimal.

The example given enables us to draw a conclusion about the structure of the mean-variance optimal hedging strategy for an arbitrary contingent claim f_T in the case of a complete market.

Indeed, as follows from Chapter 3, the optimal value (rational cost) of a contingent claim f_T is

$$\mathbb{C}_0 = B_0 \, \mathbf{E}_{\mathbf{P}^*}[B_T^{-1} f_T].$$

Correspondingly, we get from (6.7) that

$$\begin{aligned} G_T^{\pi^*} &= \left(\frac{f_T}{B_T} - \frac{x}{B_0}\right) - \frac{Z_T^*}{\mathbf{E}_{\mathbf{P}}[(Z_T^*)^2]}\left(\mathbf{E}_{\mathbf{P}^*}\left[\frac{f_T}{B_T}\right] - \frac{x}{B_0}\right) \\ &= \left(\frac{f_T}{B_T} - \frac{\mathbb{C}_0}{B_0}\right) + \frac{\mathbb{C}_0 - x}{B_0}\left(1 - \frac{Z_T^*}{\mathbf{E}_{\mathbf{P}}[(Z_T^*)^2]}\right). \end{aligned} \tag{6.14}$$

We denote by $\widehat{\gamma}_t$ the integrand in the representation

$$\left(\frac{f_T}{B_T} - \frac{\mathbb{C}_0}{B_0}\right) = \int_0^T \widehat{\gamma}_t \, d\frac{S_t}{B_t}, \tag{6.15}$$

which exists and determines a component of the strategy hedging f_T ($\mathbf{P}$-a.s.). Next, defining $\widetilde{\gamma}_t$ by

$$\widetilde{\gamma}_t = \frac{\mathbb{C}_0 - x}{B_0} \frac{Z_t^*}{\mathbf{E_P}[(Z_t^*)^2]} \frac{\mu - r}{\sigma^2} \frac{B_t}{S_t},$$

we get from (6.14) and (6.15) that

$$G_T^{\pi^*} = \int_0^T \widehat{\gamma}_t \, d\frac{S_t}{B_t} + \int_0^T \widetilde{\gamma}_t \, d\frac{S_t}{B_t} = \int_0^T \gamma_t^{\pi^*} d\frac{S_t}{B_t},$$

which confirms the optimality of the choice $\gamma_t^{\pi^*} = \widehat{\gamma}_t + \widetilde{\gamma}_t$.

The second component $\beta_t^{\pi^*}$ of the optimal strategy π^* can be obtained from the balance equation

$$X_t^{\pi^*} = B_t Y_t^{\pi^*} = B_t(x + G_t^{\pi^*}) = B_t\bigg(\beta_t^{\pi^*} + \int_0^t \gamma_u^{\pi^*} dS_u\bigg).$$

It follows immediately from (6.14) that for an initial value x the risk is determined by the formula

$$r(x) = R\big(x, \pi^*(x)\big) = \mathbf{E_P}\left[\left(\frac{x - \mathbb{C}_0}{B_0} \frac{Z_T}{\mathbf{E_P}[(Z_T)^2]}\right)^2\right] = \frac{(x - \mathbb{C}_0)^2}{B_0^2 \, \mathbf{E_P}[(Z_T)^2]}.$$

The Merton model.

$$dB_t = rB_t \, dt, \quad B_0 = 1,$$
$$dS_t = S_{t-}(\mu \, dt - \nu \, d\Pi_t),$$

where $\mu, r \in \mathbb{R}_+^1$, $\nu < 1$, and $\Pi = (\Pi_t)_{t \geqslant 0}$ is a Poisson process with intensity $\lambda > 0$ that generates the filtration $\mathbb{F}$.

This (B, S)-market is a complete no-arbitrage market, and the density of the martingale measure $\mathbf{P}^*$ has the form

$$Z_t^* = \frac{d\mathbf{P}_t^*}{d\mathbf{P}_t} = \mathcal{E}_t(N) = \exp\big\{(\lambda - \lambda^*)t + (\ln \lambda^* - \ln \lambda)\Pi_t\big\},$$

where $\lambda^* = (\mu - r)/\nu$ is the intensity of Π_t with respect to the new measure $\mathbf{P}^*$.

We consider a deterministic contingent claim f_T. Then it follows from (6.7) that

$$G_T^{\pi^*} = \widehat{h}\bigg(1 - \frac{Z_T^*}{\mathbf{E_P}[(Z_T^*)^2]}\bigg). \tag{6.16}$$

Applying the Kolmogorov–Itô formula to (6.16), we have

$$\begin{aligned} dG_T^{\pi^*} &= \widehat{h}\frac{Z_{t-}}{\mathbf{E_P}[(Z_t^*)^2]}\left(\frac{\lambda^* - \lambda}{\lambda\nu}\right)[\lambda^*\nu \, dt - \nu \, d\Pi_t] \\ &= \widehat{h}\frac{Z_{t-}}{\mathbf{E_P}[(Z_t^*)^2]}\left(\frac{\lambda^* - \lambda}{\lambda\nu}\right)\frac{B_t}{S_{t-}} \, d\frac{S_t}{B_t}, \end{aligned}$$

and then by (6.16),

$$dG_T^{\pi^*} = (\widehat{h} - G_{t-}^{\pi^*})\left(\frac{\lambda^* - \lambda}{\lambda\nu}\right)\frac{B_t}{S_{t-}} \, d\frac{S_t}{B_t}.$$

Thus, for a deterministic contingent claim f_T the optimal strategy π^* has the form

$$\gamma_t^* = (\widehat{h} - G_{t-}^{\pi^*})\Big(\frac{\lambda^* - \lambda}{\lambda\nu}\Big)\frac{B_t}{S_{t-}},$$
$$\beta_t^* = Y_t^{\pi^*} - \gamma_t^*\frac{S_t}{B_t},$$

where $\widehat{h}$ is determined by the formula (6.5).

Suppose now that $f_T = (S_T - K)^+$. Then it follows from (6.7) that

$$G_T^{\pi^*} = \Big(\frac{f_T}{B_T} - \frac{x}{B_0}\Big) - \frac{Z_T^*}{\mathbf{E_P}[(Z_T^*)^2]}\Big(\mathbf{E_{P^*}}\Big[\frac{f_T}{B_T}\Big] - \frac{x}{B_0}\Big)$$
$$= \Big(\frac{f_T}{B_T} - \mathbb{C}_0\Big) + (\mathbb{C}_0 - x)\Big(1 - \frac{Z_T^*}{\mathbf{E_P}[(Z_T^*)^2]}\Big). \tag{6.17}$$

As in the case of the preceding model, the first term of the sum in (6.17) is represented by a stochastic integral

$$\int_0^T \widehat{\gamma}_t\, d\frac{S_t}{B_t} = \Big(\frac{(S_T - K)^+}{B_T} - \mathbb{C}_0\Big),$$

where $\widehat{\gamma}$ is the investment strategy in the Merton model, and the second term in (6.17) can be represented in the form

$$\int_0^T \widetilde{\gamma}_t\, d\frac{S_t}{B_t} = (\mathbb{C}_0 - x)\Big(1 - \frac{Z_T^*}{\mathbf{E_P}[(Z_T^*)^2]}\Big),$$

where

$$\widetilde{\gamma}_t = (\mathbb{C}_0 - x)\frac{Z_{t-}^*}{\mathbf{E_P}[(Z_t^*)^2]}\Big(\frac{\lambda^* - \lambda}{\lambda\nu}\Big)\frac{B_t}{S_{t-}}.$$

Letting $\gamma_t^{\pi^*} = \widehat{\gamma}_t + \widetilde{\gamma}_t$, we have

$$G_T^{\pi^*} = \int_0^T \widehat{\gamma}_t\, d\frac{S_t}{B_t} + \int_0^T \widetilde{\gamma}_t\, d\frac{S_t}{B_t} = \int_0^T \gamma_t^{\pi^*}\, d\frac{S_t}{B_t},$$

from which it follows that the strategy $\gamma_t^{\pi^*}$ is optimal.

As in the Black–Scholes model, the risk $r(x)$ of the strategy π^* for the initial value x is equal to

$$R(x, \pi^*(x)) = \mathbf{E_P}\Big[\Big((x - \mathbb{C}_0)\frac{Z_T^*}{\mathbf{E_P}[(Z_T^*)^2]}\Big)^2\Big] = \frac{(x - \mathbb{C}_0)^2}{\mathbf{E_P}[(Z_T^*)^2]}.$$

The Cox–Ross–Rubinstein model.

We consider a model of a discrete market with the price of a nonrisky asset $(B_n)_{n=0,1,\dots}$ and the price of a risky asset $(S_n)_{n=0,1,\dots}$ determined by the formulas

$$\Delta B_n = B_n - B_{n-1} = rB_{n-1}, \quad B_0 = 1,$$
$$\Delta S_n = S_n - S_{n-1} = \rho_n S_{n-1}, \quad S_0 > 0,$$

where $(\rho_n)_{n=1,\dots}$ is a sequence of independent random variables taking two values $a < b$ with probabilities p and $q = 1 - p$.

Setting $B_t = B_n$ and $S_t = S_n$ for $t \in [n, n+1)$, we get a standard semimartingale model for a complete (B, S)-market, where the density Z_t^* of the unique martingale measure $\mathbf{P}^*$ is

$$Z_n^* = \mathcal{E}_n\biggl(-\frac{m-r}{\sigma^2}\sum_{k\leqslant n}(\rho_k - m)\biggr).$$

For a deterministic contingent claim f_T we get from (6.7) and the Kolmogorov–Itô formula that

$$\begin{aligned} G_T^{\pi^*} &= f_T \sum_{k\leqslant T} \frac{Z_{k-1}^*}{\mathbf{E_P}[(Z_{k-1}^*)^2]}\,\frac{m-r}{\sigma^2}\biggl(\frac{\rho_k - r}{1+\frac{(m-r)^2}{\sigma^2}}\biggr) \\ &= f_T \sum_{k\leqslant T} \frac{Z_{k-1}^*}{\mathbf{E_P}[(Z_{k-1}^*)^2]}\,\frac{(m-r)(1+r)}{\sigma^2+m-r}\,\frac{B_{k-1}}{S_{k-1}}\,\Delta\,\frac{S_k}{B_k} \\ &= \sum_{k\leqslant T}(f_T - G_{k-1}^{\pi^*})\frac{(m-r)(1+r)}{\sigma^2+m-r}\,\frac{B_{k-1}}{S_{k-1}}\,\Delta\,\frac{S_k}{B_k}, \end{aligned}$$

and hence

$$\gamma_k^* = (f_T - G_{k-1}^{\pi^*})\,\frac{(m-r)(1+r)}{\sigma^2+m-r}\,\frac{B_{k-1}}{S_{k-1}}\,.$$

For a standard European option to buy and for $f_T = (S_T - K)^+$, we have

$$\begin{aligned} G_T^{\pi^*} &= \biggl(\frac{f_T}{B_T} - \frac{x}{B_0}\biggr) - \frac{Z_T^*}{\mathbf{E_P}[(Z_T^*)^2]}\biggl(\mathbf{E_{P^*}}\biggl[\frac{f_T}{B_T}\biggr] - \frac{x}{B_0}\biggr) \\ &= \biggl(\frac{f_T}{B_T} - \mathbb{C}_0\biggr) + (\mathbb{C}_0 - x)\biggl(1 - \frac{Z_T^*}{\mathbf{E_P}[(Z_T^*)^2]}\biggr). \end{aligned}$$

As earlier, $G_T^{\pi^*}$ can be represented in the form

$$G_T^{\pi^*} = \sum_{k\leqslant T}\widehat{\gamma}_k\,\Delta\,\frac{S_k}{B_k} + \sum_{k\leqslant T}\widetilde{\gamma}_k\,\Delta\,\frac{S_k}{B_k} = \sum_{k\leqslant T}\gamma_k^*\,\Delta\,\frac{S_k}{B_k}\,,$$

where $\widehat{\gamma}$ is the investment strategy in the Cox–Ross–Rubinstein model, and $\widetilde{\gamma}$ is the strategy determined by the formula

$$\widetilde{\gamma}_k = (\mathbb{C}_0 - x)\,\frac{Z_{k-1}^*}{\mathbf{E_P}[(Z_{k-1}^*)^2]}\,\frac{(m-r)(1+r)}{\sigma^2+m-r}\,\frac{B_{k-1}}{S_{k-1}}\,.$$

Thus, the optimal strategy γ_t^* can be represented in the form $\gamma_t^* = \widehat{\gamma}_t + \widetilde{\gamma}_t$, and the risk of the strategy π^* for initial value x is equal to

$$R\bigl(x, \pi^*(x)\bigr) = \mathbf{E_P}\biggl[\biggl((x-\mathbb{C}_0)\,\frac{Z_T^*}{\mathbf{E_P}[(Z_T^*)^2]}\biggr)^2\biggr] = \frac{(x-\mathbb{C}_0)^2}{\mathbf{E_P}[(Z_T^*)^2]}\,.$$

§ 6.2. Quantile hedging

Together with mean-variance hedging we consider one more form of imperfect hedging in which the hedging is implemented with *probability less than* 1. In this case there are two forms of problems when either the hedging probability is given and it is necessary to minimize the value of a minimal hedge, or a constraint on the initial value of a minimal hedge is given and it is necessary to maximize the hedging probability.

Thus, the problem of hedging in this formulation is methodologically connected with problems of statistical confidence estimation. One of the main concepts in the general theory of estimation is the concept of *quantile*, that is, the boundary of the domain of estimation with a specific probability. The method of hedging with probability less than 1 has thus come to be called "quantile" hedging.

The Black–Scholes model.

Let us consider the Black–Scholes model (3.19). We recall that an *admissible* strategy is defined to be a self-financing strategy $\pi_t = (\beta_t, \gamma_t)$ with value

$$X_t^\pi = \beta_t B_t + \gamma_t S_t$$

such that $X_t^\pi \geqslant 0$ for all $t \geqslant 0$.

We denote the discounted value of a portfolio by $Y_t = Y_t^\pi = X_t^\pi / B_t$. Then the Kolmogorov–Itô formula gives us that (see Chapter 3 for details)

$$dY_t = \phi_t \, dW_t^*, \qquad Y_0 = X_0^\pi, \tag{6.18}$$

where $\phi_t = \sigma \gamma_t S_t / B_t$ and $W_t^* = W_t + \left(\dfrac{\mu - r}{\sigma}\right) t$ is a standard Wiener process with respect to the measure $\mathbf{P}^*$.

As earlier, by a *contingent claim* with exercise time T we mean a nonnegative $\mathcal{F}_T$-measurable random variable f.

The *successful hedging set* of a claim f for a strategy π with initial value x is defined to be

$$A = A(x, \pi, f) = \{\omega : X_T^\pi(x) \geqslant f\} = \{\omega : Y_T(x) \geqslant f/B_T\}. \tag{6.19}$$

The theory of perfect hedging enables one to find a hedging strategy with initial value $X_0 = \mathbf{E}^* f/B_T$ and $\mathbf{P}(A) = 1$. What if an investor responsible according to the claim f is not in a position to invest the initial value X_0 necessary for perfect hedging? The following problem arises naturally for him:

- *among all admissible strategies, choose a strategy minimizing the risk of losses associated with exercising the claim f,*
- *or, among all admissible strategies, choose a strategy maximizing the probability that the value X_T^π of the portfolio at the time of exercising the contingent claim is not less than the payment according to this contingent claim, that is,*

$$\mathbf{P}\big(A(x, \pi, f)\big) \to \max \tag{6.20}$$

under the budget constraint

$$x \leqslant x_0 < \mathbf{E}^* f/B_T = X_0, \tag{6.21}$$

where x_0 is the initial capital the investor has at his disposition.

The relations (6.20)–(6.21) represent a mathematical formulation of the problem of quantile hedging, which is our object of study in this section.

LEMMA 6.1. *Let $\widetilde{A} \in \mathcal{F}_T$ be a solution of the problem*

$$P(A) \to \max, \tag{6.22}$$

$$\mathbf{E}^* f/B_T \cdot I_A \leqslant x. \tag{6.23}$$

Then a perfect hedge $\widetilde{\pi}$ with initial value x for the contingent claim $\widetilde{f} = f \cdot I_{\widetilde{A}}$ is a solution of the problem (6.20)–(6.21), *and the successful hedging set $A = A(x, \widetilde{\pi}, f)$ coincides with $\widetilde{A}$.*

PROOF. 1. We consider an arbitrary admissible strategy π with initial value $x \leqslant \mathbf{E}^* f/B_T = X_0$. It follows from (6.18) that its discounted value

$$Y_t = x + \int_0^t \phi_s \, dW_s^* \tag{6.24}$$

is a nonnegative (local) martingale with respect to $\mathbf{P}^*$, and hence is a supermartingale. For the successful hedging set $A = A(x, \pi, f)$ we get that

$$Y_T \geqslant f/B_T \cdot I_A \quad (\mathbf{P}\text{-a.s.}),$$

therefore,

$$x = \mathbf{E}^* Y_T \geqslant \mathbf{E}^*(f/B_T \cdot I_A),$$

and thus the condition (6.23) holds.

Moreover, in view of (6.22) we have for $\widetilde{A}$ that $\mathbf{P}(A) \leqslant \mathbf{P}(\widetilde{A})$.

2. Suppose now that the strategy $\widetilde{\pi}$ is a perfect hedge for the contingent claim $f \cdot I_{\widetilde{A}}$ with initial value satisfying the inequality

$$\mathbf{E}^*(f/B_T \cdot I_{\widetilde{A}}) \leqslant x,$$

where x is determined from (6.21). We show that the strategy $\widetilde{\pi}$ is optimal for the problem (6.20)–(6.21). To do this we note first of all that, since

$$x + \int_0^t \widetilde{\phi}_s \, dW_s^* \geqslant \mathbf{E}^*(f/B_T \cdot I_{\widetilde{A}}) + \int_0^t \widetilde{\phi}_s \, dW_s^* = \mathbf{E}^*(f/B_T \cdot I_{\widetilde{A}} \,|\, \mathcal{F}_t) \geqslant 0,$$

the strategy $\widetilde{\pi}$ is admissible.

Let

$$A' = \left\{ x + \int_0^T \widetilde{\phi}_s \, dW_s^* \geqslant f/B_T \right\}$$

be the successful hedging set for $\widetilde{\pi}$ in the problem (6.20)–(6.21). Because $\widetilde{\pi}$ is "perfect" for the contingent claim $f \cdot I_{\widetilde{A}}$, we get that

$$A' \supseteq \{f \cdot I_{\widetilde{A}} \geqslant f\} \supseteq \widetilde{A},$$

and hence $\mathbf{P}(A') \geqslant \mathbf{P}(\widetilde{A})$.

3. Using the definition of the sets $A' = A$, we have from items 1 and 2 that $A = \widetilde{A}$ ($\mathbf{P}$-a.s.). But $\widetilde{A}$ is the successful hedging set for the strategy $\widetilde{\pi}$, and therefore $\widetilde{\pi}$ is an optimal strategy for the problem (6.20)–(6.21).

Thus, if the equivalence (6.20)–(6.21)$\Longleftrightarrow$(6.22)–(6.23) has been proved, then for constructing the maximal success set it is natural to turn to the Neyman–Pearson fundamental lemma, which allows us to devise a most powerful test for the problem of distinguishing between two simple hypotheses under restriction to an error of the first kind.

Below it will be assumed that the distribution $\mathbf{Q}^*$ corresponds to the hypothesis H_0 and the distribution $\mathbf{P}$ corresponds to the hypothesis H_1. Moreover, let $\alpha = \mathbf{E}_{\mathbf{Q}^*}\phi$ be the probability of an error of the first kind, and $\beta = \mathbf{E}_{\mathbf{P}}\phi$ the power of the test corresponding to the *critical function* ϕ. The Neyman–Pearson test having the structure

$$\phi = \begin{cases} 1, & \dfrac{d\mathbf{P}}{d\mathbf{Q}^*} > c, \\ 0, & \dfrac{d\mathbf{P}}{d\mathbf{Q}^*} < c, \end{cases} \tag{6.25}$$

is most powerful, that is, maximizes β under the condition that the probability of an error of the first kind does not exceed the given level α. Here c is a constant, and the value of the critical function ϕ (1 or 0) shows which of the hypotheses H_1 or H_0 should be accepted.

Let us return to the problem (6.22)–(6.23). We introduce a probability measure $\mathbf{Q}^*$ by the relation

$$\frac{d\mathbf{Q}^*}{d\mathbf{P}^*} = \frac{f}{B_T \cdot \mathbf{E}^* f/B_T} = \frac{f}{\mathbf{E}^* f}. \tag{6.26}$$

Then the restriction (6.23) takes the form

$$\mathbf{Q}^*(A) = \int_A \frac{d\mathbf{Q}^*}{d\mathbf{P}^*}\, d\mathbf{P}^* \leqslant \frac{x}{\mathbf{E}^* f/B_T} = \alpha. \tag{6.27}$$

LEMMA 6.2. *The solution of the problem* (6.22)–(6.23) *is the set*

$$\widetilde{A} = \left\{\omega : \frac{d\mathbf{P}}{d\mathbf{Q}^*} > c\right\} = \left\{\omega : \frac{d\mathbf{P}}{d\mathbf{P}^*} > c\,\frac{f}{\mathbf{E}^* f}\right\}, \tag{6.28}$$

where $c = \inf\left\{a : \mathbf{Q}^*\left(\dfrac{d\mathbf{P}}{d\mathbf{Q}^*} > a\right) \leqslant \alpha\right\}$.

The proof follows from the Neyman–Pearson fundamental lemma in view of the equalities $\alpha = \mathbf{E}_{\mathbf{Q}^*}\phi = \mathbf{Q}^*(\widetilde{A})$ and $\beta = \mathbf{E}_{\mathbf{P}}\phi = \mathbf{P}(\widetilde{A}) = \max$.

We get the following theorem from Lemmas 6.1 and 6.2.

THEOREM 6.1. *An optimal strategy π for the problem* (6.20)–(6.21) *coincides with a perfect hedge for the contingent claim $f \cdot I_{\widetilde{A}}$, where the maximal success set $\widetilde{A}$ is defined in* (6.28).

REMARK 6.1. If $\mathbf{Q}^*(d\mathbf{P}/d\mathbf{Q}^* = c) \neq 0$, then for the equality $\alpha = \mathbf{E}_{\mathbf{Q}^*}\phi$ to hold for an error of the first kind, it is necessary to use the *randomized* critical function ϕ, $0 \leqslant \phi \leqslant 1$, of the form

$$\phi = I_{\{d\mathbf{P}/d\mathbf{Q}^* > c\}} + \gamma I_{\{d\mathbf{P}/d\mathbf{Q}^* = c\}}, \qquad \gamma = \frac{\alpha - \mathbf{Q}^*(d\mathbf{P}/d\mathbf{Q}^* > c)}{\mathbf{Q}^*(d\mathbf{P}/d\mathbf{Q}^* = c)}.$$

In this case the optimal strategy π coincides with a perfect hedge for the contingent claim $f \cdot \phi$.

REMARK 6.2. $\mathbf{Q}^*(d\mathbf{P}/d\mathbf{Q}^* = c) = 0$ in the Black–Scholes model by the continuity of the distribution.

The dual problem.

Let a number $\varepsilon \in (0,1)$ be fixed. We consider a situation in which the investor agrees to undertake a certain risk in exercising a contingent claim. Namely, we shall solve the problem of finding the smallest initial value necessary for exercising a contingent claim with probability at least $1-\varepsilon$.

This problem is *dual* to (6.20)–(6.21) and can be expressed mathematically in the form

$$x \to \min, \tag{6.29}$$

$$\mathbf{P}\big(A(x,\pi,f)\big) \geqslant 1-\varepsilon, \tag{6.30}$$

or, equivalently,

$$\mathbf{E}^*[f/B_T \cdot I_A] \to \min, \tag{6.31}$$

$$\mathbf{P}(A) \geqslant 1-\varepsilon. \tag{6.32}$$

We denote by $\bar{A} = \Omega \setminus A$ the complement of the set A:

$$\bar{A} = \{\omega : Y_T(x) < f/B_T\}.$$

Then we can write the problem dual to (6.31)–(6.32) as follows:

$$\mathbf{E}^*[f/B_T \cdot I_{\bar{A}}] \to \max, \tag{6.33}$$

$$\mathbf{P}(\bar{A}) \leqslant \varepsilon. \tag{6.34}$$

By (6.26), we can rewrite (6.33) in the form

$$\mathbf{E}^* f/B_T \cdot \mathbf{Q}^*(\bar{A}) \to \max,$$

and hence the problem (6.33)–(6.34) is equivalent to the relations

$$\mathbf{Q}^*(\bar{A}) \to \max, \tag{6.35}$$

$$\mathbf{P}(\bar{A}) \leqslant \varepsilon. \tag{6.36}$$

Thus, the problem (6.29)–(6.30) reduces to finding the critical set $\bar{A}$, and this problem has already been solved. Using the Neyman–Pearson fundamental lemma, we can show that the solution in this case is (cf. Lemma 6.2)

$$\bar{A} = \left\{\omega : \frac{d\mathbf{Q}^*}{d\mathbf{P}} > a\right\},$$

where $a = \inf\left\{s : \mathbf{P}\left(\frac{d\mathbf{Q}^*}{d\mathbf{P}} > s\right) \leqslant \varepsilon\right\}$. Here an optimal strategy in the problem (6.35)–(6.36) is a perfect hedge for the contingent claim $f \cdot I_{\bar{A}}$.

We remark also that a perfect hedge for the contingent claim $f \cdot I_A$, where $A = \Omega \setminus \bar{A}$, is an optimal strategy in the problem (6.29)–(6.30), which in turn is dual to the problem (6.35)–(6.36).

Let us now consider the problem of quantile hedging of a standard option to buy with payment function $f = (S_T - K)^+$ in the classical Black–Scholes model. As shown in Chapter 3, the initial value of a perfect hedge in this case is

$$X_0 = S_0\Phi(d_+) - Ke^{-rT}\Phi(d_-), \tag{6.37}$$

where $d_{\pm} = \dfrac{\ln(S_0/K) + T(r \pm \sigma^2/2)}{\sigma\sqrt{T}}$, and $\Phi(x)$ is the normal distribution function

$$\Phi(x) = \frac{1}{\sqrt{2\pi}} \int_{-\infty}^{x} e^{-x^2/2}\, dx.$$

Assume that the investor has initial capital x with $x < X_0$. By Theorem 6.1, an optimal strategy in the problem of quantile hedging coincides with a perfect hedge of the contingent claim $f \cdot I_A$, where the set A has the form

$$A = \left\{\omega : \frac{d\mathbf{P}}{d\mathbf{Q}^*} > c\right\} = \left\{\omega : \frac{d\mathbf{P}}{d\mathbf{P}^*} > \text{const} \cdot f\right\}. \tag{6.38}$$

Using the fact that the process equal to the density of the martingale measure $\mathbf{P}^*$ with respect to $\mathbf{P}$ is

$$\frac{d\mathbf{P}^*_T}{d\mathbf{P}_T} = \exp\left\{-\frac{\mu - r}{\sigma} W^*_T + \frac{1}{2}\left(\frac{\mu - r}{\sigma}\right)^2 T\right\}, \tag{6.39}$$

we can rewrite (6.38) in the form

(6.40)

$$\begin{aligned}
A &= \left\{\exp\left(\frac{\mu - r}{\sigma} W^*_T - \frac{1}{2}\left(\frac{\mu - r}{\sigma}\right)^2 T\right) > \text{const} \cdot (S_T - K)^+\right\} \\
&= \left\{\exp\left(\frac{\mu - r}{\sigma^2}\left(\ln S_0 + \left(r - \frac{\sigma^2}{2}\right)T + \sigma W^*_T\right)\right)\right. \\
&\qquad \times \exp\left(-\frac{\mu - r}{\sigma^2}\left(\ln S_0 + \left(r - \frac{\sigma^2}{2}\right)\right)T - \frac{1}{2}\left(\frac{\mu - r}{\sigma}\right)^2 T\right) \\
&\qquad\qquad \left. > \text{const} \cdot (S_T - K)^+\right\} \\
&= \left\{S_T^{\frac{\mu - r}{\sigma^2}} \exp\left(-\frac{\mu - r}{\sigma^2}\left(\ln S_0 + \frac{\mu + r - \sigma^2}{2}\right)T\right) > \text{const} \cdot (S_T - K)^+\right\}.
\end{aligned}$$

Here two cases have to be considered separately.

1. Assume that $\dfrac{\mu - r}{\sigma^2} \leqslant 1$.

In Figure 6.1 the shaded region is the solution domain of the inequality in (6.40). Thus, in this case the set A can be written as follows:

$$A = \{S_T < d\} = \{W^*_T < b\} = \left\{S_T < S_0 \exp\left(\frac{r - \sigma^2}{2} T + b\sigma\right)\right\} \tag{6.41}$$

under the restriction

$$\mathbf{E}^*\left(\frac{f}{B_T} \cdot I_A\right) = x_0. \tag{6.42}$$

We have from (6.41) that

$$\mathbf{P}(A) = \Phi\left(\frac{b - \frac{\mu - r}{\sigma} T}{\sqrt{T}}\right),$$

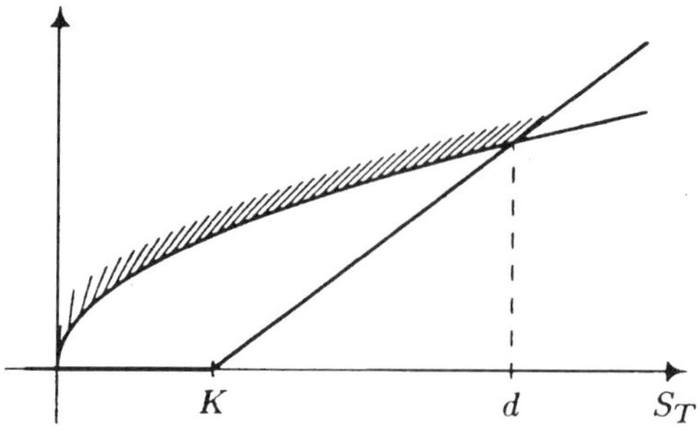

FIGURE 6.1. Structure of the hedging set.

and to find the constant b we have the condition

$$x_0 = \mathbf{E}^*(e^{-rT} f \cdot I_A) = e^{-rT} F_T^*(S_T), \tag{6.43}$$

where

$$\begin{aligned} F_T^*(S_T) &= \frac{1}{\sqrt{2\pi}} \int_{-\infty}^{b/\sqrt{T}} f\biggl(S_0 \exp\Bigl\{\sigma\sqrt{T}\,y + \frac{r-\sigma^2}{2}\,T\Bigr\}\biggr) e^{-y^2/2}\,dy \\ &= \frac{1}{\sqrt{2\pi}} \int_{d_0}^{b/\sqrt{T}} \biggl(S_0 \exp\Bigl\{\sigma\sqrt{T}\,y + \frac{r-\sigma^2}{2}\,T\Bigr\} - K\biggr) e^{-y^2/2}\,dy, \end{aligned}$$

$$d_0 = \frac{\ln(K/S_0) - \dfrac{r-\sigma^2}{2}T}{\sigma\sqrt{T}}.$$

Consequently,

(6.44)

$$\begin{aligned} x_0 &= S_0\Bigl[\Phi(\sigma\sqrt{T} - d_0) - \Phi\Bigl(\sigma\sqrt{T} - \frac{b}{\sqrt{T}}\Bigr)\Bigr] - Ke^{-rT}\Bigl[\Phi(-d_0) - \Phi\Bigl(-\frac{b}{\sqrt{T}}\Bigr)\Bigr] \\ &= S_0\Bigl[\Phi(d_+) - \Phi\Bigl(\sigma\sqrt{T} - \frac{b}{\sqrt{T}}\Bigr)\Bigr] - Ke^{-rT}\Bigl[\Phi(d_-) - \Phi\Bigl(-\frac{b}{\sqrt{T}}\Bigr)\Bigr]. \end{aligned}$$

We remark that in the case of the dual problem with the constraint $\mathbf{P}(A) = 1 - \varepsilon$, $0 < \varepsilon < 1$, we have the equation

$$b = \sqrt{T}\,\Phi^{-1}(1-\varepsilon) + \frac{\mu - r}{\sigma}\,T \tag{6.45}$$

for finding the constant b, and the corresponding value of a minimal hedge can be computed from (6.44).

2. Assume that $\dfrac{\mu - r}{\sigma^2} > 1$. In this case the set A has the structure

$$A = \{W_T^* < b_1\} \cup \{W_T^* > b_2\}.$$

The solution of the problem of quantile hedging is given by

$$\mathbf{P}(A) = \Phi\left(\frac{b_1 - \dfrac{\mu - r}{\sigma}T}{\sqrt{T}}\right) + \Phi\left(\frac{-b_2 + \dfrac{\mu - r}{\sigma}T}{\sqrt{T}}\right),$$

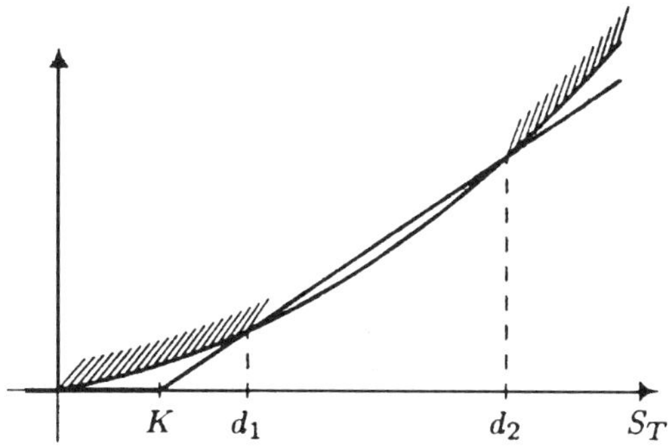

FIGURE 6.2. Structure of the hedging set.

and for determining the constants b_1 and b_2 we have the condition

$$x_0 = \mathbf{E}^*[e^{-rT} f \cdot I_A] = e^{-rT} F_T^*(S_T),$$

where

$$F_T^*(S_T) = \frac{1}{\sqrt{2\pi}} \int_{-\infty}^{b_1/\sqrt{T}} f\left(S_0 \exp\left\{\sigma\sqrt{T}\, y + \frac{r-\sigma^2}{2} T\right\}\right) e^{-y^2/2}\, dy$$
$$+ \frac{1}{\sqrt{2\pi}} \int_{b_2/\sqrt{T}}^{\infty} f\left(S_0 \exp\left\{\sigma\sqrt{T}\, y + \frac{r-\sigma^2}{2} T\right\}\right) e^{-y^2/2}\, dy.$$

By analogy with the case $\frac{\mu - r}{\sigma^2} \leqslant 1$, we get that

$$x_0 = S_0\left[\Phi(d_+) - \Phi\left(\sigma\sqrt{T} - \frac{b_1}{\sqrt{T}}\right) + \Phi\left(\sigma\sqrt{T} - \frac{b_2}{\sqrt{T}}\right)\right]$$
$$- Ke^{-rT}\left[\Phi(d_-) - \Phi\left(-\frac{b_1}{\sqrt{T}}\right) + \Phi\left(-\frac{b_2}{\sqrt{T}}\right)\right].$$

The Merton model.
In the Merton model (3.31) there are two assets:

$$dB_t = rB_t\, dt, \qquad B_0 = 1,$$
$$dS_t = S_{t-}(\mu\, dt - \nu\, d\Pi_t), \quad S_0 > 0,$$

where μ, $r \in \mathbb{R}_+$, $\nu < 1$, and $\Pi = (\Pi_t)_{t \geqslant 0}$ is a Poisson process with intensity $\lambda > 0$.

As shown in Chapter 3, there is a unique martingale measure $\mathbf{P}^*$ for the discounted stock price process S_t/B_t in this model, and its density with respect to the initial measure $\mathbf{P}$ is

$$Z_t^* = \frac{d\mathbf{P}_t^*}{d\mathbf{P}_t} = \exp\bigl((\lambda - \lambda^*)t + (\ln\lambda^* - \ln\lambda)\Pi_t\bigr),$$

where $\lambda^* = \frac{\mu - r}{\nu}$ is the intensity of Π_t with respect to $\mathbf{P}^*$.

By the Kolmogorov–Itô formula,

$$S_t = S_0 \exp\bigl(\Pi_t \ln(1-\nu) + \mu t\bigr).$$

Using this equality, we can rewrite the density $d\mathbf{P}_T^*/d\mathbf{P}_T$ as

$$\begin{aligned}\frac{d\mathbf{P}_T^*}{d\mathbf{P}_T} &= \exp\bigl(\Pi_T(\ln\lambda^* - \ln\lambda) + (\lambda-\lambda^*)T\bigr)\\ &= \exp\biggl(\frac{\ln\lambda^*-\ln\lambda}{\ln(1-\nu)}\bigl(\Pi_T\ln(1-\nu) + \ln S_0 + \mu T\bigr)\biggr)\\ &\quad\times\exp\biggl((\lambda-\lambda^*)T + \frac{\ln\lambda^*-\ln\lambda}{\ln(1-\nu)}(-\ln S_0 - \mu T)\biggr)\\ &= S_T^{\frac{\ln\lambda^*-\ln\lambda}{\ln(1-\nu)}}\exp\biggl((\lambda-\lambda^*)T + \frac{\ln\lambda^*-\ln\lambda}{\ln(1-\nu)}(-\ln S_0 - \mu T)\biggr).\end{aligned}$$

We consider the situation when the initial value x_0 of the investor's capital is $< \mathbf{E}^* f/B_T = X_0$. By analogy with the case of the Black–Scholes model, the maximal successful hedging set for a contingent claim f has the structure

$$A = \biggl\{\frac{d\mathbf{P}}{d\mathbf{P}^*} > \mathrm{const}\cdot f\biggr\},$$

where const is chosen so that $\mathbf{E}^* f/B_T\cdot I_A = x_0$. In our case the set A has the form

(6.46)

$$\begin{aligned}A = \biggl\{S_T^{-\frac{\ln\lambda^*-\ln\lambda}{\ln(1-\nu)}}\exp\biggl(-(\lambda-\lambda^*)T + \frac{\ln\lambda^*-\ln\lambda}{\ln(1-\nu)}(\ln S_0 + \mu T)\biggr)\\ > \mathrm{const}\cdot(S_T-K)^+\biggr\}\end{aligned}$$

$$= \bigl\{S_T^\alpha > \mathrm{const}\cdot(S_T-K)^+\bigr\},$$

where $\alpha = -\dfrac{\ln\lambda^*-\ln\lambda}{\ln(1-\nu)}$.

We consider the following variants for the value of the quantity α.

1. Let $\alpha \leqslant 1$. In this case we have a situation analogous to that in Figure 6.1, that is, there is a constant b such that the set A can be written as

$$A = \{S_T < b\} = \{\Pi_T < d\} = \bigl\{S_T < S_0\exp\bigl(d\ln(1-\nu) + \mu T\bigr)\bigr\}.$$

For finding b we have the condition

$$x_0 = \mathbf{E}^*(e^{-rT} f\cdot I_A) = \mathbf{E}^*\bigl(e^{-rT}(S_T-K)^+\cdot I_{\{S_T<b\}}\bigr).$$

Note that

$$(S_T-K)^+\cdot I_{\{S_T<b\}} = (S_T-K)^+ - (S_T-b)^+ - (b-K)\cdot I_{\{S_T\geqslant b\}}.$$

Consequently,

$$\begin{aligned}\mathbf{E}^*(e^{-rT} f\cdot I_A) &= \mathbf{E}^* e^{-rT}(S_T-K)^+ - \mathbf{E}^* e^{-rT}(S_T-b)^+\\ &\quad - \mathbf{E}^* e^{-rT}(b-K)\cdot I_{\{S_T\geqslant b\}}\\ &= \mathbf{E}^* e^{-rT}(S_T-K)^+ - \mathbf{E}^* e^{-rT}(S_T-b)^+\\ &\quad - e^{-rT}(b-K)\mathbf{P}^*(S_T\geqslant b).\end{aligned}$$

Here the first two terms are the costs of standard options to buy with exercise prices K and b, respectively, and with exercise time T. According to (3.37),

$$\mathbf{E}^* e^{-rT}(S_T - K)^+ = S_0 \Psi(n_0, \lambda^*(1-\nu)T) - Ke^{-rT}\Psi(n_0, \lambda^* T),$$
$$\mathbf{E}^* e^{-rT}(S_T - b)^+ = S_0 \Psi(n_1, \lambda^*(1-\nu)T) - be^{-rT}\Psi(n_1, \lambda^* T),$$

where the constants n_0 and n_1 are determined from the relations

$$n_0 = \left[\frac{\ln K/S_0 - \mu T}{\ln(1-\nu)}\right], \qquad n_1 = \left[\frac{\ln b/S_0 - \mu T}{\ln(1-\nu)}\right],$$

and the function $\Psi(x,y)$ is $\sum_{n=x}^{\infty} e^{-y}\frac{y^n}{n!}$.

For the third term we have

$$\mathbf{P}^*(S_T \geqslant b) = \mathbf{P}^*(\Pi_T \geqslant d) = \sum_{n \geqslant [d]}^{\infty} e^{-\lambda^* T}\frac{(\lambda^* T)^n}{n!}.$$

2. Suppose that $\alpha > 1$. We have a situation analogous to that in Figure 6.2. Thus, the set A has the structure

$$A = \{S_T < b_1\} \cup \{S_T > b_2\} = \{\Pi_T < d_1\} \cup \{\Pi_T > d_2\},$$

where $b_i = S_0 \exp\big(d_i \ln(1-\nu) + \mu T\big)$, $i = 1, 2$.

The constants b_1 and b_2 are found as in the preceding case, with use of the relation

$$\begin{aligned} f \cdot I_A &= (S_T - K)^+ \cdot I_{\{S_T < b_1\} \cup \{S_T > b_2\}} \\ &= (S_T - K)^+ - (S_T - b_1)^+ + (S_T - b_2)^+ \\ &\quad - (b_1 - K) \cdot I_{\{S_T \geqslant b_1\}} + (b_2 - K) \cdot I_{\{S_T \geqslant b_2\}}. \end{aligned}$$

The jump-diffusion model.

In this model we consider three assets

$$\begin{aligned} dB_t &= rB_t\,dt, & B_0 &= 1, \\ dS_t^i &= S_{t-}^i(\mu^i\,dt + \sigma^i\,dW_t - \nu^i\,d\Pi_t), & S_0^i &> 0,\ i = 1, 2. \end{aligned}$$

Here μ^i, $r \in \mathbb{R}_+$ are constants, $\sigma^i > 0$, $\nu^i < 1$, $W = (W_t)_{t \geqslant 0}$ is a Wiener process, and $\Pi = (\Pi_t)_{t \geqslant 0}$ is a Poisson process with intensity $\lambda > 0$ and is independent of W.

In the case $\sigma^2\nu^1 - \sigma^1\nu^2 \neq 0$ there is a unique martingale measure $\mathbf{P}^*$ for the discounted stock price processes $(S_t^1/B_t)_{t \geqslant 0}$ and $(S_t^2/B_t)_{t \geqslant 0}$, and its density has the form

$$Z_t^* = \frac{d\mathbf{P}_t^*}{d\mathbf{P}_t} = \exp\left\{\phi W_t - \frac{\phi^2}{2}t + (\lambda - \lambda^*)t + \Pi_t(\ln\lambda^* - \ln\lambda)\right\}, \tag{6.47}$$

where

$$\phi = \frac{(\mu^2 - r)\nu^1 - (\mu^1 - r)\nu^2}{\sigma^2\nu^1 - \sigma^1\nu^2}, \qquad \lambda^* = \frac{(\mu^1 - r)\sigma^2 - (\mu^2 - r)\sigma^1}{\sigma^2\nu^1 - \sigma^1\nu^2}.$$

Furthermore, $W_t^* = W_t - \phi t$ is a Wiener process with respect to the measure $\mathbf{P}^*$, and Π_t is a Poisson process with intensity λ^*.

We next consider a European contingent claim on one of the assets, say S^1: $f = (S^1 - K)^+$.

Using the Kolmogorov–Itô formula, we have

$$S_T^1 = S_0 \exp\biggl(\Bigl(\mu^1 - \frac{(\sigma^1)^2}{2}\Bigr)T + \sigma^1 W_T + \Pi_T \ln(1-\nu^1)\biggr).$$

Let $\psi = \lambda^*/\lambda - 1$. Then for the density process in (6.47),

$$\begin{aligned}\frac{d\mathbf{P}_T^*}{d\mathbf{P}_T} &= \exp\Bigl\{\phi W_T - \Bigl(\frac{\phi^2}{2} + \lambda\psi\Bigr)T + \Pi_T \ln(1+\psi)\Bigr\}\\ &= \exp\Bigl\{\frac{\phi}{\upsilon^1}\Bigl(\ln S_0^1 + \sigma^1 W_T + \Bigl(\mu^1 - \frac{(\sigma^1)^2}{2}\Bigr)T + \Pi_T \ln(1-\nu^1)\Bigr)\Bigr\}\\ &\quad\times \exp\Bigl\{-\frac{\phi}{\sigma^1}\ln S_0^1 - \frac{\phi}{\sigma^1}\Bigl(\mu^1 - \frac{(\sigma^1)^2}{2}\Bigr)T - \Bigl(\frac{\phi^2}{2} + \lambda\psi\Bigr)T\\ &\qquad\qquad + \Pi_T \ln\frac{1+\psi}{(1-\nu^1)^{\phi/\sigma^1}}\Bigr\}\\ &= (S_T^1)^{\phi/\sigma^1}\cdot C_1 \cdot C_2^{\Pi_T},\end{aligned}$$

where C_1 and C_2 denote the corresponding expressions.

In our case the critical set A has the form

$$A = \bigl\{(S_T^1)^{-\phi/\sigma^1} > \text{const}\cdot C_1 \cdot C_2^{\Pi_T}\cdot f\bigr\}.$$

Since W^* and Π are independent with respect to $\mathbf{P}^*$,

$$\mathbf{E}^*(e^{-rT} f\cdot I_A) = \sum_{n=0}^{\infty}\mathbf{E}^*(e^{-rT} f\cdot I_{A|\Pi_T=n})\cdot \mathbf{P}^*(\Pi_T = n).$$

As in the case of the previous model we must treat separately two cases.

1. Let $-\phi/\sigma^1 \leqslant 1$. For a fixed $n\in\mathbb{N}$,

$$f\cdot I_{A|\Pi_T=n} = (S_T^1 - K)^+ - (S_T^1 - b(n))^+ - (b(n) - K)\cdot I_{\{S_T^1 \geqslant b(n)\}},$$

where $b(n)$ is a constant depending only on n.

Consequently,

$$\begin{aligned}\mathbf{E}^*(e^{-rT} f\cdot I_{A|\Pi_T=n}) &= \mathbf{E}^* e^{-rT}(S_T^1 - K)^+ - \mathbf{E}^* e^{-rT}(S_T^1 - b(n))^+\\ &\quad - e^{-rT}(b(n) - K)\mathbf{P}^*(S_T^1 \geqslant b(n)).\end{aligned}$$

Moreover,

$$\mathbf{P}^*(S_T^1 \geqslant b(n)) = \mathbf{P}^*(W_T \geqslant d(n)) = \Phi\Bigl(-\frac{d(n) - \phi T}{\sqrt{T}}\Bigr),$$

where $b(n) = S_0^1 \exp\Bigl(\sigma^1 d(n) + \Bigl(\mu^1 - \frac{(\sigma^1)^2}{2}\Bigr)T + n\ln(1-\nu^1)\Bigr)$.

Introducing the notation

$$U(x,y) = \mathbf{E}^* e^{-rT}\Bigl[S_0^1 \exp\Bigl(\sigma^1 W_T + \Bigl(\mu^1 - \frac{(\sigma^1)^2}{2}\Bigr)T + x\ln(1-\nu^1)\Bigr) - y\Bigr]^+,$$

we have

$$
\begin{aligned}
(6.48) \qquad \mathbf{E}^*(e^{-rT}f\cdot I_A) &= \sum_{n=0}^{\infty}\mathbf{E}^*(e^{-rT}f\cdot I_{A|\Pi_T=n})\cdot \mathbf{P}^*(\Pi_T=n)\\
&= \sum_{n=0}^{\infty} e^{-\lambda^* T}\frac{(\lambda^* T)^n}{n!}\bigl\{U(n,K)-U(n,b(n))\\
&\qquad - e^{-rT}(b(n)-K)\cdot \mathbf{P}^*(S_T^1\geqslant b(n))\bigr\}.
\end{aligned}
$$

The constants $b(n)$ are found by solving the equation

$$x^{\phi/\sigma^1} = C_1\cdot C_2^n\cdot \mathrm{const}\cdot (x-K)^+$$

for x. We then sum over n in (6.48), and to find the unknown const we have the condition $\mathbf{E}^*(e^{-rT}f\cdot I_A)=x_0$.

2. Suppose that $-\phi/\sigma^1>1$. In this case the critical set has the structure

$$A|_{\Pi_T=n} = \{S_T^1 < b_1(n)\}\cup\{S_T^1 > b_2(n)\}.$$

Thus,

$$
\begin{aligned}
f\cdot I_{A|\Pi_T=n} &= (S_T^1-K)^+ - (S_T^1-b_1(n))^+ + (S_T^1-b_2(n))^+\\
&\quad - (b_1(n)-K)\cdot I_{\{S_T^1\geqslant b_1(n)\}} + (b_2(n)-K)\cdot I_{\{S_T^1\geqslant b_2(n)\}}.
\end{aligned}
$$

The unknown constants $b_1(n)$ and $b_2(n)$ are then found as in the first case.

The jump-diffusion model of an incomplete market.

Let us consider the following model of an incomplete market with two assets:

$$
\begin{aligned}
dB_t &= rB_t\,dt, & B_0 &= 1,\\
dS_t &= S_{t-}(\mu\,dt+\sigma\,dW_t-\nu\,d\Pi_t), & S_0 &> 0.
\end{aligned}
$$

Here, as before, μ, $r\in\mathbb{R}_+$, $\nu<1$, $W=(W_t)_{t\geqslant 0}$ is a Wiener process, and $\Pi=(\Pi_t)_{t\geqslant 0}$ is a Poisson process with intensity λ.

In this case there is a *set* $\mathcal{P}=\{\mathbf{P}^*\}$ of martingale measures, whose densities with respect to the initial measure $\mathbf{P}$ are determined by

$$
(6.49) \qquad \frac{d\mathbf{P}_t^*}{d\mathbf{P}_t} = \exp\biggl\{\phi W_t - \biggl(\frac{\phi^2}{2}+\lambda\psi\biggr)t + \Pi_t\ln(1+\psi)\biggr\},
$$

where ϕ and ψ are connected by the relation

$$\mu - r + \sigma\phi - \nu\lambda(1+\psi) = 0.$$

With respect to each such measure, $W_t^* = W_t - \phi t$ is a Wiener process, and Π is a Poisson process with intensity $\lambda^* = \lambda(1+\psi)$.

By the Kolmogorov–Itô formula,

$$S_T = S_0\exp\biggl(\biggl(\mu-\frac{\sigma^2}{2}\biggr)T + \sigma W_T + \Pi_T\ln(1-\nu)\biggr).$$

Then for the density $d\mathbf{P}_T^*/d\mathbf{P}_T$ we can write

$$\begin{aligned}\frac{d\mathbf{P}_T^*}{d\mathbf{P}_T} &= \exp\left\{\frac{\phi}{\sigma}\left(\ln S_0 + \sigma W_T + \left(\mu - \frac{\sigma^2}{2}\right)T + \Pi_T \ln(1-\nu)\right)\right\} \\ &\quad \times \exp\left\{-\frac{\phi}{\sigma}\ln S_0 - \left(\mu - \frac{\sigma^2}{2} + \frac{\phi^2}{2} + \lambda\psi\right)T + \Pi_T \ln\frac{1+\psi}{(1-\nu)^{\phi/\sigma}}\right\} \\ &= S_T^{\phi/\sigma}\cdot C_1 \cdot C_2^{\Pi_T}.\end{aligned}$$

Thus, the critical set in this case has the form

$$A = \left\{S_T^{-\phi/\sigma} > C_1 \cdot C_2^{\Pi_T}\cdot \text{const}\cdot f\right\}$$

and by analogy with the preceding model we must consider two cases.

1. Suppose that $-\phi/\sigma \leqslant 1$. For a fixed $n \in \mathbb{N}$ the critical set has the form

$$A|_{\Pi_T = n} = \{S_T < b(n)\},$$

where $b(n)$ is a constant depending only on n.

Consequently,

$$\begin{aligned}\mathbf{E}^*(e^{-rT} f \cdot I_{A|\Pi_T=n}) = \mathbf{E}^* e^{-rT}(S_T - K)^+ - \mathbf{E}^* e^{-rT}(S_T - b(n))^+ \\ - e^{-rT}(b(n) - K)\cdot \mathbf{P}^*(S_T \geqslant b(n)).\end{aligned}$$

We have

$$\mathbf{P}^*(S_T \geqslant b(n)) = \mathbf{P}^*(W_T \geqslant d(n)) = \Phi\left(-\frac{d(n) - \phi T}{\sqrt{T}}\right),$$

where the constants $b(n)$ and $d(n)$ are connected by the relation

$$b(n) = S_0 \exp\left(\sigma d(n) + \left(\mu - \frac{\sigma^2}{2}\right)T + n\ln(1-\nu)\right).$$

Let

$$U(x,y) = \mathbf{E}^* e^{-rT}\left[S_0 \exp\left(\sigma W_T + \left(\mu - \frac{\sigma^2}{2}\right)T + x\ln(1-\nu)\right) - y\right]^+.$$

Then

$$\begin{aligned}\mathbf{E}^*(e^{-rT} f \cdot I_A) &= \sum_{n=0}^{\infty} \mathbf{E}^*(e^{-rT} f \cdot I_{\{A|\Pi_T=n\}})\cdot \mathbf{P}^*(\Pi_T = n) \\ &= \sum_{n=0}^{\infty} e^{-\lambda^* T}\frac{(\lambda^* T)^n}{n!} \\ &\qquad \times \left\{U(n,K) - U(n,b(n)) - e^{-rT}(b(n) - K)\cdot \mathbf{P}^*(S_T \geqslant b(n))\right\}.\end{aligned}$$

The constants $b(n)$ are found by solving the equation

$$x^{-\phi/\sigma} = C_1 \cdot C_2^n \cdot \text{const}\cdot (x - K)^+$$

for x. Then the unknown const is found by maximizing the probability of the successful hedging set over the family $\mathcal{P} = \{\mathbf{P}^*\}$ of all martingale measures determined in (6.49):

$$\mathbf{P}(A) = \mathbf{P}\left(\omega : \frac{d\mathbf{P}}{d\mathbf{P}^*} > \text{const}\cdot f\right) \to \max_{\mathbf{P}^* \in \mathcal{P}}$$

under the constraint

$$\mathbf{E}^*(f/B_T \cdot I_A) \leqslant x_0 \quad \text{for all} \quad \mathbf{P}^* \in \mathcal{P}.$$

2. Let $-\phi/\sigma > 1$. By analogy with the preceding model we have

$$A|_{\Pi_T=n} = \{S_T < b_1(n)\} \cup \{S_T > b_2(n)\},$$

and hence

$$\begin{aligned} f \cdot I_{A|\Pi_T=n} &= (S_T - K)^+ - (S_T - b_1(n))^+ + (S_T - b_2(n))^+ \\ &\quad - (b_1(n) - K) \cdot I_{\{S_T \geqslant b_1(n)\}} + (b_2(n) - K) \cdot I_{\{S_T \geqslant b_2(n)\}}. \end{aligned}$$

The constants $b_1(n)$ and $b_2(n)$ are found by a procedure analogous to the first case.

The Cox–Ross–Rubinstein model.

Let us consider the discrete-market model (3.44):

$$\begin{aligned} \Delta B_n &= r B_{n-1}, \quad & B_0 &= 1, \\ \Delta S_n &= \rho_n S_{n-1}, \quad & S_0 &> 0, \end{aligned}$$

where $(\rho_n)_{1 \leqslant n \leqslant N}$ is a sequence of independent random variables taking two values $a < b$ with probabilities p and $q = 1 - p$.

As noted in Chapter 3, this model of a market is complete. The density of the initial measure with respect to the martingale measure can be represented in the form

$$\frac{d\mathbf{P}_N}{d\mathbf{P}_N^*} = \left(\frac{p}{p^*}\right)^{\Delta_N} \left(\frac{1-p}{1-p^*}\right)^{N-\Delta_N},$$

where $p^* = \dfrac{r-a}{b-a}$, $\Delta_N = \delta_1 + \cdots + \delta_N$, and $(\delta_n)_{1 \leqslant n \leqslant N}$ is a sequence of independent identically distributed random variables such that

$$\rho_i = a + (b-a)\delta_i \quad \text{for all} \quad 1 \leqslant i \leqslant N.$$

Of course, here

$$\rho_n = \begin{cases} b \\ a \end{cases} \iff \delta_n = \begin{cases} 1 \\ 0 \end{cases}$$

and $\mathcal{F}_n = \sigma(\rho_1, \ldots, \rho_n) = \sigma(\delta_1, \ldots, \delta_n)$.

We note that the stock price can be represented with the help of Δ_N in the form

$$S_N = S_0 \prod_{k=1}^{N} (1 + \rho_k) = S_0 (1+a)^N \left(\frac{1+b}{1+a}\right)^{\Delta_N}.$$

Thus, the critical set A can be written as

$$\begin{aligned} A &= \left\{ \left(\frac{p}{p^*}\right)^{\Delta_N} \left(\frac{1-p}{1-p^*}\right)^{N-\Delta_N} > \text{const} \cdot (S_N - K)^+ \right\} \\ &= \left\{ \left(\frac{1-p}{1-p^*}\right)^{N} \left(\frac{p}{p^*} \cdot \frac{1-p^*}{1-p}\right)^{\Delta_N} > \text{const} \cdot \left(S_0 (1+a)^N \left(\frac{1+b}{1+a}\right)^{\Delta_N} - K \right)^+ \right\} \\ &= \left\{ C_1 \left(\frac{p}{p^*} \cdot \frac{1-p^*}{1-p}\right)^{\Delta_N} > \text{const} \cdot \left(C_2 \left(\frac{1+b}{1+a}\right)^{\Delta_N} - K \right)^+ \right\}, \end{aligned}$$

where C_1 and C_2 are positive constants.

Consequently, two cases must be distinguished.

1. Suppose that $p < \dfrac{1+b}{1+r} p^*$. In this case A can be represented in the form (Figure 6.3)

$$A = \{S_N < h\} = \{\Delta_N < d\} = \big\{S_0(1+a)^{N-d}(1+b)^d < h\big\}.$$

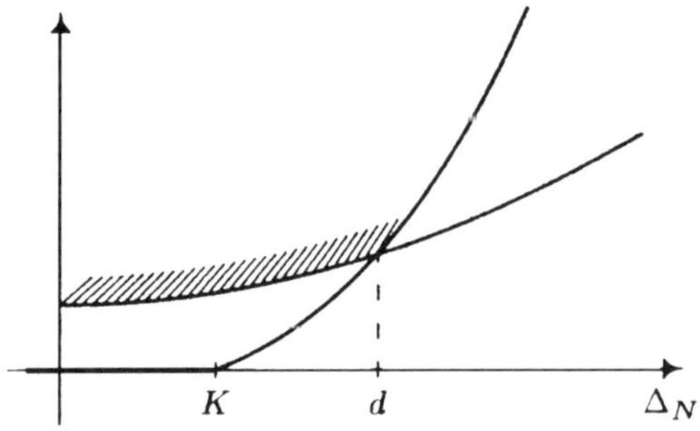

FIGURE 6.3. Structure of the hedging set.

The condition

$$\mathbf{E}^*(f/B_N \cdot I_A) = x_0 \tag{6.50}$$

can be used for finding the constant h.

We have the relation

$$(S_N - K)^+ \cdot I_{\{S_N<h\}} = (S_N - K)^+ - (S_N - h)^+ - (h-K) \cdot I_{\{S_N \geqslant h\}}.$$

Thus, (6.50) can be rewritten in the form

$$\begin{aligned}\sum_{k=0}^{N} \binom{N}{k} (p^*)^k (1-p^*)^{N-k} \Big[\big(S_0(1+a)^{N-k}(1+b)^k - K\big)^+ \\ - \big(S_0(1+a)^{N-k}(1+b)^k - h\big)^+ \\ - (h-K) \cdot I_{\{S_0(1+a)^{N-k}(1+b)^k \geqslant h\}}\Big] = x_0(1+r)^N.\end{aligned}$$

2. Let $p > \dfrac{1+b}{1+r} p^*$. In this case A has the form (Figure 6.4):

$$A = \{S_N < h_1\} \cup \{S_N > h_2\} = \{\Delta_N < d_1\} \cup \{\Delta_N > d_2\},$$

where $h_i = S_0(1+a)^{N-d_i}(1+b)^{d_i}$, $i = 1, 2$.

Since

$$\begin{aligned} f \cdot I_A = (S_N - K)^+ - (S_N - h_1)^+ + (S_N - h_2)^+ \\ - (h_1 - K) \cdot I_{\{S_N \geqslant h_1\}} + (h_2 - K) \cdot I_{\{S_N \geqslant h_2\}}, \end{aligned}$$

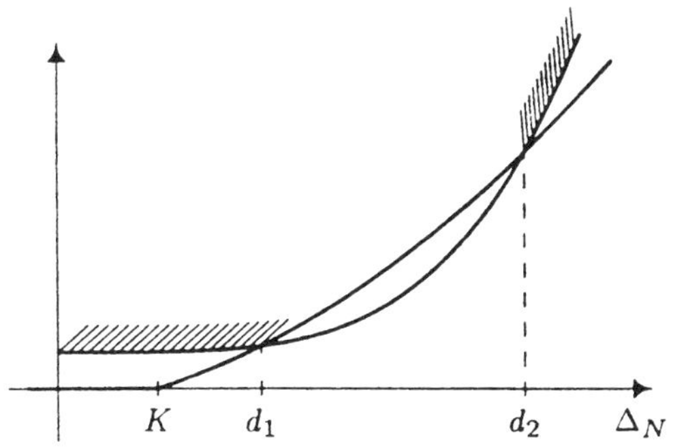

FIGURE 6.4. Structure of the hedging set.

we get the following expression for determining the constants h_i, $i = 1, 2$:

$$\sum_{k=0}^{N} \binom{N}{k} (p^*)^k (1-p^*)^{N-k} \big[(S_0(1+a)^{N-k}(1+b)^k - K)^+ \\ - \big(S_0(1+a)^{N-k}(1+b)^k - h_1\big)^+ + \big(S_0(1+a)^{N-k}(1+b)^k - h_2\big)^+ \\ - (h_1 - K) \cdot I_{\{S_0(1+a)^{N-k}(1+b)^k \geqslant h_1\}} \\ + (h_2 - K) \cdot I_{\{S_0(1+a)^{N-k}(1+b)^k \geqslant h_2\}} \big] = x_0(1+r)^N.$$

Quantile hedging and value at risk (VaR).

Quantile hedging can be regarded as a dynamic version of the well-known concept of risk management VaR. Just as in the dual problem of quantile hedging, a certain given constant ε with $0 < \varepsilon < 1$ is assumed in it, and the quantity VaR is defined as the maximum possible size of decrease of the value of the portfolio (in percent of the initial value) on the time interval $[0, T]$ for the given probability $1 - \varepsilon$:

$$\text{VaR} = \inf\big\{s : \mathbf{P}\big(\ln(X_T^\pi / X_0) < -s\big) = 1 - \varepsilon\big\}.$$

We remark that in view of the fact that the VaR strategy is static, that is, does not change in correspondence with movements of the stock price, the initial value of the VaR strategy turns out to be greater than the initial value necessary for constructing the corresponding quantile hedging strategy.

References for Chapter 6

[**25**], [**46**], [**47**], [**51**], [**56**]–[**58**], [**71**], [**84**], [**93**], [**97**], [**101**], [**108**], [**114**], [**126**], [**139**], [**141**], [**148**].

CHAPTER 7

Dynamic Contingent Claims and American Options

We study contingent claims that are random processes adapted to the price dynamics of the underlying assets of a financial market. This dynamic character of the contingent claims under consideration "changes the dimension" of the corresponding calculations, since along with the value of a hedging portfolio it presupposes the determination of a (random) moment of time for presenting the claim for payment. It is shown that an adequate solution of these problems can be attained in the framework of a study of the class of problems on optimal stopping of random processes.

§ 7.1. Pricing dynamic contingent claims and the optimal stopping problem

As in § 3.1 and § 4.1, we consider a *semimartingale* (B, S)-market given on some stochastic base $(\Omega, \mathcal{F}, \mathbb{F}, \mathbf{P})$.

A *dynamic* contingent claim with last exercise date $T < \infty$ is defined to be any nonnegative optional random *process* $f = (f_t)_{0 \leqslant t \leqslant T}$.

A strategy with consumption (π, C) is called a *hedging* strategy for f (or a *hedge* for f) on $[0, T]$ if

$$X_t^{\pi,C} \geqslant f_t \tag{7.1}$$

for any $t \leqslant T$ ($\mathbf{P}$-a.s.). We remark that the inequality (7.1) is preserved if a stopping time $\tau \in \mathcal{M}_0^T$ is taken instead of t.

A *minimal hedge* is defined to be a hedging strategy (π^*, C^*) such that

$$X_t^{\pi^*,C^*} \leqslant X_t^{\pi,C}$$

for any other hedge (π, C) and any $t \leqslant T$ ($\mathbf{P}$-a.s.). The *initial value* of a minimal hedge, if there is a minimal hedge, is denoted by $\mathbb{C}_T^*(f)$ and called the (upper) *price* of the contingent claim f.

The *mathematical problem* of the calculations under consideration here consists in finding the price $\mathbb{C}_T^*(f)$ and a minimal hedging strategy (π^*, C^*).

A somewhat different terminology is usually used in *finance* in the solution of this problem. We introduce the concept of an *American option* as a derivative security with term of validity $[0, T]$ giving the holder the right to present it at any moment of time $\tau \leqslant T$ and receive a payment in the amount f_τ. The time τ is unknown beforehand. It is determined on the basis of the current market information $\mathbb{F}$, and hence is a stopping time with respect to $\mathbb{F}$. In view of the obvious dualism between dynamic contingent claims and American options, this problem admits the equivalent formulation of finding the price $\mathbb{C}_T^*(f)$ and a minimal hedge (π^*, C^*) for an American option with *payment function* $f = (f_t)_{t \leqslant T}$.

The *general methodology* for solving this problem consists essentially in the following.

On a (B, S)-market we consider a dynamic contingent claim f such that

$$\sup_{\widetilde{\mathbf{P}} \in \mathcal{M}(S/B, \mathbf{P})} \widetilde{\mathbf{E}} B_\tau^{-1} f_\tau < \infty. \tag{7.2}$$

Under the condition (7.2) the nonnegative process

$$Y_t = \operatorname*{ess\,sup}_{\tau \in \mathcal{M}_t^T,\, \widetilde{\mathbf{P}} \in \mathcal{M}(S/B, \mathbf{P})} \widetilde{\mathbf{E}}(B_\tau^{-1} f_\tau \,|\, \mathcal{F}_t)$$

is well defined and is a nonnegative supermartingale according to (2.24).

The optional decomposition (2.23) gives us the possibility of representing Y in the form

$$Y_t = Y_0 + \int_0^t \gamma_u^* \, d\left(\frac{S}{B}\right)_u - D_t^* \tag{7.3}$$

with a predictable process γ_u and with an optional nondecreasing process D^*.

By Theorem 4.1, the supermartingale Y is the discounted value of some strategy with consumption whose structure is determined from the relation (7.3). Denoting by (π^*, C^*) the strategy with consumption found, we can see at once (see the proof of Theorem 4.2) that this is also a minimal hedge, and

$$\mathbb{C}_T^*(f) = \sup_{\widetilde{\mathbf{P}} \in \mathcal{M}(S/B, \mathbf{P}),\, \tau \in \mathcal{M}_0^T} \widetilde{\mathbf{E}} B_\tau^{-1} f_\tau, \tag{7.4}$$

$$X_t^{\pi^*, C^*} \equiv X_t^* = \operatorname*{ess\,sup}_{\widetilde{\mathbf{P}} \in \mathcal{M}(S/B, \mathbf{P}),\, \tau \in \mathcal{M}_t^T} \widetilde{\mathbf{E}}(B_\tau^{-1} f_\tau \,|\, \mathcal{F}_t), \tag{7.5}$$

$$B_t^{-1} X_t^* = \mathbb{C}_T^*(f) + \int_0^t \gamma_u^* \, d\left(\frac{S}{B}\right)_u - \int_0^t B_u^{-1} \, dC_u^*. \tag{7.6}$$

The formulas (7.4)–(7.6) obtained determine a general methodology for (super)hedging American options.

We reproduce the above in greater detail for the Black–Scholes model (3.19). In this case $\mathcal{M}(S/B, \mathbf{P})$ consists of a *unique* martingale measure $\mathbf{P}^*$ with density Z^* (see (3.21)).

It was noted above that in calculating an American option it is necessary to find both its price and its exercise time. Here the following remark is appropriate.

REMARK. Assume that an American option with payment function $f = (f_t)_{t \leqslant T}$ was purchased at the price $\mathbb{C}_T^*(f)$, and is exercised at a time $\tau \in \mathcal{M}_0^T$ when the value of the hedging strategy $X_\tau^{\pi, C}$ strictly exceeds f_τ with positive probability. The *irrationality* of this behavior on the market is clear, because for the initial value $\mathbb{C}_T^*(f)$ one should use the strategy (π, C) at once. This yields (with the same positive probability) a net gain equal to $X_\tau^{\pi, C} - f_\tau$ in comparison with the purchase of the option.

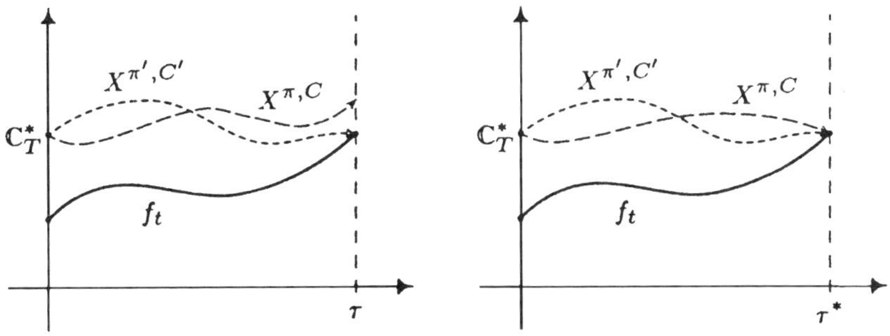

FIGURE 7.1. A rational (τ^*) and an irrational (τ) exercise time.

The foregoing justifies the following *definition* (Figure 7.1).

An exercise time τ^* is called *rational* if for any admissible strategy (π, C) with initial value $X_0^{\pi,C} = \mathbb{C}_T^*(f)$ the inequality $X_{\tau^*}^{\pi,C} \geqslant f_{\tau^*}$ (**P**-a.s.) implies the equality $X_{\tau^*}^{\pi,C} = f_{\tau^*}$ (**P**-a.s.).

THEOREM 7.1. *Suppose that on the (B,S)-market* (3.19) *a dynamic contingent claim $f = (f_t)_{t \leqslant T}$ is given such that*

$$\sup_{\tau \in \mathcal{M}_0^T} \mathbf{E}^* e^{-r\tau} f_\tau < \infty. \tag{7.7}$$

Then there exists a minimal hedging strategy $(\pi^*, C^*) = (\beta^*, \gamma^*, C^*)$ with consumption, determined by the relations

$$\mathbb{C}_T^*(f) = \sup_{\tau \in \mathcal{M}_0^T} \mathbf{E}^* e^{-r\tau} f_\tau, \tag{7.8}$$

$$X_t^* = X_t^{\pi^*, C^*} = \operatorname*{ess\,sup}_{\tau \in \mathcal{M}_t^T} \mathbf{E}^*(e^{-r(\tau - t)} f_\tau \mid \mathcal{F}_t), \tag{7.9}$$

$$e^{-rt} X_t^* = \mathbb{C}_T^*(f) + \int_0^t \gamma_u^* \, d(e^{-ru} S_u) - \int_0^t e^{-ru} \, dC_u^*, \tag{7.10}$$

$$\beta_t^* = \frac{X_t^* - \gamma_t^* S_t}{e^{rt}}.$$

The stopping time $\tau^* \in \mathcal{M}_0^T$ is rational if and only if

$$\mathbb{C}_T^*(f) = \mathbf{E}^* e^{-r\tau^*} f_{\tau^*}. \tag{7.11}$$

Further,

$$\tau^* = \inf\{t \leqslant T : X_t^* \geqslant f_t\} \tag{7.12}$$

if $\Delta f_t \geqslant 0$ for all $t \leqslant T$ and the family $\{e^{-r\tau} f_\tau\}_{\tau \in \mathcal{M}_0^T}$ is uniformly integrable.

PROOF. The relations (7.8)–(7.10) follow from (7.4)–(7.6), since the condition (7.7) ensures that (7.2) holds.

Further, if τ^* is a rational exercise time, then the difference $c = \mathbb{C}_T^*(f) - \mathbf{E}^* e^{-r\tau^*} f_{\tau^*}$ is nonnegative. We define the nonnegative supermartingale

$$Y_t = c + \mathbf{E}^*(e^{-r\tau^*} f_{\tau^*} \mid \mathcal{F}_t).$$

By Theorem 3.1, $Y_t = e^{-rt}X_t^{\pi,C}$ for some strategy (π, C) with consumption, and $X_0^{\pi,C} = Y_0 = \mathbb{C}_T^*(f)$. Further, the "rationality" of τ^* implies that

$$X_{\tau^*}^{\pi,C} = e^{r\tau^*}Y_{\tau^*} = f_{\tau^*} + ce^{r\tau^*} = X_{\tau^*}^{\pi,C} + ce^{r\tau^*}.$$

Hence, $c = 0$ and (7.11) holds.

Using (7.11), we now prove that τ^* is a rational stopping time. Let us consider a strategy (π, C) with consumption such that $X_0^{\pi,C} = \mathbb{C}_T^*(f)$ and $X_{\tau^*}^{\pi,C} \geqslant f_{\tau^*}$. Then in view of the supermartingale property of the discounted value of this strategy we get that

$$\mathbb{C}_T^*(f) = X_0^{\pi,C} \geqslant \mathbf{E}^* e^{-r\tau^*} X_{\tau^*}^{\pi,C} \geqslant \mathbf{E}^* e^{-r\tau^*} f_{\tau^*} = \mathbb{C}_T^*(f),$$

and hence $X_{\tau^*}^{\pi,C} = f_{\tau^*}$.

To prove (7.12) we establish first that for $\varepsilon > 0$

$$\mathbf{E}^* e^{-t\tau_\varepsilon} f_{\tau_\varepsilon} \geqslant \mathbb{C}_T^*(f) - \varepsilon, \tag{7.13}$$

where $\tau_\varepsilon = \inf\{t : f_t \geqslant X_t^* - \varepsilon\}$.

The inequality (7.13) can be proved using the following arguments. By the definition of the least upper bound, there is a sequence of stopping times $\sigma_n \in \mathcal{M}_0^T$ such that $\mathbf{E}^* e^{-r\sigma_n} f_{\sigma_n} \uparrow \mathbb{C}_T^*(f)$ as $n \uparrow \infty$. Further, from the chain of obvious relations

$$\begin{aligned}
\mathbf{E}^* e^{-r\sigma_n} f_{\sigma_n} &= \mathbf{E}^* e^{-r\sigma_n} f_{\sigma_n} I_{\{\sigma_n < \tau_\varepsilon\}} + \mathbf{E}^* e^{-r\sigma_n} f_{\sigma_n} I_{\{\sigma_n \geqslant \tau_\varepsilon\}} \\
&\leqslant \mathbf{E}^* e^{-r\sigma_n} (X_{\sigma_n}^* - \varepsilon) + \mathbf{E}^* e^{-r\sigma_n} X_{\sigma_n}^* I_{\{\sigma_n \geqslant \tau_\varepsilon\}} \\
&\leqslant \mathbf{E}^* e^{-r\sigma_n} X_{\sigma_n}^* - \varepsilon\, \mathbf{P}^*\{\sigma_n < \tau_\varepsilon\} \\
&\leqslant \mathbb{C}_T^*(f) - \varepsilon\, \mathbf{P}^*\{\sigma_n < \tau_\varepsilon\}
\end{aligned}$$

and the definition of $(\sigma_n)_{n\geqslant 1}$ it follows that $\mathbf{P}^*\{\sigma_n < \tau_n\} \to 0$ as $n \uparrow \infty$.

We define $\tau_n = \sigma_n \wedge \tau_\varepsilon$ and note that for any $n \geqslant 1$

$$\begin{aligned}
\mathbf{E}^* e^{-r\tau_n} f_{\tau_n} &= \mathbf{E}^* e^{-r\sigma_n} f_{\sigma_n} I_{\{\sigma_n \leqslant \tau_\varepsilon\}} + \mathbf{E}^* e^{-r\tau_\varepsilon} f_{\tau_\varepsilon} I_{\{\sigma_n > \tau_\varepsilon\}} \\
&\geqslant \mathbf{E}^* e^{-r\sigma_n} f_{\sigma_n} I_{\{\sigma_n \leqslant \tau_\varepsilon\}} + \mathbf{E}^* e^{-r\tau_\varepsilon} X_{\tau_\varepsilon}^* I_{\{\sigma_n > \tau_\varepsilon\}} - \varepsilon \\
&\geqslant \mathbf{E}^* e^{-r\sigma_n} f_{\sigma_n} I_{\{\sigma_n \leqslant \tau_\varepsilon\}} + \mathbf{E}^* e^{-r\sigma_n} X_{\sigma_n}^* I_{\{\sigma_n > \tau_\varepsilon\}} - \varepsilon \\
&\geqslant \mathbf{E}^* e^{-r\sigma_n} f_{\sigma_n} I_{\{\sigma_n \leqslant \tau_\varepsilon\}} + \mathbf{E}^* e^{-r\sigma_n} f_{\sigma_n} I_{\{\sigma_n > \tau_\varepsilon\}} - \varepsilon \\
&= \mathbf{E}^* e^{-r\sigma_n} f_{\sigma_n} - \varepsilon.
\end{aligned} \tag{7.14}$$

From (7.14),

$$\begin{aligned}
\mathbf{E}^* e^{-r\tau_\varepsilon} f_{\tau_\varepsilon} &= \mathbf{E}^* e^{-r\tau_n} f_{\tau_n} I_{\{\tau_\varepsilon < \sigma_n\}} + \mathbf{E}^* e^{-r\tau_\varepsilon} f_{\tau_\varepsilon} I_{\{\tau_\varepsilon \geqslant \sigma_n\}} \\
&= \mathbf{E}^* e^{-r\tau_n} f_{\tau_n} - \mathbf{E}^* e^{-r\tau_n} f_{\tau_n} I_{\{\tau_\varepsilon \geqslant \sigma_n\}} + \mathbf{E}^* e^{-r\tau_\varepsilon} f_{\tau_\varepsilon} I_{\{\tau_\varepsilon > \sigma_n\}} \\
&\geqslant \mathbf{E}^* e^{-r\sigma_n} f_{\sigma_n} - \varepsilon - \mathbf{E}^* e^{-r\tau_n} f_{\tau_n} I_{\{\tau_\varepsilon > \sigma_n\}}.
\end{aligned} \tag{7.15}$$

We get (7.13) from (7.15) by passing to the limit as $n \uparrow \infty$ and using the uniform integrability of $\{e^{r\tau} f_\tau\}_{\tau \in \mathcal{M}_0^T}$.

Assume now that $\varepsilon_n \downarrow 0$ $(n \uparrow \infty)$ and $\widetilde{\tau}_n \equiv \tau_{\varepsilon_n} \uparrow \widetilde{\tau}$, where $\widetilde{\tau}$ is some stopping time with $\widetilde{\tau} \leqslant \tau^*$. Since $\Delta f_t \geqslant 0$, it follows that $f_{\widetilde{\tau}} \geqslant \lim_{n\to\infty} f_{\widetilde{\tau}_n}$. Then

$$\sup_{\tau \in \mathcal{M}_0^T} \mathbf{E}^* e^{-r\tau} f_\tau = \mathbb{C}_T^*(f) \leqslant \limsup_{n\to\infty} \mathbf{E}^* e^{-r\widetilde{\tau}_n} f_{\widetilde{\tau}_n} \leqslant \mathbf{E}^* e^{-r\widetilde{\tau}} f_{\widetilde{\tau}}$$

by Fatou's lemma and the condition of uniform integrability in Theorem 7.1. Hence, $\mathbb{C}^*_T(f) = \mathbf{E}^* e^{-r\widetilde{\tau}} f_{\widetilde{\tau}}$, and the equality $X^*_{\widetilde{\tau}} = f_{\widetilde{\tau}}$ follows from the fact that $X^*_{\widetilde{\tau}} \geqslant f_{\widetilde{\tau}}$. Then the definition of the stopping time τ^* implies that $\widetilde{\tau} = \tau^*$, so $X^*_{\tau^*} = f_{\tau^*}$ and τ^* is rational.

Theorem 7.1 is proved.

It should be noted that the assertion in the theorem about the existence of a rational stopping time is untrue in general without the indicated conditions of uniform integrability and nonnegativity of the jumps Δf_t.

Theorem 7.1 reduces the problem of finding the price $\mathbb{C}^*_T(f)$ and exercise time τ^* of an American option with payment function f to the extremal problem (7.8), (7.12), which is called an *optimal stopping problem*.

The essence of this problem is to find, for the nonnegative random process $(X_t)_{t\geqslant 0}$ on $(\Omega, \mathcal{F}, \mathbb{F}, \mathbf{P})$, the supremum $\sup_\tau \mathbf{E}X_\tau$ and an *optimal* stopping time τ^* at which this supremum is attained.

One of the interesting and effective approaches to the solution is to find another random process $(Y_t)_{t\geqslant 0}$ such that

$$X_t = g(Y_t)M_t, \tag{7.16}$$

where g is a function with a unique maximum, and $(M_t)_{t\geqslant 0}$ is a nonnegative martingale with $M_0 = 1$.

If a representation (7.16) has been found, then $X_t = g(Y_t)M_t \leqslant g(y^*)M_t$, where $y^* = \operatorname{Arg\,max}_y g(y)$, and for any stopping time τ

$$\mathbf{E}X_\tau \leqslant g(y^*). \tag{7.17}$$

The representation (7.16) and the inequality (7.17) lead to a natural determination of the stopping time,

$$\tau^* = \inf\{t : Y_t \geqslant y^*\}.$$

The optimality of τ^* for continuous processes in this problem follows from the equalities

$$\mathbf{E}X_{\tau^*} = \mathbf{E}g(Y_{\tau^*})M_{\tau^*}I_{\{\tau^*<\infty\}} = g(y^*)\mathbf{E}M_{\tau^*}I_{\{\tau^*<\infty\}} = g(y^*).$$

Let us use this approach in the framework of the Black–Scholes model (3.19) to calculate the price of an American option (to buy), regarded on the infinite time interval $[0,\infty)$.

To simplify the computations we assume that the rate of yield μ and the interest rate r coincide. Consequently, the martingale measure $\mathbf{P}^*$ coincides with the original probability $\mathbf{P}$, and the dynamic behavior of the stock prices S_t is determined by the formula

$$S_t = \exp\left\{\left(r - \frac{\sigma^2}{2}\right)t + \sigma W_t\right\}, \qquad S_0 = 1. \tag{7.18}$$

Let the payment function of the option be $f_t = e^{-\delta t}(S_t - K)^+$, where the constants K, δ are > 0. According to Theorem 7.1,

$$\mathbb{C}^*_\infty(f) = \sup_{\tau\in\mathcal{M}_0^\infty} \mathbf{E}e^{-\lambda\tau}(S_\tau - K)^+,$$

where $\lambda = r + \delta$.

The martingale M_t in the decomposition (7.16) will be found in the form $e^{-\lambda t}S_t^\alpha$, $\alpha > 0$. Applying the Kolmogorov–Itô formula to $e^{-\lambda t}S_t^\alpha$ and using (7.18) and the martingale property of this product, we get that

$$\frac{(\alpha\sigma)^2}{2} + \alpha\left(r - \frac{\sigma^2}{2}\right) = \lambda, \tag{7.19}$$

$$\alpha = -\left(\frac{r}{\sigma^2} - \frac{1}{2}\right) + \left(\frac{2\lambda}{\sigma^2} + \left(\frac{r}{\sigma^2} - \frac{1}{2}\right)^2\right)^{1/2},$$

and thus

$$M_t = \exp\left\{\alpha\sigma W_t - \frac{(\alpha\sigma)^2}{2}t\right\}.$$

Further,

$$e^{-\lambda t}(S_t - K)^+ = (S_t - K)^+ S_t^{-\alpha} M_t = g(S_t)M_t,$$

where $g(y) = (y - K)^+ y^{-\alpha}$.

Direct calculations show that $g(y)$ has a unique maximum point $y^* = \dfrac{\alpha}{\alpha - 1}K$, and $g(y^*) = (\alpha - 1)^{-1}\left(\dfrac{\alpha - 1}{\alpha}\right)^\alpha K^{1-\alpha}$.

To prove the optimality of the stopping time $\tau^* = \inf\{t : S_t = y^*\}$ it suffices to show that

$$\mathbf{E}M_{\tau^*}I_{\{\tau^* < \infty\}} = 1.$$

Denote by $\mathbf{Q}$ the measure with local density $\dfrac{d\mathbf{Q}}{d\mathbf{P}} = M_t$. With respect to this measure, $\widetilde{W}_t = W_t - \alpha\sigma t$ is a Wiener process.

We get from (7.18) that with respect to the measure $\mathbf{Q}$ the drift $\log S_t$ is equal to

$$r - \frac{\sigma^2}{2} + \sigma^2\alpha = \sigma^2\left(\frac{2\lambda}{\sigma^2} + \left(\frac{r}{\sigma^2} - \frac{1}{2}\right)^2\right)^{1/2} > 0,$$

and $\mathbf{Q}(\tau^* < \infty) = 1$ and $\mathbf{P}(\tau^* < \infty) = 1$ because $\log y^* > 0$.

As a result we arrive at the following formulas for the price and exercise time of this option:

$$\mathbb{C}^*_\infty(e^{-\delta t}(S_t - K)^+) = (\alpha - 1)^{-1}\left(\frac{\alpha - 1}{\alpha}\right)^\alpha K^{1-\alpha},$$

$$\tau^* = \inf\left\{t : S_t \geqslant \frac{\alpha}{\alpha - 1}K\right\},$$

where α is defined by (7.19).

§ 7.2. Concretization of option calculations and closed analytic formulas for prices and strategies

The concept of a *dual martingale measure* turns out to be useful for many practical calculations of options. Namely, let $\widehat{\mathbf{P}}$ be a probability measure on $(\Omega, \mathcal{F}, \mathbb{F}, \mathbf{P})$ such that:

1) $\mathbf{P}$ and $\widehat{\mathbf{P}}$ are mutually equivalent;
2) the process $(B_t/S_t)_{0\leqslant t\leqslant T}$ is a local martingale with respect to $\widehat{\mathbf{P}}$.

It can be shown that the local density of the measure $\widehat{\mathbf{P}}$ with respect to $\mathbf{P}$ is

$$\widehat{Z}_t = \exp\left(\frac{r - \mu + \sigma^2}{\sigma} W_t - \frac{(r - \mu + \sigma^2)^2}{2\sigma^2} t\right), \quad 0 \leqslant t \leqslant T,$$

and

$$\widehat{W}_t = W_t - \frac{r - \mu + \sigma^2}{\sigma}$$

is a Wiener process with respect to $\widehat{\mathbf{P}}$. We remark also that the local density of $\widehat{\mathbf{P}}$ with respect to the *martingale* measure $\mathbf{P}^*$ defined in (3.21) is

$$\frac{d\widehat{\mathbf{P}}_t}{d\mathbf{P}^*_t} = \exp\left(\sigma W_t - \left(r - \mu + \frac{\sigma^2}{2}\right)t\right) = \frac{S_t}{S_0} e^{-rt}. \tag{7.20}$$

Making a change of probability measures, we arrive at the following analogue of Theorem 4.1. This analogue gives a *dual* representation for a minimal hedging portfolio and the exercise time of the option.

THEOREM 7.2. *Suppose that on a (B, S)-market (3.19) a dynamic contingent claim $f = (f_t)_{t \leqslant T}$ is given such that*

$$\sup_{\tau \in \mathfrak{M}_0^T} \widehat{\mathbf{E}} \frac{f_\tau}{S_\tau} < \infty. \tag{7.21}$$

Then there exists a minimal hedging strategy $(\widehat{\pi}, \widehat{C}) = (\widehat{\beta}, \widehat{\gamma}, \widehat{C})$ with consumption, given by

$$\widehat{\mathbb{C}}_T(f) = S_0 \sup_{\tau \in \mathfrak{M}_0^T} \widehat{\mathbf{E}} \frac{f_\tau}{S_\tau}, \tag{7.22}$$

$$\widehat{X}_t = X_t^{\widehat{\pi}, \widehat{C}} = S_t \operatorname*{ess\,sup}_{\tau \in \mathfrak{M}_t^T} \widehat{\mathbf{E}}\left(\frac{f_\tau}{S_\tau} \,\Big|\, \mathcal{F}_t\right), \tag{7.23}$$

$$\frac{\widehat{X}_t}{S_t} = \widehat{\mathbb{C}}_T(f) + \int_0^t \widehat{\beta}_u \, d\left(\frac{e^{ru}}{S_u}\right) - \int_0^t \frac{d\widehat{C}_u}{S_u}, \tag{7.24}$$

$$\widehat{\gamma}_t = \frac{\widehat{X}_t - \widehat{\beta}_t e^{rt}}{S_t}. \tag{7.25}$$

The stopping time $\widehat{\tau} \in \mathfrak{M}_0^T$ is rational if and only if

$$\widehat{\mathbb{C}}_T(f) = S_0 \widehat{\mathbf{E}} \frac{f_{\widehat{\tau}}}{S_{\widehat{\tau}}}. \tag{7.26}$$

Furthermore,

$$\widehat{\tau} = \inf\{t \leqslant T : \widehat{X}_t \geqslant f_t\} \tag{7.27}$$

if $\Delta f_t \geqslant 0$ for all $t \leqslant T$ and the family $\{f_\tau / S_\tau\}_{\tau \in \mathfrak{M}_0^T}$ is uniformly integrable.

Using (7.20), we find that

$$\widehat{\mathbb{C}}_T(f) = S_0 \sup_{\tau \in \mathfrak{M}_0^T} \widehat{\mathbf{E}} \frac{f_\tau}{S_\tau} = \sup_{\tau \in \mathfrak{M}_0^T} \mathbf{E}^* e^{-r\tau} f_\tau = \mathbb{C}_T^*(f).$$

We present some concrete examples of calculating American options in which it is possible to get an exact analytic solution.

Let $\gamma_1 < \gamma_2$ be the roots of the equation

$$\frac{\sigma^2}{2}\gamma^2 - \left(\frac{\sigma^2}{2} + r\right)\gamma - \lambda = 0.$$

THEOREM 7.3. *The fair (rational) price* $\mathbb{C}^*_\infty(f)$ *of an American option with payment function* $f_t = e^{-\lambda t}(S_t - K)^+$ *and exercise times with values in the interval* $[0, \infty)$ *is given by the formula*

$$\mathbb{C}^*_\infty(f) = S_0 \begin{cases} \left(\dfrac{\psi}{\psi^*}\right)^{\gamma_1}(1 - K\psi^*), & \psi > \psi^*, \\ 1 - K\psi, & \psi \leqslant \psi^*, \end{cases} \tag{7.28}$$

where

$$\psi^* = \frac{\gamma_1}{K(\gamma_1 - 1)}.$$

The rational exercise time of the option is

$$\tau^* = \inf\{t \geqslant 0 : \psi_t \leqslant \psi^*\},$$

where $\psi_t = \dfrac{1}{S_t}$.

THEOREM 7.4. *The fair (rational) price* $\mathbb{C}^*_\infty(f)$ *of an American option with payment function* $f_t = e^{-\lambda t}(K - S_t)^+$ *and exercise times with values in the interval* $[0, \infty)$ *is given by the formula*

$$\mathbb{C}^*_\infty(f) = S_0 \begin{cases} \left(\dfrac{\psi}{\psi^*}\right)^{\gamma_2}(K\psi^* - 1), & \psi < \psi^*, \\ K\psi - 1, & \psi \geqslant \psi*, \end{cases} \tag{7.29}$$

where

$$\psi^* = \frac{\gamma_2}{K(\gamma_2 - 1)}.$$

The rational exercise time of the option is

$$\tau^* = \inf\{t \geqslant 0 : \psi_t \geqslant \psi^*\},$$

where $\psi_t = \dfrac{1}{S_t}$.

THEOREM 7.5. *The fair (rational) price* $\mathbb{C}^*_\infty(f)$ *of an American option with payment function*

$$f_t = e^{-\lambda t}\Big(aS_t - \min\Big[\min_{u \leqslant t} S_u, s_0\psi_0\Big]\Big)^+,$$

$a > 0$, $0 < \psi_0 \leqslant 1$, $r + \lambda > 0$, *and exercise times with values in the interval* $[0, \infty)$ *is given by the formula*

$$\mathbb{C}^*_\infty(f) = S_0 \begin{cases} a - \psi_0, & \psi_0 \leqslant \psi^*, \\ (\gamma_1\psi^{\gamma_2} - \gamma_2\psi^{\gamma_1})\dfrac{a - \psi^*}{\gamma_1\psi^{*\gamma_2} - \gamma_2\psi^{*\gamma_1}}, & \psi_0 > \psi^*, \end{cases} \tag{7.30}$$

where the boundary value ψ^* *is the solution of the equation*

$$\psi^{\gamma_2}\left(1 - \frac{1}{\gamma_2} - \frac{a}{\psi}\right) = \psi^{\gamma_1}\left(1 - \frac{1}{\gamma_1} - \frac{a}{\psi}\right), \quad \psi < a.$$

The rational exercise time of the option is

$$\tau^* = \inf\{t \geqslant 0 : \psi_t \geqslant \psi^*\},$$

where

$$\psi_t = \frac{1}{S_t} \cdot e^{-\lambda t}\Big(aS_t - \min\Big[\min_{u \leqslant t} S_u, s_0\psi_0\Big]\Big).$$

THEOREM 7.6. *The fair (rational) price $\mathbb{C}^*_\infty(f)$ of an American option with payment function*

$$f_t = e^{-\lambda t}\Big(\max\Big[\max_{u \leqslant t} S_u, s_0\psi_0\Big] - aS_t\Big)^+,$$

$a > 0$, $1 \leqslant \psi_0 < \infty$, $r + \lambda > 0$, *and exercise times with values in the interval* $[0, \infty)$ *is given by the formula*

$$\mathbb{C}^*_\infty(f) = S_0 \begin{cases} \psi_0 - a, & \psi_0 \geqslant \psi^{**}, \\ (\gamma_1\psi^{\gamma_2} - \gamma_2\psi^{\gamma_1}) \dfrac{\psi^{**} - a}{\gamma_1\psi^{**\gamma_2} - \gamma_2\psi^{**\gamma_1}}, & \psi_0 < \psi^{**}, \end{cases} \tag{7.31}$$

where the boundary value ψ^{**} is the solution of the equation

$$\psi^{\gamma_2}\Big(1 - \frac{1}{\gamma_2} - \frac{a}{\psi}\Big) = \psi^{\gamma_1}\Big(1 - \frac{1}{\gamma_1} - \frac{a}{\psi}\Big), \quad \psi > a.$$

The rational exercise time of the option is

$$\tau^* = \inf\{t \geqslant 0 : \psi_t \geqslant \psi^{**}\},$$

where

$$\psi_t = \frac{1}{S_t} \cdot e^{-\lambda t}\Big(\max\Big[\max_{u \leqslant t} S_u, s_0\psi_0\Big] - aS_t\Big).$$

THEOREM 7.7. *The fair (rational) price $\mathbb{C}^*_\infty(f)$ of an American option with payment function*

$$f_t = e^{-\lambda t}\Big(\int_0^t S_u^n\, du + (s_0\psi_0)^n\Big)^{1/n}, \tag{7.32}$$

$r + \lambda > 0$, *and exercise times with values in the interval* $[0, \infty)$ *is given by the formula*

$$\mathbb{C}^*_\infty(f) = \begin{cases} \psi_0, & \psi_0 \geqslant \widetilde{\psi}, \\ u(\psi_0)\dfrac{\widetilde{\psi}}{u(\widetilde{\psi})}, & \psi_0 < \widetilde{\psi}, \end{cases} \tag{7.33}$$

where the function $u(\psi)$ admits an integral representation

$$u(\psi) = \int_0^\infty \exp\Big(-\frac{2y}{n^2\sigma^2}\Big) y^{-\gamma_1 - 1}(1 + y\psi^n)^{\gamma_2}\, dy$$

with $\gamma_1 < \gamma_2$ the roots of the equation

$$\frac{\sigma^2}{2}n^2\gamma^2 - \Big(\frac{\sigma^2}{2} + r\Big)n\gamma - \lambda = 0,$$

and $\widetilde{\psi}$ is determined from the relation

$$u(\psi) = \psi u'(\psi).$$

The rational exercise time of the option is

$$\tau^* = \inf\{t \geqslant 0 : \psi_t \geqslant \widetilde{\psi}\},$$

where

$$\psi_t = \frac{1}{S_t}\left(\int_0^t S_u^n \, du + (s_0\psi_0)^n\right)^{1/n}. \tag{7.34}$$

THEOREM 7.8. *The fair (rational) price $\mathbb{C}^*_\infty(f)$ of an American option with payment function*

$$f_t = e^{-\lambda t}\left(\int_0^t \frac{1}{S_u^n} \, du + \frac{1}{(s_0\psi_0)^n}\right)^{-1/n}, \tag{7.35}$$

$r + \lambda > 0$, and exercise times with values in the interval $[0, +\infty)$ is given by the following formulas.

I. *If one of the conditions*

1) $\lambda < 0$,

2) $\lambda = 0$, $r \geqslant \sigma^2/2$,

3) $0 < \lambda \leqslant \dfrac{\sigma^2}{2}\left(\dfrac{1}{2} + \dfrac{r}{\sigma^2}\right)$, $r > \sigma^2/2$

holds, then $\mathbb{C}^*_\infty(f) = 0$.

II. *If $r \leqslant \sigma^2/2$, then*

$$\mathbb{C}^*_\infty(f) = S_0 \cdot \begin{cases} \psi_0, & \psi_0 < \widetilde{\psi}, \\ \dfrac{\widetilde{\psi}}{m(\widetilde{\psi})} m(\psi_0), & \psi_0 \geqslant \widetilde{\psi}, \end{cases}$$

where the function $m(\psi)$ admits an integral representation depending on the value of the parameter λ:

1) *if $\lambda < \dfrac{\sigma^2}{8}\left(1 + \dfrac{2r}{\sigma^2}\right)$, then*

$$m(\psi) = x^{n\gamma_2} \int_0^1 \exp\left(\frac{2t\psi^n}{n^2\sigma^2}\right) t^{\gamma_2 - 1}(1-t)^{-\gamma_1} \, dt,$$

where γ_1 and γ_2 are the roots of the equation

$$\frac{\sigma^2}{2}n^2\gamma^2 - \left(\frac{\sigma^2}{2} + r\right)n\gamma + \lambda = 0; \tag{7.36}$$

2) *if $\lambda = \dfrac{\sigma^2}{8}\left(1 + \dfrac{2r}{\sigma^2}\right)$, then*

$$m(\psi) = x^{n\gamma_2} \int_0^1 \exp\left(\frac{2t\psi^n}{n^2\sigma^2}\right) t^{\frac{1}{n}(\frac{r}{\sigma^2}+\frac{1}{2})-1}(1-t)^{-\frac{1}{n}(\frac{r}{\sigma^2}+\frac{1}{2})} \, dt;$$

3) *if $\lambda > \dfrac{\sigma^2}{8}\left(1 + \dfrac{2r}{\sigma^2}\right)$, then*

$$m(\psi) = x^{n\gamma_2} \int_0^1 \exp\left(\frac{2t\psi^n}{n^2\sigma^2}\right) t^{\gamma - 1}(1-t)^{-\overline{\gamma}} \, dt,$$

where γ and $\overline{\gamma}$ are the complex roots of the equation (7.36).

Here the constant $\widetilde{\psi}$ is determined from the relation

$$m(\psi) = \psi\, m'(\psi).$$

The rational exercise time of the option is

$$\tau^* = \inf\{t \geqslant 0 : \psi_t \leqslant \widetilde{\psi}\},$$

where

$$\psi_t = \frac{1}{S_t} \cdot \left(\int_0^t \frac{1}{S_u^n}\, du + \frac{1}{(s_0\psi_0)^n} \right)^{-1/n}.$$

REMARK. We remark that

$$\left(\int_0^t S_u^n\, du + (s_0\psi_0)^n \right)^{1/n} \to \max\Big[\max_{u \leqslant t} S_u, s_0\psi_0\Big], \tag{7.37}$$

$$\left(\int_0^t \frac{1}{S_u^n}\, du + \frac{1}{(s_0\psi_0)^n} \right)^{-1/n} \to \min\Big[\min_{u \leqslant t} S_u, s_0\psi_0\Big] \tag{7.38}$$

as $n \to \infty$, where the "limit" payment functions define a *Russian option* to sell and buy, respectively. In the case $n = 1$ we get from (7.19) an *integral* option.

We present the basic ideas of the proofs of these theorems by using Theorem 7.7 as an example. First of all, note that with respect to the measure $\widehat{\mathbf{P}}$ the process $\psi = (\psi_t)_{t \geqslant 0}$ defined in (7.34) is a Markov diffusion process satisfying the following stochastic differential equation:

$$d\psi_t = -\psi_t(r\, dt + \sigma\, d\widehat{W}_t) + \frac{1}{n\psi_t^{n-1}}\, dt.$$

Using Theorem 7.2 on a characterization of a minimal hedging portfolio, we arrive at the standard problem on optimal stopping of the diffusion process $\psi = (\psi_t)_{t \geqslant 0}$:

$$\mathbb{C}_\infty^*(f) = S_0 \sup_{\tau \leqslant \infty} \widehat{\mathbf{E}}\, e^{-\lambda\tau} \psi_\tau.$$

According to the general theory for solving such problems, an optimal time τ^* has the structure

$$\tau^* = \inf\{t \geqslant 0 : \psi_t \geqslant \widetilde{\psi}\},$$

where $\widetilde{\psi}$ is a constant.

We let $\widetilde{V}(\psi) = \sup_\tau \widehat{\mathbf{E}}\, e^{-\lambda\tau}\psi_\tau$ and assume that the function $\widetilde{V}(\psi)$ is sufficiently smooth (for example, in the class $C^2(\mathbb{R}_+)$). According to the general theory, the equation

$$L\widetilde{V}(\psi) = \lambda \widetilde{V}(\psi) \tag{7.39}$$

must hold in the region $\widetilde{C} = \{\psi : \psi_t < \widetilde{\psi}\}$ of "continuing the observations", where

$$L = \left(-r + \frac{1}{n\psi^n} \right) \psi \frac{\partial}{\partial \psi} + \frac{\sigma^2}{2} \psi^2 \frac{\partial^2}{\partial \psi^2}$$

is the generating operator of the process ψ.

The general solution of (7.22) will be found in the form

$$V(\psi) = x^{n\gamma_1}\big(C_1 M(a, b, z) + C_2 U(a, b, z)\big),$$

where $a = -\gamma_1$, $b = 1 - \gamma_1 + \gamma_2$, $z = \dfrac{2}{n^2\sigma^2 x^n}$, $\gamma_1 < \gamma_2$ are the roots of the equation

$$\frac{\sigma^2}{2} n^2\gamma^2 - \left(\frac{\sigma^2}{2} + r\right) n\gamma - \lambda = 0,$$

C_1, C_2 are unknown constants, and $M(a,b,z)$ and $U(a,b,z)$ are so-called *Kummer functions of confluent type*. They have the integral representations

$$M(a,b,z) = \frac{\Gamma(b)}{\Gamma(a)\Gamma(b-a)} \int_0^1 \exp(zt) t^{a-1} (1-t)^{b-a-1}\, dt,$$
$$U(a,b,z) = \frac{1}{\Gamma(a)} \int_0^\infty \exp(-zt) t^{a-1} (1+t)^{b-a-1}\, dt.$$

In this case the constant C_1 is 0, since otherwise $\widetilde{V}(\psi) \to \infty$ as $\psi \to 0$, which can be excluded, because according to the meaning of the problem the solution of interest to us must be bounded.

Thus, we have two unknown parameters $C_2 = C$ and $\widetilde{\psi}$, and the following conditions are used to determine them:

1) $\widetilde{V}(\widetilde{\psi}) = \widetilde{\psi}$, which means that on the boundary separating the regions of "continuing" and "stopping" the observations the gain from continuing the observations coincides with the gain from stopping them;
2) $\widetilde{V}'(\widetilde{\psi}) = 1$, or the "smooth pasting" condition on the boundary separating the regions.

Under these two additional conditions the solution of the above problem is given by the formula (7.23) in Theorem 7.2. Of course, we must still prove that the solution found coincides with $\mathbb{C}^*_\infty(f)$. To this end it suffices to prove the following two properties:

A1) for any finite ($\widehat{\mathbf{P}}$-a.s.) Markov time τ

$$\widehat{\mathbf{E}} e^{-\lambda\tau} e^{-\lambda\tau} \psi_\tau \leqslant \widetilde{V}(\psi), \qquad \psi \geqslant 0$$

(that is, $\mathbb{C}^*_\infty \leqslant \widetilde{V}(\psi)$);

A2) the time $\tau^* = \inf\{t \geqslant 0 : \psi_t \geqslant \widetilde{\psi}\}$ is finite $\widehat{\mathbf{P}}$-a.s., and

$$\widehat{\mathbf{E}} e^{-\lambda\tau^*} \psi_{\tau^*} = \widetilde{V}(\psi).$$

These assertions can be proved by applying the Kolmogorov–Itô formula to the process $(e^{-\lambda t}\widetilde{V}(\psi_t))_{t\geqslant 0}$.

§ 7.3. Quantile hedging of dynamic contingent claims

In this section we show how quantile hedging is realized in its dynamic variant for American options in the framework of the Black–Scholes model (3.19).

Let $SF(x_0)$ denote the set of self-financing admissible strategies $\pi = (\beta, \gamma)$ with initial value $x_0 > 0$. According to (3.2), (3.20)–(3.21), for any such strategy we have

$$Y_t^\pi = \frac{X_t^\pi}{B_t} = x_0 + \int_0^t \gamma_u\, d\left(\frac{S}{B}\right)_u = x_0 + \int_0^t \alpha_u\, dW_u^*, \quad t \geqslant 0, \tag{7.40}$$

where $\alpha_t = \sigma\gamma_t \dfrac{S_t}{B_t}$, and $W_t^* = W_t + \dfrac{(\mu - r)}{\sigma} t$ is a Wiener process with respect to the unique martingale measure $\mathbf{P}^*$ with density

$$Z_t^* = \exp\biggl\{ -\Bigl(\frac{\mu - r}{\sigma}\Bigr) W_t - \frac{1}{2}\Bigl(\frac{\mu - r}{\sigma}\Bigr)^2 t \biggr\}. \tag{7.41}$$

It is necessary to find the function

$$U(x_0) = \sup_{\tau\in\mathcal{M}(T),\, \pi\in SF(x_0)} \mathbf{P}\{Y_\tau^\pi \geqslant 1\}, \tag{7.42}$$

where $\mathcal{M}(T) = \{\tau \in \mathcal{M}_0^\infty : \mathbf{E}\tau \leqslant T\}$, $0 < x_0 < 1$, and to find a stopping time $\tau^* \in \mathcal{M}(T)$ and a strategy $\pi^* \in SF(x_0)$ for which the supremum in (7.42) is attained.

The *financial* interpretation of this problem amounts to obtaining a discounted value exceeding a given level at the exercise time τ^* of the option.

We remark that the "unitness" of the contingent claim in (7.42) does not involve a loss of generality, in view of the following considerations.

If g is a (discounted) contingent claim ($g_0 = 1$ for simplicity), then in the framework of the "coordinate" Black–Scholes model (3.19) (see Chapter 2) it can be represented in the form

$$g_t = 1 + \int_0^t \phi_u^g \, dW_u^*.$$

Hence, $\{Y_\tau^\pi \geqslant g_\tau\} = \{x_0 + \int_0^t (\phi_u - \phi_u^g)\, dW_u^* \geqslant 1\}$, and the problem reduces to (7.42) with ϕ_u replaced by $\phi_u - \phi_u^g$.

The basic result with regard to the problem (7.42) is the following.

THEOREM 7.9. *There exist constants a and b such that $x_0 < a < 1 < b$,*

$$U(x_0) = \frac{b-1}{b-a}, \tag{7.43}$$

$$\tau^* = \inf\{t : Z_t^* \notin (a,b)\}, \tag{7.44}$$

where

$$Z_t^* = \Bigl(\frac{S_t}{S_0}\Bigr)^{-\frac{\mu - r}{\sigma^2}} \exp\biggl\{ -\frac{1}{2}\frac{\mu - r}{\sigma^2}(\sigma^2 - \mu - r)t \biggr\}.$$

The optimal portfolio $\pi^* = (\beta^*, \gamma^*)$ has the structure

$$\gamma_t^* = \frac{ab(\mu - r)B_t}{(b-a)\sigma^2 Z_t^* S_t}, \qquad \beta_t^* = \frac{X^{\pi^*}}{B_t} - \gamma_t^* S_t. \tag{7.45}$$

The constants a and b can be expressed in terms of the original parameters of the model:

$$a = \frac{x_0}{1-\beta}, \qquad b = \frac{1-x_0}{\beta}, \tag{7.46}$$

where β is the unique solution of the equation

$$\frac{1}{2}\Bigl(\frac{\mu - r}{\sigma}\Bigr)^2 T = (1-\beta)\ln\frac{1-\beta}{x_0} + \beta\ln\frac{\beta}{1-x_0}. \tag{7.47}$$

PROOF. A predictable function $\phi_t(\omega)$ will be called an *x_0-admissible strategy* if $x_0 + \int_0^t \alpha_s \, dW_s^* \geqslant 0$ (a.s.) for all $t \geqslant 0$. We denote the corresponding set by $\mathcal{A}(x_0)$, and by the definition of $SF(x_0)$ we get from (7.42) that

$$U(x_0) = \sup_{\tau \in \mathcal{M}(T),\, \phi \in \mathcal{A}(x_0)} \mathbf{P}\Big\{ x_0 + \int_0^\tau \alpha_s \, dW_s^* \geqslant 1 \Big\}. \tag{7.48}$$

Further, we introduce for a stopping time $\tau \in \mathcal{M}(T)$ the following class of sets:

$$\mathcal{A}(x_0, \tau) = \Big\{ A \in \mathcal{F}_\tau : A = \Big\{ \omega : v + \int_0^\tau \phi_s \, dW_s^* \geqslant 1 \Big\} \\ \text{for some } v \in [0, x_0] \text{ and some } \phi \in \mathcal{A}(v) \Big\}.$$

For a set $A \in \mathcal{F}_\tau$ we have the equivalence

$$A \in \mathcal{A}(x, \tau) \iff \mathbf{P}^*(A) \leqslant x_0. \tag{7.49}$$

The implication ($\Rightarrow$) follows from the Chebyshev inequality in view of the supermartingale property of the process $v + \int_0^t \phi_s \, dW_s^*$:

$$\mathbf{P}^*(A) = \mathbf{P}^* \Big\{ v + \int_0^\tau \phi_s \, dW_s^* \geqslant 1 \Big\} \leqslant \mathbf{E}^* \Big\{ v + \int_0^\tau \phi_s \, dW_s^* \Big\} \leqslant v \leqslant x_0.$$

To verify the converse implication ($\Leftarrow$) in (7.49) we consider the nonnegative martingale $m_t = \mathbf{P}^*(A \,|\, \mathcal{F}_t)$ and observe that

$$m_t = \mathbf{P}^*(A) + \int_0^t \phi_s \, dW_s^*$$

for some predictable ϕ (martingale representation).

Further, $m_\tau = I_A = v + \int_0^\tau \phi_s \, dW_s^*$ with $v = \mathbf{P}^*(A) \leqslant x_0$, and hence $A \in \mathcal{A}(x_0, \tau)$.

It is clear that the relation (7.49) enables us to give the problem (7.42) the form

$$U(x_0) = \sup_{\tau \in \mathcal{M}(T),\, A \in \mathcal{A}(x_0, \tau)} \mathbf{P}(A), \qquad \mathbf{P}^*(A) \leqslant x_0. \tag{7.50}$$

It is natural to approach (7.50) like the *Wald problem* of sequential distinction between two simple hypotheses about the mean value of a Wiener process:

$$H_0\colon\ d\zeta_t = \frac{\mu - r}{\sigma} \, dt + dW_t, \quad \zeta_0 = 0,$$
$$H_1\colon\ d\zeta_t = dW_t, \quad \zeta_0 = 0,$$

in which A is the *critical set* for the hypothesis H_0, x_0 is the *probability of an error of the first kind*, and $\mathbf{P}(A)$ is the *power* of the sequential test.

Let $\lambda \stackrel{\text{def}}{=} \ln Z$. According to the general theory of sequential analysis, the "region of continuing the observations" in the problem (7.50) has the following structure.

There are numbers $\widetilde{a}$ *and* $\widetilde{b}$ *with* $\widetilde{a} < 0 < \widetilde{b}$ *having the property that the decision rule with time*

$$\tau^* = \inf\{t \geqslant 0 : \lambda_t \notin (\widetilde{a}, \widetilde{b})\} \tag{7.51}$$

for "stopping the observations" and with critical set A *has the form*

$$I_A = \begin{cases} 1, & \lambda_{\tau^*} \geqslant \widetilde{b}, \\ 0, & \lambda_{\tau^*} \leqslant \widetilde{a}, \end{cases} \tag{7.52}$$

and the probability of an error of the first kind is equal to x. *The indicated rule is optimal in the sense that for any other rule with time* $\widetilde{\tau}$

$$\mathbf{E}\tau^* \leqslant \mathbf{E}\widetilde{\tau}.$$

We turn to the problem of finding the thresholds $\widetilde{a}$, $\widetilde{b}$ and the mathematical expectation $\mathbf{E}\tau^*$ of the duration of the observations. For this we need some auxiliary assertions.

LEMMA 7.1. *For the decision rule* (7.51)–(7.52)

$$\mathbf{P}(\tau^* < \infty) = \mathbf{P}^*(\tau^* < \infty) = 1.$$

PROOF. We define the stopping times

$$\sigma_n = \inf\left\{t \geqslant 0 : \int_0^t \frac{(\mu - r)^2}{\sigma^2}\,dt \geqslant n\right\} = \frac{\sigma^2}{(\mu - r)^2}\,n.$$

Then

$$\lambda_{\tau^* \wedge \sigma_n} = -\int_0^{\tau^* \wedge \sigma_n} \frac{\mu - r}{\sigma}\,dW_t - \frac{1}{2}\int_0^{\tau^* \wedge \sigma_n} \frac{(\mu - r)^2}{\sigma^2}\,dt$$

and

$$\widetilde{a} \leqslant \lambda_{\tau^* \wedge \sigma_n} \leqslant \widetilde{b}.$$

Consequently,

$$\mathbf{E}\int_0^{\tau^* \wedge \sigma_n} \frac{(\mu - r)^2}{\sigma^2}\,dt \leqslant 2(\widetilde{b} - \widetilde{a}) < \infty,$$

and by passing to the limit as $n \to \infty$, we get that

$$\mathbf{E}\int_0^{\tau^*} \frac{(\mu - r)^2}{\sigma^2}\,dt \leqslant 2(\widetilde{b} - \widetilde{a}) < \infty.$$

Since

$$\mathbf{E}\int_0^{\tau^*} \frac{(\mu - r)^2}{\sigma^2}\,dt \geqslant \mathbf{E}I_{\{\tau^* = \infty\}}\int_0^{\infty} \frac{(\mu - r)^2}{\sigma^2}\,dt \quad \text{and} \quad \int_0^{\infty} \frac{(\mu - r)^2}{\sigma^2}\,dt = \infty,$$

we conclude that $\mathbf{P}(\tau^* < \infty) = 1$.

Similarly, noting that $\lambda_t = -\dfrac{\mu - r}{\sigma} W_t^* + \dfrac{(\mu - r)^2}{2\sigma^2} t$, we prove the equality $\mathbf{P}^*(\tau^* < \infty) = 1$.

REMARK. As a consequence of the above arguments we get that the random variable λ_{τ^*} takes only the two values $\widetilde{a}$ and $\widetilde{b}$.

LEMMA 7.2. $\mathbf{P}^*(A) = \dfrac{e^{\widetilde{a}}(e^{\widetilde{b}} - 1)}{e^{\widetilde{b}} - e^{\widetilde{a}}}$ *is the probability of an error of the first kind, and* $1 - \mathbf{P}(A) = \dfrac{1 - e^{\widetilde{a}}}{e^{\widetilde{b}} - e^{\widetilde{a}}}$ *is the probability of an error of the second kind.*

PROOF. Let $\alpha(x)$ and $\beta(x)$ be the solutions of the differential equations

$$\alpha''(x) + \alpha'(x) = 0, \qquad \beta''(x) + \beta'(x) = 0$$

satisfying the boundary conditions

$$\alpha(\widetilde{a}) = 1, \quad \alpha(\widetilde{b}) = 0, \quad \beta(\widetilde{a}) = 0, \quad \beta(\widetilde{b}) = 1.$$

We show that

$$\mathbf{P}^*(A) = \alpha(0), \qquad \mathbf{P}(A) = U(x) = \beta(0).$$

We have

$$\alpha(x) = \frac{e^{\widetilde{a}}(e^{\widetilde{b}-x} - 1)}{e^{\widetilde{b}} - e^{\widetilde{a}}}, \qquad \beta(x) = \frac{e^{x} - e^{\widetilde{a}}}{e^{\widetilde{b}} - e^{\widetilde{a}}}.$$

On the other hand, by the Kolmogorov–Itô formula,

$$\begin{aligned} \alpha(\lambda_{\tau^*\wedge\sigma_n}) &= \alpha(0) + \int_0^{\tau^*\wedge\sigma_n} \alpha'(\lambda_t)\,\frac{\mu - r}{\sigma}\,dW_t^* \\ &\quad + \frac{1}{2}\int_0^{\tau^*\wedge\sigma_n} [\alpha'(\lambda_t) + \alpha''(\lambda_t)]\,\frac{(\mu - r)^2}{\sigma^2}\,dt \\ &= \alpha(0) + \int_0^{\tau^*\wedge\sigma_n} \alpha'(\lambda_t)\,\frac{\mu - r}{\sigma}\,dW_t^*. \end{aligned} \tag{7.53}$$

Consequently, since

$$\mathbf{E}^* \int_0^{\tau^*\wedge\sigma_n} \left(\alpha'(\lambda_t)\,\frac{\mu - r}{\sigma}\right)^2 dt \leqslant n \sup_{\widetilde{a}\leqslant x\leqslant\widetilde{b}} [\alpha'(x)]^2 < \infty,$$

we have

$$\mathbf{E}^* \int_0^{\tau^*\wedge\sigma_n} \alpha'(\lambda_t)\,\frac{\mu - r}{\sigma}\,dW_t^* = 0.$$

Taking the expectation with respect to $\mathbf{P}^*$ of both sides of (7.53), we get that $\mathbf{E}^*(\alpha(\lambda_{\tau^*\wedge\sigma_n})) = \alpha(0)$. By the dominated convergence theorem, $\mathbf{E}^*(\alpha(\lambda_{\tau^*})) = u(0)$ results when we pass to the limit as $n \to \infty$. On the other hand, by the remark after Lemma 7.1,

$$\mathbf{E}^*\alpha(\lambda_{\tau^*}) = 1 \cdot \mathbf{P}^*(\lambda_{\tau^*} = \widetilde{a}) + 0 \cdot \mathbf{P}^*(\lambda_{\tau^*} = \widetilde{b}) = \mathbf{P}^*(A).$$

It can be proved similarly that

$$1 - \mathbf{P}(A) = 1 - \mathbf{P}(\lambda_{\tau^*} = \widetilde{a}) = \mathbf{P}(\lambda_{\tau^*} = \widetilde{b}) = \beta(0).$$

LEMMA 7.3. *The expectation of the duration of observations is*

$$\mathbf{E}\tau^* = \frac{2\sigma^2}{(\mu - r)^2} = \frac{2\sigma^2}{(\mu - r)^2}\left\{\frac{(e^{\widetilde{b}} - 1)(\widetilde{b} - \widetilde{a})}{e^{\widetilde{b}} - e^{\widetilde{a}}} - \widetilde{b}\right\}. \tag{7.54}$$

PROOF. Denote by $g(x)$ the solution of the differential equation

$$g''(x) - g'(x) = -2$$

with boundary condition

$$g(\widetilde{b}) = g(\widetilde{a}) = 0.$$

Direct computations show that

$$g(x) = 2\left\{ \frac{(e^{\widetilde{b}} - e^x)(\widetilde{b} - \widetilde{a})}{e^{\widetilde{b}} - e^{\widetilde{a}}} - \widetilde{b} + x \right\}.$$

Applying the Kolmogorov–Itô formula to the process $g(\lambda_{\tau^* \wedge \sigma_n})$, we see that

$$\mathbf{E} g(\lambda_{\tau^* \wedge \sigma_n}) = g(0) + \mathbf{E} \int_0^{\tau^* \wedge \sigma_n} \frac{(\mu - r)^2}{\sigma^2}\, dt.$$

Consequently, passing to the limit as $n \to \infty$, we get that

$$\frac{(\mu - r)^2}{\sigma^2} \mathbf{E}\tau^* = g(0).$$

REMARK. According to Lemma 7.2, $x_0 = \dfrac{e^{\widetilde{a}}(e^{\widetilde{b}} - 1)}{e^{\widetilde{b}} - e^{\widetilde{a}}}$. Let $\beta \stackrel{\text{def}}{=} \dfrac{1 - e^{\widetilde{a}}}{e^{\widetilde{b}} - e^{\widetilde{a}}}$. Then we can rewrite (7.54) in the form

$$\frac{(\mu - r)^2}{\sigma^2} \mathbf{E}\tau^* = \frac{\widetilde{b}(e^{\widetilde{a}} - 1) + \widetilde{a}(1 - e^{\widetilde{b}})}{e^{\widetilde{b}} - e^{\widetilde{a}}} = (1 - \beta) \ln \frac{1 - \beta}{x_0} + \beta \ln \frac{\beta}{1 - x_0}.$$

Let us proceed to the search for an optimal strategy in the problem (7.50). We have shown that the critical set A on which the hypothesis H_0 is rejected has the structure

$$A = \{\omega : Z^*_{\tau^*}(\omega) = a\}$$

with optimal time

$$\tau^* = \inf\{t \geqslant 0 : Z^*_t \notin (a, b)\}$$

for stopping the observations, where $a = e^{\widetilde{a}}$ and $b = e^{\widetilde{b}}$. Since the terminal value of the optimal strategy is $Y_{\tau^*}(\omega) = I_A(\omega)$, the value of the strategy at a time t with $0 \leqslant t \leqslant \tau^*$ is

$$Y^*_t = \mathbf{E}^*[I_A \,|\, \mathcal{F}_t] = \mathbf{P}^*(Z^*_{\tau^*} = a \,|\, \mathcal{F}_t).$$

It follows from the properties of the filtration and the Markov property of the process Z_t that the conditional probability in the last equality must be found in the form

$$\mathbf{P}^*(Z^*_{\tau^*} = a \,|\, \mathcal{F}_t) = \phi(Z^*_{t \wedge \tau^*}). \tag{7.55}$$

According to the "telescope" property of conditional expectations, $\phi(Z_{t\wedge\tau^*})$ is a martingale with respect to the measure $\mathbf{P}^*$; therefore, use of the Kolmogorov–Itô formula necessarily leads to the differential equation

$$\frac{1}{2}\phi''(z)z + \phi'(z) = 0$$

in the function $\phi(z)$ with boundary conditions $\phi(a) = 1$ and $\phi(b) = 0$. Solving this, we get

$$\phi(z) = \frac{a}{b-a}\left(\frac{b}{z} - 1\right).$$

On the other hand, it follows from the definition of self-financing strategies that

$$\mathbf{P}^*(Z^*_{\tau^*} = a \mid \mathcal{F}_t) = v + \int_0^{t\wedge\tau^*} \frac{\sigma\gamma^*_u S_u}{B_u}\, dW^*_u. \tag{7.56}$$

Applying the Kolmogorov–Itô formula to the process ϕ, we get from (7.55)–(7.56) that

$$\frac{\sigma\gamma^*_u S_u}{B_u} = -\phi'(Z^*_u) Z^*_u \frac{\mu - r}{\sigma},$$

and we arrive at the expression (7.45) for the number of shares and the number of units in the bank account in an optimal portfolio.

REMARK. We compare the mean time $\mathbf{E}\tau^*$ defined in (7.44) for stopping the observations with the mean time $\mathbf{E}\widetilde{\tau}$ corresponding to the optimal decision rule with fixed observation time $t(x_0, \beta)$ needed for distinguishing between the hypotheses about the mean value of a Wiener process with the use of the Neyman–Pearson most powerful classical rule. Suppose that the probabilities of errors of the first and second kinds do not exceed x_0 and β, respectively. The Neyman–Pearson optimal test proposes an answer on the finite time interval $t(x_0, \beta)$ in the form of the critical set

$$A = \left\{ \frac{d\mathbf{P}^*}{d\mathbf{P}}\bigg|_{t(x_0,\beta)} \geqslant h(x_0, \beta) \right\},$$

where $h(x_0, \beta)$ is determined from the equality $\mathbf{P}^*(A) = x_0$. It can be shown that

$$t(x_0, \beta) = (C_{x_0} + C_\beta)^2 \left(\frac{\sigma}{\mu - r}\right)^2, \qquad h(x_0, \beta) = \frac{1}{2}(C^2_\beta - C^2_{x_0}),$$

where C_γ is the root of the equation

$$\frac{1}{\sqrt{2\pi}} \int_{C_\gamma}^\infty e^{-x^2/2}\, dx = \gamma, \qquad 0 \leqslant \gamma \leqslant 1.$$

Thus, for given x_0 and β we get from (7.47) that

$$\frac{\mathbf{E}\tau^*}{t(x_0, \beta)} = 2\frac{\omega(x_0, \beta)}{(C_{x_0} + C_\beta)^2},$$

where $\omega(x, y) = (1 - y)\ln\dfrac{1-y}{x} + y\ln\dfrac{y}{1-x}$. Moreover, if $x_0 = \beta$, then

$$\lim_{x_0\downarrow 0} \frac{\mathbf{E}\tau^*}{t(x_0, x_0)} = \frac{1}{2}\lim_{x_0\downarrow 0}\frac{\omega(x_0, x_0)}{C^2_{x_0}} = \frac{1}{4}.$$

References for Chapter 7

[**19**], [**46**], [**51**], [**87**], [**90**], [**96**], [**98**], [**126**], [**132**], [**142**], [**143**], [**155**], [**156**], [**158**], [**162**].

CHAPTER 8

Analysis of "Bond" Contingent Claims

Bonds are a significant part of the financial market. This chapter is devoted to an analysis of the structure of bond prices, or the term structure of interest rates, and to the pricing of contingent claims on these markets.

§ 8.1. Models of the term structure of interest rates

A *bond* is a security issued by a government of other issuer, with an obligation to pay a certain sum (the *face value*) to the owner at the expiration of the established time T until maturity.

During the period before a bond matures, intermediate payments can be made, called coupon payments or simply *coupons*. Correspondingly, bonds are divided into zero-coupon (discount) bonds and bonds with coupon payments.

Let us consider a (B, P)-*market* consisting of a bank account B and zero-coupon bonds $P(T)$ with different redemption dates T. The value at time t of a unit of the bank account and the value of a bond with redemption date $T \geqslant t$ are denoted by B_t and $P_t(T)$, respectively. We assume that the face value $P_T(T)$ of the bond is equal to 1 and the prices of the zero-coupon (discount) bonds satisfy the relation $0 < P_t(T) \leqslant 1$ in view of the no-arbitrage requirement.

Since the bonds have to be redeemed by the issuer at a fixed price, the investor receives a *deterministic* return at the time of redemption of a bond, the amount being determined by the price $P_t(T)$ for acquiring the bond. From this viewpoint, bonds can be regarded as nonrisky assets.

At the same time, the price $P_t(T)$ of the bond can vary on the time interval (t, T), and an investor selling it on this time interval does not know the amount of return (or loss). This feature makes investing in bonds and the associated contingent claims risky, and gives rise to problems of hedging and investment on a bond market.

The *interest rate structure* or evolution of bond prices can often be represented in the form

$$P_t(T) = P_0(T)B(t)\frac{\exp\bigl(\int_0^t \sigma(s, T)\,dM_s\bigr)}{\mathbf{E}_{\mathbf{P}^*}\bigl[\exp\bigl(\int_0^t \sigma(s, T)\,dM_s\bigr)\bigr]}, \tag{8.1}$$

where M_t is a random process with independent increments, and $B(t)$ describes the dynamics of the spot market,[1] that is, it characterizes the movement of the bank

[1]The spot market, in this case, is the market of very "short" funds, when resources are made available for a period of at most 24 hours, like overnight credit, when funds of European financial instruments are made available in credit to far-Asian banks for playing the stock markets of Singapore and Tokyo in the course of a night while the European markets are closed.

account and (or) other nonrisky assets. The exact form of $B(t)$ is determined by the boundary condition $P_t(t) = 1$.

From the practical point of view, the formula (8.1) asserts that the dynamic behavior of the bond market is determined by the movement of the spot market $B(t)$, and "on the average" the resources invested in bonds must bring a return according to how the price of the nonrisky assets has changed. At the same time, if the resources are put into bonds whose redemption period has not yet expired, then the bond prices (and hence also the return) can deviate from the movement of the spot market, which is taken into account in the second factor

$$\frac{\exp\left(\int_0^t \sigma(s,T)\,dM_s\right)}{\mathbf{E}_{\mathbf{P}^*}\left[\exp\left(\int_0^t \sigma(s,T)\,dM_s\right)\right]} \tag{8.2}$$

of the formula (8.1).

The structure (8.2) is determined by the following: the factor cannot be negative or zero, the bond prices must "remember" the past, and deviations from the movement of the spot market must not allow arbitrage opportunities. Consequently, the factor must be represented in the form $\exp(X_t(T))C(t,T)$, where $C(t,T) = \mathbf{E}_{\mathbf{P}^*}\left[\exp(X_t(T))\right]^{-1}$, and $\mathbf{P}^*$ is a certain measure equivalent to the original measure. Among the different processes $X_t(T)$ an integral form seems convenient that gives the possibility of constructing processes "remembering" the past but having independent increments.

The amount of possible deviation of bond prices from the movement of the spot market is determined both by the properties of the process M and by the "volatility" $\sigma(s,t)$. In each concrete case we must impose on the process M, the function σ, and the process $B(t)$ restrictions of a technical nature that render the model noncontradictory and make it possible to use stochastic integrals and other methods of stochastic analysis.

The Ho–Lee model.

The Ho–Lee model has both a discrete and a continuous form which is a special case of a more general model of Heath, Jarrow, and Morton. In both cases the main distinguishing feature of the model is the form of the function σ, which is represented as a linear function $\sigma(s,t) = \delta(t-s)$ of the time left before the bond is redeemed. In the discrete case $t = 0, 1, \dots$ we should set

$$M_t = \sum_{i=1}^{t} \xi_i,$$

where $\xi = (\xi_t)_t$ is a sequence of independent identically distributed random variables taking the values 0 and 1 with probabilities p and $1-p$, respectively.

Specifying a measure $\mathbf{P}^*$ equivalent to the original one and substituting in the formula (8.1) the explicit expressions for M_t and $\sigma(s,T)$, we get that

$$P_t(T) = P_0(T)B(t)\frac{\exp\left(\sum\limits_{i=1}^{t} \delta(T-i)\xi_i\right)}{\mathbf{E}_{\mathbf{P}^*}\left[\exp\left(\sum\limits_{i=1}^{t} \delta(T-i)\xi_i\right)\right]}.$$

Note that

$$\begin{aligned}\text{(8.3)}\qquad \frac{P_t(T)}{B(t)} &= \frac{P_{t-1}(T)}{B(t-1)}\,\frac{\exp(\delta(T-t)\xi_t)}{p^*+(1-p^*)\exp(\delta(T-t))}\\ &= \frac{P_{t-1}(T)}{B(t-1)}\,h(t,T,\xi_t),\end{aligned}$$

where $h(t,T,\xi_t)=\exp\bigl(\delta(T-t)\xi_t\bigr)\bigl(p^*+(1-p^*)\exp(\delta(T-t))\bigr)^{-1}$ is often called the *perturbation function.*

In view of the condition $P_t(t)=1$ and the property $h(t,t,\xi)=1$, we get from (8.3) that

$$B(t)^{-1}=P_{t-1}(t)B(t-1)^{-1},$$

and hence

$$\text{(8.4)}\qquad B(t)=B(t-1)P_{t-1}(t)^{-1},\qquad B(t)=\prod_{i=1}^{t}P_{i-1}(i)^{-1}.$$

The expression (8.4) reflects the essence of the spot market, when money is invested for a short time with a fixed return. Indeed, investing a sum V at time $t-1$ in a bond with redemption date t, the investor receives a deterministic return in the amount

$$V\frac{P_t(t)}{P_{t-1}(t)}-V=V\biggl(1-\frac{1}{P_{t-1}(t)}\biggr).$$

Investing the sum V in this way at time 0 and continuing the process by investing resources at each moment of time t only in bonds with redemption date $t+1$, which yield a fixed (nonrisky) return at the next moment of time, the investor will have the capital

$$X_T=V\prod_{i=1}^{T}P_{i-1}(i)^{-1}=VB(T)$$

at the time T. This allows us to interpret the factor $B(t)$ as the value of a unit of the nonrisky asset (bank account) at the time t.

Thus, we get an explicit formula for the price of the bond $P(T)$ at the time t:

$$\text{(8.5)}\qquad P_t(T)=P_0(T)\prod_{i=1}^{t}P_{i-1}(i)^{-1}\frac{\prod_{i=1}^{t}\theta^{(T-i)\xi_i}}{\prod_{i=1}^{t}\bigl(p^*+(1-p^*)\theta^{(T-i)}\bigr)},$$

which depends on the structure of the bond prices $\{P_0(t)\}_{t\geqslant 1}$ at the initial moment of time and the dynamics of the stochastic sequence $\xi=\{\xi_t\}_{t\geqslant 1}$, where $\theta=\exp(\delta)$.

The deficiencies of the Ho–Lee model stem from the unnecessary simplicity of the function $\sigma(s,t)$. For large values of t, when the central limit theorem begins to "operate" in view of the linearity of $\sigma(s,T)$, there is a "disappearance" of randomness. This property is clearly evident in Figure 8.1, which is constructed on the basis of selecting random numbers. Formally, we get by using the formula (8.5) that

$$P_t(T)=\frac{P_0(T)}{P_0(t)}\exp\biggl(\delta(T-t)\sum_{i=1}^{t}\xi_i\biggr)\prod_{i=1}^{t}\frac{p^*+(1-p^*)\theta^{t-i}}{p^*+(1-p^*)\theta^{T-i}}.$$

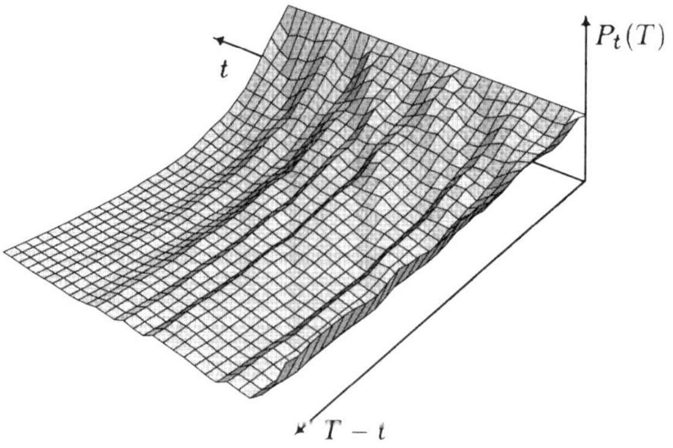

FIGURE 8.1. Example of the interest rate structure for the Ho–Lee model.

Hence, for sufficiently large t the value $P_t(T)$ will be determined by a random factor of the order $\theta\sqrt{tp^*(1-p^*)}$, which decreases as t increases.

On the basis of the above arguments we can construct a very broad spectrum of discrete models using the structure of the formula (8.1). For example, employing a simple autoregression scheme with $\sigma(s,t) = \delta(T-t)I_{\{T-t\leqslant K\}}$ along with a sequence $\xi = \{\xi_t\}_{t\geqslant 1}$ of independent standard normal variables, we get a model lacking the deficiency indicated above.

Moreover, we can construct not only single-factor but also multifactor discrete models of the interest rate structure, when the price dynamics is determined by formulas of the kind

$$P_t(T) = P_0(T)B(t)\frac{\exp\left(\sum_{i=1}^t (\sigma^1(i,T)\xi_i^1 + \cdots + \sigma^n(i,T)\xi_i^n)\right)}{\mathbf{E}_{\mathbf{P}^*}\left[\exp\left(\sum_{i=1}^t (\sigma^1(i,T)\xi_i^1 + \cdots + \sigma^n(i,T)\xi_i^n)\right)\right]}.$$

The independence of the elements of the sequence $\xi = \{\xi_t\}_{t\geqslant 1}$ actually ensures the absence of arbitrage in these models, because with respect to the measure $\mathbf{P}^*$ the discounted price

$$\overline{P}_t(T) = \frac{P_t(T)}{B(t)} = P_0(T)\frac{\exp\left(\sum_{i=1}^t \sigma(i,T)\xi_i\right)}{\mathbf{E}_{\mathbf{P}^*}\left[\exp\left(\sum_{i=1}^t \sigma(i,T)\xi_i\right)\right]}$$

of the bond is a martingale.

The Heath–Jarrow–Morton model.

This model is constructed on the basis of the *forward interest rates* $f_t(T)$, when under the assumption that $P_t(T)$ is continuously differentiable with respect to T the structure of the function is given by the expression

$$P_t(T) = \exp\left(-\int_t^T f_t(s)\,ds\right). \tag{8.6}$$

The quantity $f_t(T)$ is the interest rate at the time t for loans over the "future" infinitesimal interval from T to $T+dT$. In other words, this is the rate with respect to which a forward agreement can be concluded at the time t concerning a loan on the interval $[T, T + dT]$.[2]

We define $f_t(T)$ by the structural formula

$$f_t(T) = f_0(T) + \int_0^t \eta_s(T)\,ds + \int_0^t \delta_s(T)\,dw_s, \tag{8.7}$$

where $w = (w_t)_{t\geqslant 0}$ is a standard Brownian motion. Under certain additional assumptions about the processes α and σ, assumptions ensuring the existence of iterated integrals and the validity of Fubini's theorem for random processes, it can be shown that $P_t(T)$ is a solution of the stochastic differential equation

$$dP_t(T) = P_t(T)\big[(r(t) + b_t(T))\,dt + a_t(T)\,dw_t\big],$$

where

$$r(t) = f_t(t), \quad a_t(T) = -\int_t^T \delta_t(s)\,ds, \quad b_t(T) = -\int_t^T \eta_t(s)\,ds + \frac{1}{2}a_t(T)^2.$$

Assuming the existence of a solution $\gamma_t(T)$ of the equation

$$a_t(T)\zeta_t(T) = -b_t(T),$$

we get that

$$dP_t(T) = P_t(T)\big[r(t)\,dt + a_t(T)\,d\widetilde{w}_t\big].$$

Here $\widetilde{w}_t = w_t - \zeta_t(T)$ is a Wiener process with respect to the measure $\mathbf{P}^*$ with density

$$dZ_t^*(T) = \exp\bigg\{\int_0^t \zeta_s(T)\,dw_s - \frac{1}{2}\int_0^t \zeta_s(T)^2\,ds\bigg\}.$$

Setting $B(t) = \exp\Big(\int_0^t r(t)\,dt\Big)$, we arrive at a multiplicative representation of $P_t(T)$ in terms of the stochastic exponential $\mathcal{E}$:

$$P_t(T) = P_0(T)B(t)\,\mathcal{E}_t\bigg(\int a_s(T)\,d\widetilde{w}_s\bigg). \tag{8.8}$$

The process $B(t)$ can be interpreted as the value of a unit of the bank account at the moment of time t. The discounted value of the price $P_t(T)$ has the form

$$\overline{P}_t(T) = \frac{P_t(T)}{B(t)} = P_0(T)\,\mathcal{E}\bigg(\int_0^t a_s(T)\,d\widetilde{w}_s\bigg),$$

and hence is a martingale with respect to the measure $\mathbf{P}^*$. If the values of the process $\gamma_t(T)$ do not depend on the value of T, then the discounted price of an arbitrary bond is a martingale with respect to $\mathbf{P}^*$, and this (B, P)-market is a no-arbitrage market.

On the other hand, defining by

$$P_t(T) = P_0(T)B(t)\frac{\exp\big(\int_0^t \sigma_s(T)\,d\widetilde{w}_s\big)}{\mathbf{E}_{\mathbf{P}^*}\big[\exp\big(\int_0^t \sigma_s(T)\,d\widetilde{w}_s\big)\big]} \tag{8.9}$$

the price process $P_t(T)$ from the start with the help of (8.1) for a bond with redemption date T, where $\widetilde{w}$ is a Wiener process with respect to a measure $\mathbf{P}^*$

[2] That is, a contract can be concluded at the time t, where a rate $R_T = f_t(T)$ is stipulated for which one of the parties agrees to extend credit to the other party on the interval $[T, T + dT]$.

equivalent to the initial measure $\mathbf{P}$, we get a model of the interest rate structure. In view of the martingale property and the positivity of the factor

$$\frac{\exp\left(\int_0^t \sigma_s(T)\,d\widetilde{w}_s\right)}{\mathbf{E}_{\mathbf{P}^*}\left[\exp\left(\int_0^t \sigma_s(T)\,d\widetilde{w}_s\right)\right]}$$

the equality (8.9) can be written in the form

$$\begin{aligned} P_t(T) &= P_0(T)B(t)\exp\left(-\frac{1}{2}\int_0^t \sigma_s(T)^2\,ds + \int_0^t \sigma_s(T)\,d\widetilde{w}_s\right) \\ &= P_0(T)B(t)\,\mathcal{E}_t\left(\int \sigma_s(T)\,d\widetilde{w}_s\right), \end{aligned}$$

which coincides with (8.8). The boundary condition $P_t(t) = 1$ makes it possible to define $B(t)$ by the formula

$$B(t) = P_0(t)^{-1}\exp\left(\frac{1}{2}\int_0^t \sigma_s(t)^2\,ds - \int_0^t \sigma_s(t)\,d\widetilde{w}_s\right).$$

Various *special cases* of the Heath–Jarrow–Morton model are obtained in dependence on the specific form of the function $\sigma_t(T)$ in (8.9) or $\delta_t(T)$ in (8.7). For example, for

$$\sigma_t(T) = \widehat{\sigma}(T-t) \quad \text{or, equivalently,} \quad \delta_t(T) = \widehat{\sigma}$$

we get a *diffusion analogue of the Ho–Lee model*, while for

$$\sigma_t(T) = \frac{\widehat{\sigma}}{a}\left(1 - e^{-a(T-t)}\right) \quad \text{or} \quad \delta_t(T) = \widehat{\sigma}e^{-a(T-t)}$$

we get a diffusion analogue of the *Bachelier model.*

Just as in the discrete case, the Heath–Jarrow–Morton model can be made a multifactor model. For example, in place of the equality (8.7) we can give the dynamics of the forward rates by the expression

$$f_t(T) = f_0(T) + \int_0^t \eta_s(T)\,ds + \sum_{i=1}^{n}\int_0^t \delta_s^i(T)\,dw_s^i,$$

where $w = (w_t^1, \dots, w_t^n)_{t\geqslant 0}$ is a multidimensional Wiener process with respect to $\mathbf{P}$. In the subsequent exposition we use the one-dimensional case of the Heath–Jarrow–Morton model in order to facilitate the reading of the text, although most of the results can be formulated also in the multidimensional case with the appropriate refinements.

§ 8.2. Hedging on a bond market

One can consider various forms of derivative securities on a bond market, and this leads to the problem of hedging the corresponding contingent claims.

We show how contingent claims should be calculated on a bond market. If on a given (B,P)-market there is a *unique* martingale measure $\mathbf{P}^*$, then the cost $\mathbb{C}_0$ of a contingent claim f_T presented for exercise at the time T must be equal to

$$\mathbb{C}_0 = \mathbf{E}_{\mathbf{P}^*}\left[B(t)^{-1}f_T\right].$$

But if there is a *family* $\mathbb{Q} = \mathcal{M}(\mathbf{P})$ of martingale measures, then we should calculate the ask and bid prices for a contingent claim f_T:

$$\mathbb{C}^* = \sup_{\mathbf{Q}\in\mathbb{Q}} \mathbf{E}_{\mathbf{Q}}\big[B(t)^{-1}f_T\big], \qquad \mathbb{C}_* = \inf_{\mathbf{Q}\in\mathbb{Q}} \mathbf{E}_{\mathbf{Q}}\big[B(t)^{-1}f_T\big].$$

At the same time, we must determine not only the price of an arbitrary contingent claim on the given market, but also a corresponding *hedging strategy* which enables us to *hedge* or even *replicate* f_T.

A basic feature of this market is the presence of a large (often infinite) number of bonds, including bonds with a redemption date less than T. Correspondingly, resources can be invested in these bonds only on the interval up to the moment of redemption.

The general theory of bond markets assumes the possibility of investing resources simultaneously in all bonds existing at the given time on the market. This circumstance requires a definition of an integral for processes taking values in some function space when the investment strategy forms a pair $(\beta,\gamma) = (\beta_t,\gamma_t)_{t\geqslant 0}$, where β_t is the number of units in the bank account in the portfolio, and $\gamma_t(ds)$ is a measure on $[t,T]$ determining the number of bonds with redemption dates $S\in[t,T]$ in which resources are invested at the time t.

In the present exposition we deal with a situation in which the portfolio can contain only *a finite number of bonds* at one time. This means that for any t the distribution of $\gamma_t(ds)$ is concentrated at a finite number of points.

A *portfolio* at time t is defined to be a pair (β_t,γ_t), where β_t is the number of units of a bank account at time t, and $\gamma_t = (\gamma_t(S_t^1),\dots,\gamma_t(S_t^{n_t}))$ is a vector determining the number of bonds with redemption dates $S_t^1,\dots,S_t^{n_t}$ contained in the portfolio at the time t.

An *investment strategy* π is defined to be a pair $(\beta,\gamma) = \{(\beta_t,\gamma_t)\}_{t\geqslant 0}$ of processes representing a portfolio at each moment of time t.

The *value* X_t^π of a strategy π at the time t is defined to be the quantity

$$X_t^\pi = \beta_t B(t) + \sum_{i=1}^{n_t} \gamma_t(S_t^i) P_t(S_t^i).$$

An investment strategy π is said to be *self financing* if:

a) (in the case of *discrete time*)

$$\Delta X_t^\pi = \beta_t \Delta B(t) + \sum_{i=1}^{n_t} \gamma_t(S_t^i)\, \Delta P_t(\widehat{S}_t^i),$$

where $\Delta P_t(\widehat{S}_t^i) = P_t(S_t^i) - P_{t-1}(S_t^i)$;

b) (in the case of *continuous time*)

$$dX_t^\pi = \beta_t\, dB(t) + \sum_{i=1}^{n_t} \gamma_t(S_t^i)\, \partial P_t(\widehat{S}_t^i),$$

where it is convenient to understand the differential $\partial P_t(\widehat{S}_t^i)$ as if the parameter S_t^i were fixed.[3]

[3]Such a definition is due to our trying to avoid complexities connected with an exposition of the theory of integration with respect to a measure-valued process. Anyone wishing to master this technique can turn to the paper [**22**].

Hedging in the Ho–Lee model.

Let us consider the problem of hedging a contingent claim in a model of a discrete bond market when the price dynamics is determined by the equations (8.3)–(8.5):

$$B(t) = \prod_{i=1}^{t} P_{i-1}(i)^{-1}, \qquad P_t(T) = P_0(T)B(t)\frac{\exp\big(\delta\sum_{i=1}^{t}(T-i)\xi_i\big)}{\mathbf{E}_{\mathbf{P}^*}\Big[\exp\big(\delta\sum_{i=1}^{t}(T-i)\xi_i\big)\Big]},$$

where ξ_i is a sequence of independent identically distributed random variables taking the values 0 and 1 with probabilities p and $1-p$, respectively.

Suppose that an investor takes as investment objects only a bank account B and a bond $P(T^1)$ with (fixed) redemption date T^1. Then at each moment of time $t \in [0, T^1]$ his portfolio is determined by the pair $(\beta_t, \gamma_t(T^1))$ indicating the number of units in the bank account and the number of bonds with redemption date T^1 contained in the portfolio on the interval $(t-1, t]$. The capital V^π of the investor at the time t is given by

$$V_t^\pi = \beta_t B(t) + \gamma_t(T^1)P_t(T^1).$$

A strategy $\pi = \{(\beta_t \gamma_t(T^1))\}_{t\geqslant 1}$ is said to be *self-financing* if

$$\Delta X_t^\pi = V_t^\pi - V_{t-1}^\pi = \beta_t\,\Delta B(t) + \gamma_t(T^1)\,\Delta P_t(T^1).$$

The $(B, P(T^1))$-market obtained in this way consists of two assets in all, and in this sense does not differ from a (B,S)-market. Obviously, $\mathbf{P}^*$ is a martingale measure for this market. By using the methodology for finding martingale measures it is not hard to show that $\mathbf{P}^*$ is *the unique martingale measure* equivalent to $\mathbf{P}$.

Consequently, the $(B, P(T^1))$-market is a *complete* and *no-arbitrage* market. For an arbitrary contingent claim f_T exercised at a time $T < T^1$ and measurable with respect to the σ-algebra $\mathcal{F}_T = \sigma(\xi_i,\, i \leqslant T)$ there exists a self-financing strategy π^* such that

$$X_T^{\pi^*} = f_T \quad (\mathbf{P}\text{-a.s.}), \qquad \mathbb{C} = X_0^{\pi^*} = \mathbf{E}_{\mathbf{P}^*}\big[B(t)^{-1}f_T\big].$$

It is clear that f_T can be taken to be any other bond with redemption date $T^2 \geqslant T$. The random variable $P_T(T^2)$ is $\mathcal{F}_T$-measurable, and the rational price of a contingent claim with payment function $P_T(T^2)$ is equal to

$$\mathbb{C} = X_0^{\pi^*} = \mathbf{E}_{\mathbf{P}^*}\big[B(T)^{-1}f_T\big] = \mathbf{E}_{\mathbf{P}^*}\big[B(T)^{-1}P_T(T^2)\big].$$

In view of the martingale property of discounted bond prices, the rational price of such a contingent claim is

$$\mathbb{C} = \mathbf{E}_{\mathbf{P}^*}[B(T)^{-1}P_T(T^2)] = P_0(T^2).$$

Since the discounted value of a self-financing strategy π is a martingale with respect to the measure $\mathbf{P}^*$, we can write

$$\begin{aligned} Y_t^{\pi^*} = B(t)^{-1}X_t^{\pi^*} &= \mathbf{E}_{\mathbf{P}^*}\big[B(T)^{-1}f_T \,|\, \mathcal{F}_t\big] \\ &= \mathbf{E}_{\mathbf{P}^*}\big[B(T)^{-1}P_T(T^2) \,|\, \mathcal{F}_t\big] = B(t)^{-1}P_t(T^2). \end{aligned}$$

Thus, the hedging of a contingent claim with payment function $P_T(T^2)$ reduces to the equality $X_t^{\pi^*} = P_t(T^2)$ for any time $t \in [0, T]$.

Further, the equality

$$\Delta Y_t^{\pi^*} = \gamma_t^*(T^1)\,\Delta\big(B(t)^{-1}P_t(T^1)\big)$$

holds for a self-financing strategy π^*. Therefore, using the formula (8.5), we get that

$$\gamma_t^*(T^1) = \frac{P_{t-1}(T^2)}{P_{t-1}(T^1)} \cdot \frac{p^* + (1-p^*)\theta^{T^1-t}}{p^* + (1-p^*)\theta^{T^2-t}} \cdot \frac{\theta^{\xi_t(T^2-t)} - p^* - (1-p^*)\theta^{T^2-t}}{\theta^{\xi_t(T^1-t)} - p^* - (1-p^*)\theta^{T^1-t}},$$

where the last factor, aside from dependence on the value of ξ_t, is equal to

$$\frac{\theta^{T^2-t}-1}{\theta^{T^1-t}-1}.$$

Consequently, $\gamma_t(T^1)$ is an $\mathcal{F}_{t-1}$-measurable function and takes the value

$$\gamma_t^*(T^1) = \frac{P_{t-1}(T^2)}{P_{t-1}(T^1)} \cdot \frac{p^* + (1-p^*)\theta^{T^1-t}}{p^* + (1-p^*)\theta^{T^2-t}} \cdot \frac{\theta^{(T^2-t)}-1}{\theta^{(T^1-t)}-1}.$$

The quantity β_t^* can be obtained from the balance equation

$$X_{t-1}^{\pi^*} = P_{t-1}(T^2) = \beta_t^* B(t-1) + \gamma_t^*(T^1)P_t(T^1).$$

The strategy π^* constructed is such that for any $t \in [0,T]$ its value $X_t^{\pi^*}$ coincides with $P_t(T^2)$. The following useful device is based on this result.

Let $\pi = (\beta, \gamma(T^2))$ be a self-financing strategy using a bank account $B(t)$ and a bond with redemption date T^2. Such a strategy must be implemented with the help of a bond $P(T^1)$ with redemption date T^1. Then it is natural to propose substituting in place of the bond $P(T^2)$ the value V^{π^*} of a portfolio replicating this bond:

$$\begin{aligned} X_t^\pi &= \beta_t B(t) + \gamma_t(T^2)P_t(T^2) = \beta_t B(t) + \gamma_t(T^2)X_t^{\pi^*} \\ &= (\beta_t + \gamma_t(T^2)\beta_t^*)B(t) + \gamma_t(T^2)\gamma_t^*(T^1)P_t(T^1). \end{aligned}$$

We let $\beta_t' = \beta_t + \gamma_t(T^2)\beta_t^*$ and $\gamma_t'(T^1) = \gamma_t(T^2)\gamma_t^*(T^1)$ and rewrite the value X_t^π in the form $X_t^\pi = \beta_t' B(t) + \gamma_t' P_t(T^1)$. Let us show that the strategy $\pi' = (\beta', \gamma'(T^1))$ is self-financing.

Indeed, it follows from the self-financing property of the strategies π and π^* that

$$\begin{aligned} \Delta X_t^{\pi'} = \Delta X_t^\pi &= \beta_t\,\Delta B(t) + \gamma_t(T^2)\,\Delta P_t(T^2) = \beta_t\,\Delta B(t) + \gamma_t(T^2)\,\Delta X_t^{\pi^*} \\ &= \beta_t\,\Delta B(t) + \gamma_t(T^2)\big(\beta_t\,\Delta B(t) + \gamma_t(T^1)\,\Delta P_t(T^1)\big) \\ &= \big(\beta_t + \gamma_t(T^2)\beta_t^*\big)\,\Delta B(t) + \gamma_t(T^2)\gamma_t^*(T^1)\,\Delta P_t(T^1) \\ &= \beta_t'\,\Delta B(t) + \gamma_t'\,\Delta P_t(T^1). \end{aligned}$$

Thus, the self-financing strategy using the bank account B and the bond $P(T^2)$ can be replaced by a self-financing strategy constructed using the bank account B and another bond $P(T^1)$. This makes it possible to construct a hedging strategy when there are restrictions on the bonds used in the portfolio.

As another example we consider the pricing of an option to buy for a bond with some redemption date $T^2 \geqslant T$ and payment function $(P_T(T^2) - K)^+$. Since the dynamics of the price of $P(T^2)$ is also determined by the sequence $\{\xi_t\}_{t \leqslant T}$, the function $(P_T(T^2) - K)^+$ is $\mathcal{F}_T$-measurable, and hence the rational price of the option is equal to

$$\mathbb{C} = X_0^{\pi^*} = \mathbf{E}_{\mathbf{P}^*}\big[B(T)^{-1} f_T\big] = \mathbf{E}_{\mathbf{P}^*}\big[B(T)^{-1}(P_T(T^2) - K)^+\big].$$

It follows from (8.4) and (8.5) that $P_T(T^2) \geqslant K$ in the case when among $\xi_1, \dots, \xi_T$ there are at least $k_0 = k(T, T^2, P_0(T), P_0(T^2))$ variables with value 1, where

$$k(n, N, P^1, P^2) = \inf_{0 \leqslant k \leqslant n} \left\{ k : k \geqslant \frac{1}{\delta(N-n)} \ln\left(K \frac{P^1}{P^2} \prod_{i=1}^{n} \frac{p^* + (1-p^*)\theta^{N-i}}{p^* + (1-p^*)\theta^{n-i}} \right) \right\}.$$

Let

$$\mathbb{B}(k_0, n, N, p) = \sum_{\substack{(x_1, \dots, x_n): \\ \sum x_i \geqslant k_0}} \left[\frac{\prod_{i=1}^{n} \theta^{x_i(N-i)}}{\prod_{i=1}^{n} (p^* + (1-p^*)\theta^{(N-i)})} p^{n - \sum x_i} (1-p)^{\sum x_i} \right],$$

where $(x_1, \dots, x_n)$ is a vector with components equal to 0 or 1. Then

$$\begin{aligned}
\mathbf{E}_{\mathbf{P}^*}\big[B(T)^{-1}(P_T(T^2) - K)^+\big] &= \mathbf{E}_{\mathbf{P}^*}\big[(B(T)^{-1} P_T(T^2) - K B(T)^{-1})^+\big] \\
&= \mathbf{E}_{\mathbf{P}^*}\left[\left(P_0(T^2) \frac{\prod_{i=1}^{T} \theta^{\xi_i(T^2-i)}}{\prod_{i=1}^{T} (p^* + (1-p^*)\theta^{(T^2-i)})} \right.\right. \\
&\qquad \left.\left. - K P_0(T) \frac{\prod_{i=1}^{T} \theta^{\xi_i(T-i)}}{\prod_{i=1}^{T} (p^* + (1-p^*)\theta^{(T-i)})} \right)^+ \right] \\
&= P_0(T^2)\, \mathbb{B}(k_0, T, T^2, p^*) - K P_0(T)\, \mathbb{B}(k_0, T, T, p^*),
\end{aligned}$$

and hence

$$\mathbb{C} = P_0(T^2)\, \mathbb{B}(k_0, T, T^2, p^*) - K P_0(T)\, \mathbb{B}(k_0, T, T, p^*). \tag{8.10}$$

The dynamic behavior of the value $X_t^{\pi^*}$ of the hedging strategy π^* is determined from the self-financing condition, which implies that the discounted value of π^* is a martingale:

$$B(t)^{-1} X_t^{\pi^*} = \mathbf{E}_{\mathbf{P}^*}\big[B(T)^{-1} f_T \,|\, \mathcal{F}_t\big] = \mathbf{E}_{\mathbf{P}^*}\big[B(T)^{-1}(P_T(T^2) - K)^+ \,|\, \mathcal{F}_t\big].$$

Setting $k_t = k(T-t, T^2-t, P_t(T), P_t(T^2))$, we get the following general formula for the value $X_t^{\pi^*}$:

$$\begin{aligned}
X_t^{\pi^*} &= B(t) \mathbf{E}_{\mathbf{P}^*}\big[B(T)^{-1}(P_T(T^2) - K)^+ \,|\, \mathcal{F}_t\big] \\
&= B(t)\big(\overline{P}_t(T^2)\, \mathbb{B}(k_t, T-t, T^2-t, p^*) - K \overline{P}_t(T)\, \mathbb{B}(k_t, T-t, T-t, p^*)\big) \\
&= P_t(T^2)\, \mathbb{B}(k_t, T-t, T^2-t, p^*) - K P_t(T)\, \mathbb{B}(k_t, T-t, T-t, p^*).
\end{aligned}$$

Hedging in the Heath–Jarrow–Morton model.

Suppose that the bond market given by the equations (8.6)–(8.8) consists solely of the bank account $B(t)$ and one bond $P(T^1)$. In this case the definition of a self-financing strategy $\pi = (\beta, \gamma(T^1))$ can be rewritten in the form

$$dX_t^\pi = \beta_t\, dB(t) + \gamma_t(T^2)\, dP_t(T^2).$$

The existence of a unique martingale measure $\mathbf{P}^*$ implies that under the condition $a_s(T^2) > 0$ such a $(B, P(T^2))$-market does not differ from a diffusion (B, S)-market. Therefore, for an arbitrary contingent claim f_T with exercise time $T \leqslant T^1$ the rational price is equal to

$$\mathbb{C} = \mathbf{E}_{\mathbf{P}^*}\big[B(t)^{-1} f_T\big].$$

We consider a contingent claim with payment function $P_T(T^2)$ coinciding with the price of a bond with redemption date T^2. Clearly, the rational price of such a contingent claim coincides with the initial bond price:

$$\begin{aligned}\mathbb{C} = \mathbf{E}_{\mathbf{P}^*}\big[B(t)^{-1} f_T\big] &= \mathbf{E}_{\mathbf{P}^*}\big[B(t)^{-1} P_T(T^2)\big] \\ &= \mathbf{E}_{\mathbf{P}^*}\Big[P_0(T^2)\,\mathcal{E}\Big(\int a_s(T^2)\, d\widetilde{w}_s\Big)\Big] = P_0(T^2).\end{aligned}$$

The evolution of the discounted value $Y_t^{\pi^*}$ of a hedging strategy π^*, which exists by the results in Chapter 3, is determined by the equation

$$\begin{aligned}Y_t^{\pi^*} = B(t)^{-1} X_t^{\pi^*} &= \mathbf{E}_{\mathbf{P}^*}\big[B(t)^{-1} f_T \,|\, \mathcal{F}_t\big] \\ &= \mathbf{E}_{\mathbf{P}^*}\big[B(t)^{-1} P_T(T^2) \,|\, \mathcal{F}_t\big] = B(t)^{-1} P_t(T^2).\end{aligned}$$

Since the discounted value of a self-financing strategy satisfies the stochastic differential equation $dY_t^{\pi^*} = \gamma_t^*(T^1)\, d\dfrac{P_t(T^1)}{B(t)}$, we get that

$$\begin{aligned}d\frac{P_t(T^2)}{B(t)} = \frac{P_t(T^2)}{B(t)} a_s(T^2)\, d\widetilde{w}_s &= \gamma_t^*(T^1)\, d\frac{P_t(T^1)}{B(t)} \\ &= \gamma_t^*(T^1) \frac{P_t(T^1)}{B(t)} a_s(T^1)\, d\widetilde{w}_s,\end{aligned}$$

and hence

$$\gamma_t^*(T^1) = \frac{P_t(T^2)}{P_t(T^1)} \frac{a_s(T^2)}{a_s(T^1)}. \tag{8.11}$$

The value of β_t^*, as usual, is found from the balance equation $P_t(T^2) = \beta_t^* B(t) + \gamma_t^*(T^1) P_t(T^1)$:

$$\beta_t^* = \frac{P_t(T^2)}{B(t)}\Big(1 - \frac{a_s(T^2)}{a_s(T^1)}\Big). \tag{8.12}$$

The ratio $\dfrac{a_s(T^2)}{a_s(T^1)}$ in the formula (8.12) represents the *portion* of the value that should be invested in the bond $P_t(T^1)$, while $\Big(1 - \dfrac{a_s(T^2)}{a_s(T^1)}\Big)$ represents the *portion* of the value deposited in the bank account.

Let $\pi = (\beta, \gamma(T^2))$ be a self-financing strategy operating with the bank account $B(t)$ and a bond with redemption date T^2. As earlier, we get that

$$\beta'_t = \beta_t + \gamma_t(T^2)\beta^*_t, \qquad \gamma'_t(T^1) = \gamma_t(T^2)\gamma^*_t(T^1).$$

Then

$$\begin{aligned} X^{\pi'}_t &= \beta'_t B(t) + \gamma'_t(T^1)P_t(T^1) \\ &= \big(\beta_t + \gamma_t(T^2)\beta^*_t\big)B(t) + \gamma_t(T^2)\gamma^*_t(T^1)P_t(T^1) \\ &= \beta_t B(t) + \gamma_t(T^2)\big(\beta^*_t B(t) + \gamma^*_t(T^1)P_t(T^1)\big) \\ &= \beta_t B(t) + \gamma_t(T^2)X^{\pi^*}_t = \beta_t B(t) + \gamma_t(T^2)P_t(T^2) = X^{\pi}_t, \end{aligned}$$

and hence

$$\begin{aligned} dX^{\pi'}_t &= \beta'_t\, dB(t) + \gamma'_t(T^1)\, dP_t(T^1) \\ &= \big(\beta_t + \gamma_t(T^2)\beta^*_t\big)\, dB(t) + \gamma_t(T^2)\gamma^*_t(T^1)\, dP_t(T^1) \\ &= \beta_t\, dB(t) + \gamma_t(T^2)\big(\beta^*_t\, dB(t) + \gamma^*_t(T^1)\, dP_t(T^1)\big) \\ &= \beta_t\, dB(t) + \gamma_t(T^2)\, dX^{\pi^*}_t = \beta_t\, dB(t) + \gamma_t(T^2)\, dP_t(T^2) = dX^{\pi}_t. \end{aligned}$$

Thus, *an arbitrary self-financing strategy using some bond and a bank account can (on the interval $[0, T^1]$) be replaced by a self-financing strategy operating with the bank account and the bond $P(T^1)$.*

The indicated property is an immediate consequence of the completeness and the no-arbitrage property of our market and is satisfied for an arbitrary self-financing strategy. Indeed, let $\pi = (\beta, \gamma(S^1_t), \dots, \gamma(S^{n_t}_t))$ be some self-financing strategy. Then the evolution of its value satisfies the equation

$$\begin{aligned} dX^{\pi}_t &= \beta_t dB(t) + \sum_{i=1}^{n_t} \gamma_t(S^i_t)\, \partial P_t(S^i_t) \\ &= \beta_t\, dB(t) + \sum_{i=1}^{n_t} \gamma_t(S^i_t)P_0(S^i_t)\, \mathcal{E}\Big(\int a_s(S^i_t)\, d\widetilde{w}_s\Big) dB(t) \\ &\quad + \sum_{i=1}^{n_t} \gamma_t(S^i_t)P_0(S^i_t)B(t)\, \mathcal{E}\Big(\int a_s(S^i_t)\, d\widetilde{w}_s\Big) a_s(S^i_t)\, d\widetilde{w}_s \\ &= \big(X^{\pi}_t B(t)^{-1}\big)\, dB(t) \\ &\quad + \sum_{i=1}^{n_t} \gamma_t(S^i_t)P_0(S^i_t)B(t)\, \mathcal{E}\Big(\int a_s(S^i_t)\, d\widetilde{w}_s\Big) a_s(S^i_t)\, d\widetilde{w}_s. \end{aligned} \tag{8.13}$$

Denoting by $Y^{\pi}_t = \dfrac{X^{\pi}_t}{B(t)}$ the discounted value of the strategy π, we get that

$$\begin{aligned} d(Y^{\pi}_t B(t)) &= Y^{\pi}_t dB(t) + B(t)dY^{\pi}_t \\ &= (X^{\pi}_t B(t)^{-1})\, dB(t) \\ &\quad + \sum_{i=1}^{n_t} \gamma_t(S^i_t)P_0(S^i_t)B(t)\, \mathcal{E}\Big(\int a_s(S^i_t)\, d\widetilde{w}_s\Big) a_s(S^i_t)\, d\widetilde{w}_s; \end{aligned}$$

consequently,

$$dY_t^\pi = \sum_{i=1}^{n_t} \gamma_t(S_t^i) P_0(S_t^i)\, \mathcal{E}\Big(\int a_s(S_t^i)\, d\widetilde{w}_s\Big) a_s(S_t^i)\, d\widetilde{w}_s,$$

that is, the discounted value of the self-financing strategy is a martingale with respect to the measure $\mathbf{P}^*$.

We consider the contingent claim $f_T = X_T^\pi$ on the $(B, P(T^1))$-market, where in view of completeness and the no-arbitrage property there exists a self-financing strategy π^* such that $X_T^\pi = X_T^{\pi^*}$ ($\mathbf{P}$-a.s.). By the martingale properties of the discounted value of a self-financing strategy, we get that

$$X_t^\pi = B(t)\, \mathbf{E}_{\mathbf{P}^*}\big[B(t)^{-1} X_T^\pi \,|\, \mathcal{F}_t\big] = B(t)\, \mathbf{E}_{\mathbf{P}^*}\big[B(t)^{-1} f_T \,|\, \mathcal{F}_t\big] = X_t^{\pi^*}.$$

Thus, an arbitrary self-financing strategy can be realized (on the interval $[0,T]$) with the use of only a bank account and a single bond with redemption date $T^1 \geqslant T$. There is also a *converse result*, in a certain sense.

Suppose that on the interval $[0,T]$ we are given a predictable process T_t with values at each moment of time satisfying the inequality $t < T_t \leqslant T^1$. We interpret this process as a rule indicating in which bond we can invest our resources at the time t.[4]

We consider a contingent claim $f_T = P_T(T^1)$ that should be hedged with the help of some self-financing strategy $\pi^* = (\beta^*, \gamma^*(T_t))$, that is, using a bond with redemption date T_t at each moment of time.

It follows from the equalities (8.11) and (8.12) that a hedging strategy has the form

$$\gamma_t^*(T_t) = \frac{P_t(T^1)}{P_t(T_t)} \frac{a_s(T^1)}{a_s(T_t)}, \qquad \beta_t^* = \frac{P_t(T^1)}{B(t)}\Big(1 - \frac{a_s(T^1)}{a_s(T_t)}\Big). \tag{8.14}$$

Indeed,

$$\begin{aligned} X_t^{\pi^*} &= \beta_t^* B(t) + \gamma_t^*(T_t) P_t(T_t) \\ &= P_t(T^1)\Big(1 - \frac{a_s(T^1)}{a_s(T_t)}\Big) + P_t(T^1)\frac{a_s(T^1)}{a_s(T_t)} = P_t(T^1), \end{aligned}$$

and at the same time we have the equality

$$\begin{aligned} dX_t^{\pi^*} &= dP_t(T^1) = P_t(T^1) a_s(T^1)\, d\widetilde{w}_t + \big(B(t)^{-1} P_t(T^1)\big)\, dB(t) \\ &= \gamma_t^*(T_t) P_t(T_t) a_s(T_t)\, d\widetilde{w}_t \\ &\qquad \pm \gamma_t^*(T_t)\big(B(t)^{-1} P_t(T_t)\big)\, dB(t) + \big(B(t)^{-1} P_t(T^1)\big)\, dB(t) \\ &= \gamma_t^*(T_t)\, \partial P_t(\widehat{T}_t) + \beta_t^*\, dB(t), \end{aligned}$$

which implies that the strategy π^* is self-financing.

[4] Such an assumption, perhaps seeming artificial at first glance, is of interest because in a real market there are only a limited number of bonds available at each moment of time. Indeed, if the government issues four-year bonds every two years, then to hedge a contingent claim with exercise time in 10 years the investor must exchange the bonds with which he is working at least three times, because there are no bonds with redemption date $T^1 \geq 10$ on the market. In this case one possible pricing variant is to assume that all bonds are present on the market, but one can invest only in bonds with redemption date 4 on the interval $[0,2]$, only in bonds with redemption dates 4 and 6 on the interval $[2,4]$, only in bonds with redemption dates 6 and 8 on the interval $[4,6]$, and so on.

According to arguments similar to those above, we can show that an arbitrary self-financing strategy $\pi = (\beta, \gamma(T^1))$ operating with a bank account $B(t)$ and a bond with redemption date T^1 can be replaced by a self-financing strategy $\pi' = (\beta', \gamma'(T_t))$, where

$$\beta'_t = \beta_t + \gamma_t(T^1)\beta^*_t, \qquad \gamma'_t(T_t) = \gamma_t(T^1)\gamma^*_t(T_t). \tag{8.15}$$

Thus, *on a complete $(B, P(T^1))$-market any self-financing strategy can be represented by some strategy $\pi = (\beta, \gamma(T_t))$ using only a bank account and a bond with redemption date T_t at each moment of time t.*

We apply the above methodology to a concrete model.

The Ho–Lee continuous model. Let us consider an option to buy with payment function

$$f_T = \big(P_T(T^2) - KB(T)\big)^+.$$

According to this option, if the discounted value of the bond $P_T(T^2)$ at the time T exceeds the level K, then the buyer of the option receives a payment in the amount

$$P_T(T^2) - KB(T) = B(T)\big(B(T)^{-1}P_T(T^1) - K\big).$$

The rational price $\mathbb{C}$ of this option in the case of the Ho–Lee model is equal to

$$\begin{aligned}
\mathbb{C} &= \mathbf{E}_{\mathbf{P}^*}\big[B(T)^{-1}f_T\big] \\
&= \mathbf{E}_{\mathbf{P}^*}\big[(B(T)^{-1}P_T(T^1) - K)^+\big] \\
&= \mathbf{E}_{\mathbf{P}^*}\bigg[\bigg(P_0(T^1)\,\mathcal{E}\Big(\int \delta(T^1 - t)\,d\widetilde{w}_t\Big) - K\bigg)^+\bigg] \\
&= \mathbf{E}_{\mathbf{P}^*}\bigg[\bigg(P_0(T^1)\exp\Big(-\frac{1}{2}\int_0^T \delta^2(T^1 - t)^2\,dt + \int_0^T \delta(T^1 - t)\,d\widetilde{w}_t\Big) - K\bigg)^+\bigg].
\end{aligned}$$

As earlier, let $\phi(x)$ and $\Phi(y)$ stand for the density and distribution function, respectively, of a standard normal random variable. We get that

(8.16)

$$\begin{aligned}
\mathbb{C} &= P_0(T^1)\int_{x_0}^{\infty}\exp\bigg(-\frac{1}{2}\int_0^T \delta^2(T^1 - t)^2\,dt + x\sqrt{\int_0^T \delta^2(T^1 - t)^2\,dt}\bigg)\phi(x)\,dx \\
&\quad - K\int_{x_0}^{\infty}\phi(x)\,dx \\
&= P_0(T^1)\int_{x_0}^{\infty}\phi\bigg(x - \sqrt{\int_0^T \delta^2(T^1 - t)^2 dt}\bigg)dx - K\int_{x_0}^{\infty}\phi(x)\,dx,
\end{aligned}$$

where x_0 is determined from the condition

$$P_0(T^1)\exp\Big(-\frac{1}{2}\int_0^T \delta^2(T^1 - t)^2\,dt + \int_0^T \delta(T^1 - t)\,d\widetilde{w}_t\Big) \geqslant K.$$

Since $\int_0^T \delta(T^1-t)\,d\widetilde{w}_t$ is a normal random variable (with respect to the measure $\mathbf{P}^*$) with parameters 0 and $\int_0^T \delta^2(T^1-t)^2\,dt$, this condition leads to the inequality

$$\xi \geqslant \frac{\ln\left(\frac{K}{P_0(T^1)}\right) + \frac{1}{2}\int_0^T \delta^2(T^1-t)^2\,dt}{\sqrt{\int_0^T \delta^2(T^1-t)^2\,dt}} = x_0,$$

where ξ is a standard normal random variable.

Setting

$$d_+ = \frac{\ln\left(\frac{P_0(T^1)}{K}\right) + \frac{1}{2}\int_0^T \delta^2(T^1-t)^2\,dt}{\sqrt{\int_0^T \delta^2(T^1-t)^2\,dt}},$$

$$d_- = \frac{\ln\left(\frac{P_0(T^1)}{K}\right) - \frac{1}{2}\int_0^T \delta^2(T^1-t)^2\,dt}{\sqrt{\int_0^T \delta^2(T^1-t)^2\,dt}},$$

we rewrite the formula (8.16) in the standard form

$$\mathbb{C} = P_0(T^1)\Phi(d_+) - K\Phi(d_-). \tag{8.17}$$

In the same way it can be shown that at the time t the value of a hedging strategy π^* can be represented in the form

$$\begin{aligned} X_t^{\pi^*} &= B(t)\,\mathbf{E}_{\mathbf{P}^*}\big[B(T)^{-1} f_T \,|\, \mathcal{F}_t\big] \\ &= P_t(T^1)\Phi(d_+(t)) - KB(t)\Phi(d_-(t)), \end{aligned} \tag{8.18}$$

where

$$d_+(t) = \frac{\ln\left(\frac{P_t(T^1)}{KB(t)}\right) + \frac{1}{2}\int_t^T \delta^2(T^1-s)^2\,ds}{\sqrt{\int_t^T \delta^2(T^1-s)^2\,ds}},$$

$$d_-(t) = \frac{\ln\left(\frac{P_t(T^1)}{KB(t)}\right) - \frac{1}{2}\int_t^T \delta^2(T^1-s)^2\,ds}{\sqrt{\int_t^T \delta^2(T^1-s)^2\,ds}}.$$

The formulas (8.17) and (8.18) are analogous to the Black–Scholes formulas for the (B,S)-market.

§8.3. Investing in a bond market

One can also consider optimal investment problems on a bond market. A methodology for solving such problems was presented in §3.3.

For some utility function U (see (3.51)) on a given (B,P)-market we consider the problem of maximizing the mean value of this function for the discounted value $Y_T^\pi = B(t)^{-1}X_T^\pi$ of a self-financing strategy π with initial value $v = X_0^\pi$.

Using Theorem 3.4 for a complete (B,P)-market, we can find a self-financing strategy π^* such that

$$\mathbf{E}[U(Y_T^{\pi^*})] = \sup_{\pi\in SF} \mathbf{E}[U(Y_T^\pi)]. \tag{8.19}$$

We set $U(x) = \ln(x)$ and find a strategy π^* for the Ho–Lee model and the Heath–Jarrow–Morton model.

The Heath–Jarrow–Morton model.

For the model determined by the equations (8.6)–(8.8) we consider the problem (8.19) under the condition that the self-financing strategy π uses only a bank account and a bond with redemption date T.

For the function

$$V(y) = \sup_{x>0}[U(x) - xy] = -\ln(y) - 1$$

conjugate to U we find that

$$v(y) = \mathbf{E}[V(yZ_T^*)] = \frac{1}{2}\mathbf{E}\left[\int_0^T \gamma_t^2\, dt\right] - 1 - \ln(y), \tag{8.20}$$

where

$$Z_t^* = \exp\left\{\int_0^t \zeta_s\, dw_s - \frac{1}{2}\int_0^t \zeta_s^2\, ds\right\}$$

is the density of the unique martingale measure $\mathbf{P}^*$.

From (8.20) we find the *price function* of the problem (8.19):

$$u(x) = \inf_{y>0}[v(y) + xy] = \ln(x) + \frac{1}{2}\mathbf{E}\left[\int_0^T \zeta_t^2\, dt\right].$$

By (3.54),

$$\begin{aligned}\frac{X_T^{\pi^*}}{B(T)} &= Y_T^{\pi^*} = I(yZ_T^*) = \frac{x}{Z_T^*}\\ &= x\exp\left\{-\int_0^t \zeta_s\, dw_s + \frac{1}{2}\int_0^t \zeta_s^2\, ds\right\},\end{aligned} \tag{8.21}$$

where

$$I(x) = (U'(x))^{-1} = -V'(y).$$

For a self-financing strategy $\pi = (\beta, \gamma(T))$ we let

$$\alpha_t(T) = \frac{\gamma_t(T)P_t(T)}{X_t^\pi}.$$

Since the evolution of the discounted value $Y_t^\pi = X_t^\pi B(t)$ for a self-financing strategy $\pi = (\beta, \gamma(T))$ is determined by the equation

$$dY_t^\pi = \gamma_t(T)d\frac{P_t(T)}{B(t)},$$

we get that

$$dY_t^\pi = \gamma_t(T)\, d\frac{P_t(T)}{B(t)} = \gamma_t(T)\frac{P_t(T)}{B(t)}a_t(T)\, d\widetilde{w}_t = Y_t^\pi \alpha_t(T)a_t(T)\, d\widetilde{w}_t.$$

Consequently,

$$\begin{aligned}(8.22)\qquad Y_T^\pi &= x\exp\left\{\int_0^T \alpha_t(T)a_t(T)\,d\widetilde{w}_t - \frac{1}{2}\int_0^T \big(\alpha_t(T)a_t(T)\big)^2\,dt\right\}\\ &= x\exp\left\{\int_0^T \alpha_t(T)a_t(T)\,dw_t\right.\\ &\qquad \left.-\int_0^T \left(\alpha_t(T)a_t(T)\zeta_t\,dt + \frac{1}{2}\big(\alpha_t(T)a_t(T)\big)^2\right)dt\right\}.\end{aligned}$$

Comparing (8.22) and (8.21), we get

$$(8.23)\qquad \alpha_t^*(T) = -\frac{\zeta_t}{a_t(T)} = \frac{b_t(T)}{a_t(T)^2},$$

and hence

$$(8.24)\qquad \gamma_t^*(T) = \frac{b_t(T)X_t^{\pi^*}}{a_t(T)^2P_t(T)}, \qquad \beta_t^* = \left(1-\frac{b_t(T)}{a_t(T)^2}\right)\frac{X_t^{\pi^*}}{B(t)}.$$

To clarify the formulas (8.23)–(8.24) obtained we observe that

$$b_t(T) = -\int_t^T \eta_s(T)\,ds + \frac{1}{2}\,a_t(T)^2.$$

Further, from (8.23),

$$(8.25)\qquad \alpha_t^*(T) = -\frac{1}{a_t(T)^2}\int_t^T \eta_s(T)\,ds + \frac{1}{2}.$$

We now set $\eta_s(T) = 0$. This corresponds to the situation when the forward rates oscillate around some value and do not have a "trend" in any particular direction. In this case we get from (8.25) that

$$\alpha_t(T) = \frac{1}{2}.$$

Thus, in this "indeterminate" situation an optimal strategy consists in investing half our resources at hand in a bond with redemption date T, and the other half in a nonrisky asset B. Corresponding to this,

$$\gamma_t^*(T) = \frac{X_t^{\pi^*}}{2P_t(T)}, \qquad \beta_t^* = \frac{X_t^{\pi^*}}{2B(t)}.$$

We note further that for the collection of bonds corresponding to a predictable process T_t specifying at each moment of time t a bond available for investment, we can use (8.14) and (8.15) to get from (8.24) a strategy $\pi' = (\beta', \gamma'(T_t))$ that represents a solution of the problem (8.19) in this case:

$$\begin{aligned}\beta_t' &= \frac{b_t(T)X_t^{\pi^*}}{a_t(T)^2B(t)}\left(1-\frac{a_t(T)}{a_t(T_t)}\right),\\ \gamma_t'(T_t) &= \frac{b_t(T)X_t^{\pi^*}}{a_t(T)a_t(T_t)P_t(T_t)}.\end{aligned}$$

The Ho–Lee model.

Let us consider the model (8.3)–(8.5). Setting $\xi_t \equiv \xi_i$ on the interval $[i, i+1)$, we get a model with continuous time and the possibility of using the previous arguments and results. Our methodology for finding martingale measures (see Chapter 3) gives us an expression for the density of a martingale measure:

$$Z_N = \mathcal{E}_N\left(-\frac{p^* - p}{p(1-p)}\sum(\xi_n - (1-p))\right). \tag{8.26}$$

We consider the problem (8.19) under the condition that the investor uses only a bank account $B(t)$ and a bond with redemption date T. This $(B, P(T))$-market is complete, and hence we can use Theorem 3.4.

In view of the form of the utility function and (8.26) we find the following expression for $v(y)$:

$$\begin{aligned}
v(y) &= \mathbf{E}[V(yZ_T)] = -1 - \mathbf{E}[\ln(yZ_T)] \\
&= -1 \quad \ln(y) - \sum_{i=1}^{T} \mathbf{E}\left[\ln\left(1 - \frac{p^* - p}{p(1-p)}(\xi_i - (1-p))\right)\right] \\
&= -1 - \ln(y) - \sum_{i=1}^{T}\left[p\ln\left(1 - \frac{p^* - p}{p(1-p)}(-(1-p))\right)\right. \\
&\qquad\qquad\qquad\qquad \left. + (1-p)\ln\left(1 - \frac{p^* - p}{p(1-p)}p\right)\right] \\
&\quad - 1 - \ln(y) - T\left(p\ln\left(\frac{p^*}{p}\right) + (1-p)\ln\left(\frac{1-p^*}{1-p}\right)\right).
\end{aligned}$$

Consequently, the price function for the problem (8.19) is equal to

$$u(x) = \ln(x) - \mathbf{E}[\ln(Z_T)] = \ln(x) - T\left(p\ln\left(\frac{p^*}{p}\right) + (1-p)\ln\left(\frac{1-p^*}{1-p}\right)\right).$$

Using (3.53), we get a formula for the discounted optimal value:

$$\frac{X_t^*(x)}{B(t)} = \frac{x}{Z_T} = x\mathcal{E}_T^{-1}\left(-\frac{p^* - p}{p(1-p)}\sum(\xi_n - (1-p))\right). \tag{8.27}$$

For an optimal strategy $\pi^* = (\beta^*, \gamma^*(T))$ let

$$\alpha_n^*(T) = \frac{\gamma_n^*(T)P_{n-1}(T)}{B(n-1)}.$$

Using the self-financing property of the strategy π^*, we have

$$\Delta\frac{X_n^*}{B(n)} = \gamma_n^*(T)\,\Delta\frac{P_n(T)}{B(n)} = \frac{X_{n-1}^*\alpha_n^*(T)}{P_{n-1}(T)}\,\Delta\frac{P_n(T)}{B(n)},$$

and by (8.5),

$$\Delta\frac{X_n^*}{B(n)} = \frac{X_{n-1}^*\alpha_n^*(T)}{B(n-1)}\left(\frac{\theta^{(T-n)\xi_n}}{p^* + (1-p^*)\theta^{T-n}} - 1\right).$$

Consequently, the discounted terminal value of the strategy π^* has the form

$$\frac{X_T^*}{B(T)} = x\,\mathcal{E}_T\left(\sum\alpha_n^*(T)\left(\frac{\theta^{(T-n)\xi_n}}{p^* + (1-p^*)\theta^{T-n}} - 1\right)\right). \tag{8.28}$$

From the formulas (8.27) and (8.28) we get that

$$\mathcal{E}_T\left(\sum \alpha_n^*(T)\left(\frac{\theta^{(T-n)\xi_n}}{p^*+(1-p^*)\theta^{T-n}}-1\right)\right)\times\mathcal{E}_T\left(-\frac{p^*-p}{p(1-p)}\sum(\xi_n-(1-p))\right)=1.$$

By an induction argument analogous to § 3.3 we prove that

$$\alpha_n^*(T)=\frac{p^*-p}{p^*(1-p^*)}\,\frac{\Theta^{T-n}-1}{p^*+(1-p^*)\Theta^{T-n}}\,.$$

The arguments given above for the Heath–Jarrow–Morton model enable us to construct an optimal strategy with a predictable process T_n specifying a bond available to the investor on the interval $[n-1, n)$.

References for Chapter 8

[**22**], [**46**], [**64**], [**74**], [**75**], [**79**], [**108**], [**126**], [**129**], [**133**], [**135**], [**141**].

CHAPTER 9

Economics of Insurance and Finance: Convergence of Quantitative Methods of Calculations

In this chapter we study problems involved with calculating premiums and reserves in insurance. It is shown how traditional actuarial methods of calculation in property insurance can be derived from the financial principle of no arbitrage. A general method is given for estimating the ruin probability as a risk characteristic of the solvency of an insurance company.

In life insurance we focus on innovation systems of "flexible" insurance, where a connection is indicated between calculations of the corresponding premiums and reserves, and the Black–Scholes formula and equation.

We present a new approach to the insurance and reinsurance of catastrophe risks by means of diversification of them on financial markets with the help of insurance derivative instruments.

§ 9.1. "Non-life" insurance. Traditional actuarial principles for calculating premiums and the financial no-arbitrage principle in a model of collective risk

The concept of *insurance* usually involves a contract whereby one party (the *insurant*) pays a certain amount of money (a *premium*) to buy from a second party (an insurance company, or *insurer*) a service consisting in indemnification by the insurer of possible losses, described in the contract, of the insurant. Here the party subject to the *risk* tries to ensure for himself a certain amount of protection, while the second party provides such protection. An insurance contract against this or that risk is called a *policy*. It specifies the conditions of the contract: the time it is in effect, the method of indemnification against damage, and so on. If during the indicated periods a corresponding risky situation arose, and the policy was presented for payment, then the insurer is said to have received a *claim*. But if no risky situation arose during the periods indicated in the policy, then the insurant loses the premium.

Thus, one of the fundamental problems of an insurer is *the calculation of the premium* for the risk-coverage service offered. For the investigation of this problem relating to "the economics of insurance" we introduce the following basic objects:

v is the *initial* value (capital) of the company;

$(\tau_n)_{n\geqslant 0}$, $\tau_0 = 0$, is a nondecreasing sequence of random variables—the *times when claims are received*;

$N_t = \sup\{n : \tau_n \leqslant t\}$ is the total *number of claims* up to the time $t \geqslant 0$;

$(Y_k)_{k\geqslant 1}$ is a sequence of nonnegative random variables determining the *size of the claims* at the times $\tau_1, \tau_2, \dots$.

The *cumulative payments* of the insurer on the time interval $[0, t]$ are called the *risk process*

$$X_t = \sum_{k=1}^{N_t} Y_k,$$

which together with the *cumulative premiums* $c(t)$ and the initial value v of the insurance company determines its *value* V at the time t (Figure 9.1):

$$V_t = v + c(t) - X_t. \tag{9.1}$$

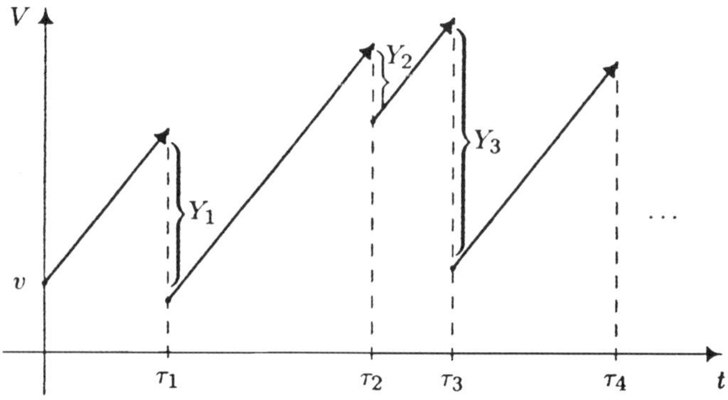

FIGURE 9.1. Typical form of the value V with linear premiums $c(t) = c \cdot t$.

Risk is understood to mean the possibility of various losses, and often the word "risk" is identified with the losses themselves. The model (9.1) introduced for the formation of capital of an insurance company is adapted to the insurance of *collective risks*, when several payments are possible according to a single policy. Therefore, it is called a *collective risk model.*

To solve the problem of "correct" calculation of premiums in insurance it is common to start from the *equivalence principle*, which amounts to the requirement that the cumulative premiums must be equal to the payments on the average:

$$\mathbf{E}c(t) = \mathbf{E}X_t. \tag{9.2}$$

It is natural to regard the process $c(t)$ as *deterministic*, and even linear: $c(t) = c \cdot t$. This reduces the problem of calculating a *linear premium* to finding the density c.

A solution of the equivalence equation (9.2) is the so-called *net premium*

$$c = T^{-1}\mathbf{E}X_T,$$

which balances the *claim* expenditures on some time interval $[0, T]$.

However, a company also has to have extra resources to ensure its proper functioning (for example, worker salaries, taxes, and so on). Therefore, a "load" on the net premium is introduced in the insurance or *actuarial* calculations, and the corresponding amount is called a *loaded premium* or *gross premium*. There are a number of traditional *actuarial principles* for determining such premiums.

1. Principles involving the concept of net premium:
   (a) *the mathematical expectation principle*,
$$c = T^{-1}(\mathbf{E}X_T + \alpha\mathbf{E}X_T), \qquad \alpha > 0;$$
   (b) *the variance principle*,
$$c = T^{-1}(\mathbf{E}X_T + \alpha\mathbf{D}X_T), \qquad \alpha > 0;$$
   (c) *the standard deviation principle*,
$$c = T^{-1}(\mathbf{E}X_T + \alpha\sqrt{\mathbf{D}X_T}\,), \qquad \alpha > 0.$$
2. *The Esscher principle*,
$$c = T^{-1}\mathbf{E}X_T e^{\alpha X_T}/\mathbf{E}e^{\alpha X_T}, \qquad \alpha > 0.$$
3. *The principle of calculating premiums on the basis of the concept of utility*,
$$U(v - cT) = \mathbf{E}U(v - X_T),$$
   where U is a convex utility function.

In the third case we get
$$U(v - cT) = \mathbf{E}U(v - X_T) \leqslant U(\mathbf{E}(v - X_T)) \implies cT \geqslant \mathbf{E}X_T$$
by Jensen's inequality, and hence the loaded premium is correctly determined. Furthermore, in the first case the choice of load is completely obvious and natural: the load is determined either as the αth portion of the mean claim, or as the αth portion of the variance and mean-square deviation, which are often convenient quantitative measures of risk. In the second case we set
$$Z = e^{\alpha X_T}/\,\mathbf{E}e^{\alpha X_T}$$
and note that
$$cT = \mathbf{E}X_T Z = \mathbf{E}_{\mathbf{Q}} X_T,$$
where $\mathbf{Q}$ is the measure with density Z. Hence, the Esscher principle is also an equivalence principle, but with respect to a specially chosen probability measure $\mathbf{Q}$.

As is clear from the preceding, there is no unified method for calculating premiums in traditional insurance. This is explained by the history and practice of insurance. Another explanation can be found in the theory of financial markets, in the interconnection between finance and insurance. Central in financial economics is the concept of a financial market with a chain of "bought-sold" actions, and so on. Although initially this concept was really untypical for insurance, in the course of time many of its forms have been characterized by a large number of assets and competitions, and this forms a certain model of the insurance market on the basis of a *system of reinsurance* with a natural chain of "insured-reinsured" actions, and so on.

Reinsurance appeared in connection with the insurance of very large risks. A basic and essentially systematizing form of reinsurance of risks is the so-called quota-share *proportional reinsurance*, when the αth portion of risk is assumed by an "initial" company (the *cedent*), with the remaining $(1-\alpha)$th portion of the risk passed on to a reinsurer and with the premiums divided in the same proportion. Another representative of proportional reinsurance is *excess of surplus*: the insurer assumes the risk of total losses up to a certain limit A (*the responsibility level of the company*), the reinsurer assumes the risk from A up to a certain level B, and all losses above B revert again to the insurer. In this form of reinsurance the *layer* A/B is reinsured. A traditional *nonproportional reinsurance* contract is an *excess of loss* contract, in which one considers not the total of payments according to the contract as above, but the individual payments according to each insurance event. Here A is called the *priority* and B the *ceiling* of guarantees.

In nonproportional reinsurance the interrelation between the premiums and the payments is more complicated. We mention also *stop loss*, when insurance protection is implemented not according to each claim, but according to the results of activities over, for example, a year. There are also other methods of reinsurance, but they all can be derived from those above (and even from quota-share proportional reinsurance) with the help of various recurrence relations.

Suppose that the sphere of insurance is characterized by the possibility at any time of concluding a new contract involving insurance and reinsurance, and cancelling the old contract, which means the existence of a certain price of the risk and its liquidity.

Let $(\Omega, \mathcal{F}, \mathbb{F} = (\mathcal{F}_t)_{t\geqslant 0}, \mathbf{P})$ be a standard stochastic base on which we are given a predictable process C_t and an adapted process X_t characterizing the premiums and payments of an insurance company up to a time $t \leqslant T$. The risk connected with the process $(X_t)_{t\in[0,T]}$ is the object of the insurance (the premiums C_t are paid for this) and reinsurance.

In connection with the latter we take as given a bounded predictable process ϕ_t characterizing the portion of risk $(X_u)_{u\in[t,T]}$. At any time t the insurance company can decide to buy or sell this portion of the risk. The process $(\phi_t)_{t\leqslant T}$ will be called a reinsurance *policy*, which determines the following flow of capital:

$$C_t(\phi) = \int_0^t \phi_s\, dX_s - \int_0^t \phi_s\, dC_s, \qquad t \leqslant T. \tag{9.3}$$

It is assumed that the corresponding integrals in (9.3) are defined.

The policy ϕ is called an *arbitrage* policy if $C_T(\phi) \geqslant 0$ and $\mathbf{P}(C_T(\phi) > 0) > 0$. The processes (C, X, ϕ) form a *reinsurance market.*

As in the case of financial markets, we have the following characterization.

On a reinsurance market there do not exist arbitrage policies $\Longleftrightarrow$ *there exists a measure* $\widetilde{\mathbf{P}}$ *equivalent to* $\mathbf{P}$ *such that the difference* $X - C$ *is a martingale with respect to* $\widetilde{\mathbf{P}}$.

The above analogy between reinsurance markets and financial markets allows us to use *martingale* arguments in the calculus of premiums (for risk) in insurance. We look for a martingale measure $\widetilde{\mathbf{P}}$ converting the difference $X - C$ into a martingale. Then averaging with respect to $\widetilde{\mathbf{P}}$ leads to the determination of the premium itself. To implement this approach we assume that in the collective risk model (9.1) $(N_t)_{t\geqslant 0}$ is a standard Poisson process with intensity $\lambda > 0$, the sequence $(Y_k)_{k\geqslant 1}$ of independent nonnegative random variables is independent of the Poisson process and has distribution function $F(x)$, and $c(t) = c \cdot t$. Under these conditions (9.1) is commonly called the *Cramér–Lundberg model.* In this framework we give a characterization of all martingale measures $\widetilde{\mathbf{P}}$ with respect to which $(N_t)_{t\geqslant 0}$ remains a Poisson process, perhaps with a different intensity.

Let $\beta\colon \mathbb{R}_+ \to \mathbb{R}$ be a measurable function such that $\mathbf{E}e^{\beta(Y_1)} < \infty$. We define

$$M_t^\beta = \exp\biggl\{\sum_{k=1}^{N_t} \beta(Y_k) - \lambda t\mathbf{E}[e^{\beta(Y_1)} - 1]\biggr\}, \qquad t \leqslant T.$$

LEMMA 9.1. *The positive process M_t^β is a martingale with respect to the original measure* $\mathbf{P}$, *and* $\mathbf{E}M_t^\beta = 1$, $t \leqslant T$.

PROOF. Let $X_t^\beta = \sum_{k=1}^{N_t} \beta(Y_k)$ and observe that X_t^β is a process with independent increments. Consequently, for any $0 \leqslant s \leqslant t$ we get that

$$\mathbf{E_P}\bigl[M_t^\beta \mid \mathcal{F}_s\bigr] = M_s^\beta \mathbf{E_P} \exp\bigl\{X_t^\beta - X_s^\beta\bigr\} \cdot \exp\bigl\{-\lambda(t-s)\mathbf{E_P}[e^{\beta(Y_1)} - 1]\bigr\} = M_s^\beta,$$

which proves the lemma.

LEMMA 9.2. *If* $\mathbf{Q}$ *and* $\mathbf{P}$ *are two equivalent probability measures on* $(\Omega, \mathcal{F}, \mathbb{F})$, *then the corresponding probability distributions* $\mathbf{Q}_{Y_1}$ *and* $\mathbf{P}_{Y_1}$ *of the random variable* Y_1 *are equivalent.*

PROOF. Let A be a Borel set in $\mathbb{R}_+$ such that $\mathbf{P}_{Y_1}(A) = 0$. Then

$$\mathbf{Q}_{Y_1}(A) = \mathbf{Q}\bigl(X_{\tau_1}^{-1}(A)\bigr) = \lim_{n\to\infty} \mathbf{Q}\bigl(X_{\tau_1}^{-1}(A) \cap (\tau_1 \leqslant n)\bigr),$$

since τ_1 is finite. Further, $X_{\tau_1}^{-1}(A) \cap (\tau_1 \leqslant n) \in \mathcal{F}_n$ and $\mathbf{Q}(X_{\tau_1}^{-1}(A) \cap (\tau_1 \leqslant n)) = 0$, and hence $\mathbf{Q}_{Y_1}(A) = 0$.

The next lemma gives a characterization of all martingale measures in the Cramér–Lundberg model.

LEMMA 9.3. *Let* $\mathbf{Q}$ *be a measure on* $(\Omega, \mathcal{F}, \mathbb{F})$ *such that*:

1) $\mathbf{Q}$ *and* $\mathbf{P}$ *are locally equivalent*;

2) $(X_t)_{t\geqslant 0}$ *is a* $\mathbf{Q}$-*compound Poisson process.*

Then there is a measurable function $\beta\colon \mathbb{R}_+ \to \mathbb{R}$ *such that* $\mathbf{E}e^{\beta(Y_1)} < \infty$ *and*

$$\mathbf{Q}(A) = \int_A M_t^\beta \, d\mathbf{P} \quad \textit{for all} \quad A \in \mathcal{F}_s, \quad 0 \leqslant s \leqslant t. \tag{9.4}$$

Conversely, if there exists such a measurable function $\beta\colon \mathbb{R}_+ \to \mathbb{R}$, *then there exists a unique probability measure* $\mathbf{Q}$ *that is defined by* (9.4) *and has the properties* 1) *and* 2).

PROOF. Consider a measure $\mathbf{Q}$ satisfying the conditions 1) and 2). It follows from 2) that $(N_t)_{t\geqslant 0}$ is a Poisson process on $(\Omega, \mathcal{F}, \mathbb{F}, \mathbf{Q})$ with some intensity $\lambda' > 0$. Setting $\alpha = \log \lambda' - \log \lambda$, we have

$$\mathbf{Q}(N_t = n) = \mathbf{P}(N_t = n)e^{n\alpha - \lambda(e^\alpha - 1)t}.$$

By Lemma 9.2, $\mathbf{P}_{Y_1}$ and $\mathbf{Q}_{Y_1}$ are equivalent. Therefore, there is a strictly positive density function f such that

$$\mathbf{Q}_{Y_1}(A) = \int_A f \, d\mathbf{P}_{Y_1} \quad \text{for all} \quad A \in \mathcal{F}.$$

Let $\gamma(x) = \log f(x)$.

If $s > 0$, then for $n \in \mathbb{N}_0 = \{0, 1, \dots\}$

$$Y_n \cap (N_s = n) \subset \sigma(Y_1, \dots, Y_n) \cap (N_s = n).$$

For $n = 0$ the right-hand side of this imbedding is trivial: $\{\varnothing, \Omega\} \cap (N_s = 0)$. Let $A \in \mathcal{F}_s$ be arbitrary. Then for each $n \in \mathbb{N}_0$ there exist $B_n \in \sigma(Y_1, \dots, Y_n)$ such that

$$\begin{aligned}\mathbf{Q}(A) &= \sum_{n\geqslant 0} \mathbf{Q}\big(A \cap (N_s = n)\big) \\ &= \sum_{n\geqslant 0} \mathbf{Q}\big(B_n \cap (N_s = n)\big) \\ &= \sum_{n\geqslant 0} \mathbf{Q}(B_n)\mathbf{Q}(N_s = n).\end{aligned}$$

Let $B_n = Y_1^{-1}(C_1) \cap \dots \cap Y_n^{-1}(C_n)$, where $C_1, \dots, C_n$ are Borel subsets of $\mathbb{R}_+$. Then

$$\begin{aligned}\mathbf{Q}(B_n) &= \mathbf{E_Q}\big[I_{C_1}(Y_1) \cdots I_{C_n}(Y_n)\big] = \mathbf{E_Q}[I_{C_1}(Y_1)] \cdots \mathbf{E_Q}[I_{C_n}(Y_n)] \\ &= \mathbf{E_P}\big[I_{C_1}(Y_1)e^{\gamma(Y_1)}\big] \cdots \mathbf{E_P}\big[I_{C_n}(Y_n)e^{\gamma(Y_n)}\big] \\ &= \mathbf{E_P}\big[I_{B_n} \exp\{\gamma(Y_1) + \dots + \gamma(Y_n)\}\big].\end{aligned}$$

The latter can be extended in the standard way to any $B_n \in \sigma(Y_1, \dots, Y_n)$. Consequently,

$$\begin{aligned}Q(A) &= \sum_{n\geqslant 0} \mathbf{E_P}\big[I_{B_n} \exp\{\gamma(Y_1) + \dots + \gamma(Y_n)\}\big] \times \mathbf{Q}(N_s = n) \\ &= \sum_{n\geqslant 0} \mathbf{E_P}\big[I_{B_n} \exp\{\gamma(Y_1) + \dots + \gamma(Y_n)\}\big] \\ &\qquad \times \mathbf{P}(N_s = n) \exp\{\alpha n - \lambda(e^\alpha - 1)t\} \\ &= \sum_{n\geqslant 0} \mathbf{E_P}\big[I_{B_n \cap (N_s = n)} \exp\big\{\gamma(Y_1) + \dots + \gamma(Y_n) + \alpha n - \lambda(e^\alpha - 1)t\big\}\big] \\ &= \sum_{n\geqslant 0} \mathbf{E_P}\left[I_{A \cap (N_s = n)} \exp\left\{\sum_{i=1}^{N_s} (\alpha + \gamma(Y_i)) - \lambda(e^\alpha - 1)t\right\}\right] \\ &= \int_A \exp\left\{\sum_{i=1}^{N_s} (\alpha + \gamma(Y_i)) - \lambda(e^\alpha - 1)t\right\} d\mathbf{P}.\end{aligned}$$

With the notation $\beta(x) = \alpha + \gamma(x)$ we have that $\mathbf{E_P}[e^{\beta(Y_1)}] = e^\alpha$ and $\mathbf{Q}(A) = \int_A M_s^\beta \, d\mathbf{P}$. Using the fact that $(M_t^\beta)_{t\geqslant 0}$ is a martingale, we obtain the equality

$$\mathbf{Q}(A) = \int_A M_t^\beta \, d\mathbf{P} \quad \text{for} \quad 0 \leqslant s \leqslant t,$$

which proves the direct implication.

To prove the reverse implication we assume that $\beta(x)$ is a real function on $\mathbb{R}_+$ with $\mathbf{E_P}[\exp\{\beta(Y_1)\}] < \infty$. Since $(M_t^\beta)_{t\geqslant 0}$ is a martingale, the equality (9.4) defines a measure $\mathbf{Q}$ on the algebra $\bigcup_{t\geqslant 0} \mathcal{F}_t$, and its restriction to each $\mathcal{F}_t$ is a probability measure.

We show that for any fixed $t > 0$ the random process $(X_u)_{u \leqslant t}$ is a stationary process with independent increments. To prove this we verify that for any $r > 0$ and $0 \leqslant u \leqslant s \leqslant t$ the conditional expectation $\mathbf{E_Q}[e^{-r(X_s - X_u)} \mid \mathcal{F}_u]$ is deterministic and depends only on the difference $s - u$.

Let $A \in \mathcal{F}_u$. Then

$$\begin{aligned}
&\mathbf{E_Q}\big[I_A \exp\{-r(X_s - X_u)\}\big] \\
&\quad = \mathbf{E_P}\big[I_A \exp\{-r(X_s - X_u)\} M_s^\beta\big] \\
&\quad = \mathbf{E_P}\big[I_A \exp\{-r(X_s - X_u) + X_s^\beta\}\big] \times \exp\{-\lambda s \mathbf{E_P}[e^{\beta(Y_1)} - 1]\} \\
&\quad = \mathbf{E_P}\big[I_A \exp\{X_u^\beta\}\big] \mathbf{E_P}\big[I_A \exp\{-r(X_s - X_u) + (X_s^\beta - X_u^\beta)\}\big] \\
&\qquad\qquad \times \exp\{-\lambda s \mathbf{E_P}[e^{\beta(Y_1)} - 1]\} \\
&\quad = \mathbf{E_P}\big[I_A M_u^\beta\big] \mathbf{E_P}\big[I_A \exp\{-r(X_s - X_u) + (X_s^\beta - X_u^\beta)\}\big] \\
&\qquad\qquad \times \exp\{-\lambda(s-u) \mathbf{E_P}[e^{\beta(Y_1)} - 1]\} \\
&\quad = \mathbf{Q}(A) \mathbf{E_P}\big[\exp\{-r X_{s-u} + X_{s-u}^\beta\}\big] \times \exp\{-\lambda(s-u) \mathbf{E_P}[e^{\beta(Y_1)} - 1]\}.
\end{aligned}$$

We consider two probability measures μ and ν on $\mathbb{R}_+$ defined on Borel sets $A \in \mathbb{R}_+$ by

$$\begin{aligned}
\mu(A) &= \int_A \frac{\exp\{\beta(x)\}}{\mathbf{E_P}[\exp\{\beta(Y_1)\}]}\, d\mathbf{P}_{Y_1}, \\
\nu(A) &= \lambda' \int_A e^{-\lambda' x}\, dx,
\end{aligned}$$

where $\lambda' = \lambda \mathbf{E_P}[\exp\{\beta(Y_1)\}]$.

Let $\widetilde{\mathbf{Q}} = \nu^{\otimes N} \otimes \mu^{\otimes N}$, as a product measure on $\Omega = \mathbb{R}_+^N \times \mathbb{R}_+^N$. It follows at once from the structure of $\widetilde{\mathbf{Q}}$ that $(X_t)_{t \geqslant 0}$ is a compound Poisson process with respect to $\widetilde{\mathbf{Q}}$. We show that the measures $\mathbf{Q}$ and $\widetilde{\mathbf{Q}}$ coincide on $\mathcal{F}_t$ for all $t \in \mathbb{R}_+$. Since $(X_s)_{s \leqslant t}$ has independent increments on the spaces $(\Omega, \mathcal{F}_t, \mathbf{Q})$ and $(\Omega, \mathcal{F}_t, \widetilde{\mathbf{Q}})$, it suffices to establish that for all Borel sets $A \in \mathbb{R}_+$ and all $s \leqslant t$

$$\mathbf{Q}_{X_s}(A) = \widetilde{\mathbf{Q}}_{X_s}(A).$$

To this end we show that for any $r > 0$

$$\mathbf{E_Q}[e^{-rX_s}] = \mathbf{E}_{\widetilde{\mathbf{Q}}}[e^{-rX_s}]. \tag{9.5}$$

From the equalities

$$\begin{aligned}
\mathbf{E}_{\widetilde{\mathbf{Q}}}[e^{-rY_1}] &= \frac{\mathbf{E_P}[\exp\{-rY_1 + \beta(Y_1)\}]}{\mathbf{E_P}[\exp\{\beta(Y_1)\}]}, \\
\widetilde{\mathbf{Q}}(N_s = n) &= \mathbf{E_P}[\exp\{\beta(Y_1)\}]^n \\
&\qquad \times \exp\big\{-\lambda s \mathbf{E_P}[\exp\{\beta(Y_1)\} - 1]\big\} \times \mathbf{P}(N_s = n)
\end{aligned}$$

it follows that

$$\begin{aligned}
\mathbf{E}_{\widetilde{\mathbf{Q}}}[e^{-rX_s}] = \sum_{n \geqslant 0} & \mathbf{E_P}[\exp\{-rY_1 + \beta(Y_1)\}]^n \\
& \times \exp\big\{-\lambda_s \mathbf{E_P}[\exp\{\beta(Y_1)\} - 1]\big\} \times \mathbf{P}(N_s = n).
\end{aligned}$$

We observe now that

$$\mathbf{E}_{\mathbf{Q}}[e^{-rX_s}] = \sum_{n\geqslant 0} \mathbf{E}_{\mathbf{P}}[\exp\{-rY_1 + \beta(Y_1)\}]^n \times \exp\bigl\{-\lambda_s \mathbf{E}_{\mathbf{P}}[\exp\{\beta(Y_1)\} - 1]\bigr\} \times \mathbf{P}(N_s = n),$$

and we get (9.5).

Further, the measures $\mathbf{Q}$ and $\widetilde{\mathbf{Q}}$ coincide on $\bigcup_{t\geqslant 0} \mathcal{F}_t$, and hence $\widetilde{\mathbf{Q}}$ is the unique extension of $\mathbf{Q}$ to the σ-algebra $\mathcal{F}$.

Lemma 9.3 is proved.

We turn to the calculation of premiums.

THEOREM 9.1. *The martingale approach to calculating the premium in the Cramér–Lundberg model leads to the following family of premiums*:

$$c = \lambda \mathbf{E}_{\mathbf{P}}[Y_1 \exp\{\beta(Y_1)\}],$$

where β is the function in Lemma 9.3.

PROOF. From $\beta(\,\cdot\,)$ we construct a measure $\mathbf{Q}$ with respect to which $X_t - c\cdot t$ is a martingale. Then it is clear that $c = \mathbf{E}_{\mathbf{Q}} X_1$. With respect to $\mathbf{Q}$ the process $(X_t)_{t\geqslant 0}$ is again a composite Poisson process; therefore, the expectation factors into the product $\mathbf{E}_{\mathbf{Q}} X_T = \mathbf{E}_{\mathbf{Q}} N_1 \cdot \mathbf{E}_{\mathbf{Q}} Y_1$. As is clear from the proof of Lemma 9.3,

$$\mathbf{E}_{\mathbf{Q}} N_1 = \lambda \mathbf{E}_{\mathbf{P}}[\exp\{\beta(Y_1)\}] \quad \text{and} \quad \mathbf{E}_{\mathbf{Q}} Y_1 = \frac{\mathbf{E}_{\mathbf{P}}[Y_1 \exp\{\beta(Y_1)\}]}{\mathbf{E}_{\mathbf{P}}[\exp\{\beta(Y_1)\}]}.$$

Consequently, their product is equal to $c = \lambda \mathbf{E}_{\mathbf{P}}[Y_1 \exp\{\beta(Y_1)\}]$, which concludes the proof of the theorem.

From the martingale principle we can get all the traditional methods for calculating premiums in insurance with the help of the function $\beta(x)$.

1. *Mathematical expectation principle.* We let $\beta(x) = \ln(1+\alpha)$ and get that

   $$c = \lambda \mathbf{E}_{\mathbf{P}}[Y_1 e^{\beta(Y_1)}] = \lambda \mathbf{E}_{\mathbf{P}}[Y_1(1+\alpha)] = \mathbf{E}_{\mathbf{P}} X_1 (1+\alpha),$$

   and hence

   $$c\cdot T = \mathbf{E}_{\mathbf{P}} X_T (1+\alpha).$$

2. *Variance principle.* We consider $\beta(x) = \ln(a + bx)$, $b > 0$, and $a = 1 - b\mathbf{E}_{\mathbf{P}} Y_1 > 0$ and get that

   $$c = \lambda \mathbf{E}_{\mathbf{P}}[Y_1(a + bY_1)] = \lambda \mathbf{E}_{\mathbf{P}} Y_1 a + \lambda \mathbf{E}_{\mathbf{P}} Y_1^2 b = a\mathbf{E}_{\mathbf{P}} X_1 + b\mathbf{D}_{\mathbf{P}} X_1.$$

   Hence,

   $$c\cdot T = a\mathbf{E}_{\mathbf{P}} X_T + b\mathbf{D}_{\mathbf{P}} X_T,$$

   which for $a = 1$ and $b = \alpha$ reduces to the variance principle.

3. *Esscher principle.* For $\beta(x) = \alpha x - \ln \mathbf{E}_{\mathbf{P}} e^{\alpha Y_1}$ we have that

   $$c = \lambda \mathbf{E}_{\mathbf{P}} \frac{Y_1 e^{\alpha Y_1}}{\mathbf{E}_{\mathbf{P}} e^{\alpha Y_1}},$$

   and hence

   $$c\cdot T = \lambda T \mathbf{E}_{\mathbf{P}} \frac{Y_1 e^{\alpha Y_1}}{\mathbf{E}_{\mathbf{P}} e^{\alpha Y_1}} = \mathbf{E}_{\mathbf{P}} N_T \mathbf{E}_{\mathbf{P}} \frac{Y_1 e^{\alpha Y_1}}{\mathbf{E}_{\mathbf{P}} e^{\alpha Y_1}}.$$

On the other hand, the "Esscher" premium has the form

$$c \cdot T = \frac{\mathbf{E_P} X_T e^{\alpha X_T}}{\mathbf{E_P} e^{\alpha X_T}} = \mathbf{E_P} \sum_{k=1}^{N_T} \frac{Y_k e^{\alpha Y_k}}{\mathbf{E_P} e^{\alpha Y_k}} = \mathbf{E_P} N_T \mathbf{E_P} \frac{Y_1 e^{\alpha Y_1}}{\mathbf{E_P} e^{\alpha Y_1}}.$$

Comparison of these two formulas leads to the required assertion.

Thus, it is established that in the Cramér–Lundberg model there is a whole series of martingale measures, which means, on the one hand, the no-arbitrage property and, on the other hand, incompleteness of the corresponding market, if we may use the "financial" terminology. This is a theoretical explanation for the nonuniqueness of the methods for determining insurance premiums. The nonuniqueness leads also to the important practical conclusion that there are risks not suitable for insurance.

§ 9.2. Life insurance. Mortality tables. Calculation of premiums and reserves in traditional and innovation insurance schemes

Along with property insurance, or "non-life" insurance, a second very important component is *life insurance*, where the corresponding specifics of the calculus of premiums and reserves stem from the *mortality* of the company's insured clients. All calculations in traditional life insurance are carried out on the basis of the "demographic history" of the insurance event: sickness, death, and so on. For these purposes one uses *mortality tables* giving a general idea about the probabilities of such events.

Mortality tables are statistical data about the longevity of a population group, categorized according to different criteria (gender, region, profession, etc.). They contain a number of variables and indicators characterizing the decrease in numbers of the group under observation and the mortality level for different periods. Tables based only on age give an idea of the mortality dynamics as a whole and are thus called *general* tables. If other parameters besides age are taken into account, then the tables are called *special* (*selective*) tables.

Key indicators of mortality tables are the probability q_x of someone of age x dying within a year and the probability p_x of someone of age x living through the year. They are often determined using the *direct method*, when q_x is taken as the ratio of the number of members of an observed group who died during a year to the number who were alive at the beginning of the year. However, the difficulties of selecting a homogeneous group and compiling data for 100 years as it evolves (and this is the approximate size of the table) lead to essential distortions of the reality. In practice, therefore, statistical data are gathered about a group of contemporaries including people of all ages and giving indicators for a complete age range (*cross-sectional analysis*). The composition of the age groups is very dynamic, while in a mortality table the number of people must decrease monotonically, and thus the tables are smoothed and completed to fill out a whole category of ages.

The mortality table obtained accumulates in itself the general probabilistic reality in which insurance companies operate, and it provides a certain physical measure of mortality. It is natural to call such "first approximation" tables *tables of the first kind*. However, real calculations of premiums and reserves done by qualified actuaries are carried out according to different, *actuarial*, tables which will be called *tables of the second kind*. The fact of the matter is that each insurance company, using its own experience, specifics, statistics, and so on, transforms the original mortality table. It is usually taken into account that not all people are insured in a given category (an age category, for example) for which the mortality is well known from tables of the first kind, but only people with a larger "data base" (security, passing a medical examination, and so on). As a result, a different more suitable mortality table emerges that provides the corresponding probability with risk-neutral properties.

In practice one can often pass from one table to another by a linear transformation of the form $q^{(\mathrm{II})} = Aq^{(\mathrm{I})} + B$. For example, $A = 0.7$ for Moscow, and hence the premium calculated from the probability measure induced by such a table will be roughly 30% less than that calculated from the original measure. Thus, in both life insurance and financial economics there is the same process of looking for measures more "inconvenient" for risk and using them in actuarial calculations.

In the traditional approach the problem of insurance is to determine the premium at the conclusion of the contract (a *one-time* premium if it is to be paid immediately upon concluding the contract, and a *periodic* premium if it is to be paid according to a chosen scheme over the period when the contract is in effect). Calculation of a one-time premium uses the equivalence principle, when the premiums are equated to the future payments according to the insurance contract, on the average with respect to the "physical" probability given by the mortality table of the first kind.

Suppose that the insurance company concludes a "*pure endowment*" contract with a group of ℓ_x clients of age x for a time T. The corresponding policy gives the holder the right to receive a payment in the amount X upon living to the date T.

To calculate a one-time individual premium ${}_TU_x$ according to such an agreement we denote by $T_1, \dots, T_{\ell_x}$ the remaining lifetimes of the members of this group as independent identically distributed variables. Then for each $i = 1, \dots, \ell_x$ the payments of the company are $X \cdot I_{\{T_i > T\}}$.

By the equivalence principle,

$$ {}_TU_x = \mathbf{E}X \cdot I_{\{T_i > T\}} = X \cdot {}_Tp_x, \tag{9.6} $$

where ${}_Tp_x = \mathbf{P}(T_i > T)$, $i = 1, \dots, \ell_x$.

Taking into account the nonzero interest rate r leads to the discounted size of the payments and to the corresponding formula

$$ {}_TU_x = e^{-rT} X \cdot {}_Tp_x. \tag{9.7} $$

For a *term insurance* agreement, when a payment of amount X is made in case of the death of the client before the date T, the corresponding formula for the premium $U(x, T)$ follows from (9.6)–(9.7):

$$ U(x,T) = (1 - {}_Tp_x)X \quad \text{and} \quad U(x,T) = (1 - {}_Tp_x)e^{-rT}X. \tag{9.8} $$

The price formulas (9.6)–(9.8) were obtained under the assumption that X is deterministic, and they take into account the unique randomness factor stemming from the mortality of people in the selected group. We show how they can be transformed for contracts with payments depending on the prices S_t of a risky asset in a Black–Scholes financial market (3.19) given, as usual, on a coordinate space $(\Omega^1, \mathcal{F}^1, \mathbb{F}^1, \mathbf{P}^1)$.

Let $N_t = \sum_{i=1}^{\ell_x} I_{\{T_i \leqslant t\}}$ be a *mortality process* on another space $(\Omega^2, \mathcal{G}^2, \mathbb{F}^2, \mathbf{P}^2)$ with $\mathbb{F}^2 = (\mathcal{F}_t^2)$ and $\mathcal{F}_t^2 = \sigma(N_u,\, u \leqslant t)$. We assume that

$$_tp_x = \mathbf{P}^2(T_i > t) = \exp\left\{-\int_0^t \mu_{x+s}\,ds\right\}$$

with some density μ_{x+s} (*the mortality rate*).

It is reasonable to assume that the two randomness factors are independent of each other. Then the whole model belongs on the probability space $(\Omega, \mathcal{F}, \mathbb{F}, \mathbf{P}) = (\Omega^1 \times \Omega^2, \mathcal{F}^1 \otimes \mathcal{F}^2, \mathbb{F}^1 \otimes \mathbb{F}^2, \mathbf{P}^1 \times \mathbf{P}^2)$.

Let us consider the pure endowment model with payment $X = \max\{S_T, K\}$, where K is a nonnegative constant that is the distinctive *guaranteed* payment for the insurant. We find a formula for the premium $_TU_x$ starting from the equivalence principle with respect to the martingale measure $\mathbf{P}^*$ in the Black–Scholes model. Such a change of measure in the equivalence principle corresponds completely to the actuarial practice described above when one goes from general mortality tables to actuarial tables.

From the property of independence we have that

$$_TU_x = {_Tp_x} \cdot \mathbf{E}^* e^{-rT} \max\{S_T, K\}. \tag{9.9}$$

Since

$$\max\{S_T, K\} = K + \max\{S_T - K, 0\}, \tag{9.10}$$

we get from (9.9) and the Black–Scholes formula (3.23) that

$$_TU_x = {_Tp_x}\left[Ke^{-rT} + S_0\Phi(d_+(0)) - Ke^{-rT}\Phi(d_-(0))\right], \tag{9.11}$$

where $d_\pm(t) = \dfrac{\ln S_0/K + (T-t)(r \pm \sigma^2/2)}{\sigma\sqrt{T-t}}$. Similarly, in the case of a term contract the individual premium is

$$U(x, T) = \int_0^T \left[S_0\Phi(d_+(t)) - Ke^{-rt}\Phi(d_-(t))\right] {_tp_x}\mu_{x+t}\,dt. \tag{9.12}$$

A periodic premium is characterized by its density $p(t)$. Balancing premiums leads to the relation

$$_TU_x = \int_0^T p(t)e^{-rt} \cdot {_tp_x}\,dt. \tag{9.13}$$

It is clear that at each moment of time t the insurance company must have some *reserve capital* to cover claims that arise. We define the *reserves* $V(t)$ to be the conditional expectation of the difference between the discounted (up to the time t) payments and the future premiums. Conditional *averaging* is necessary because neither future claims nor future receipts are known.

In the case of a pure endowment contract the payment for a claim is $X = \max\{S_T, K\}$. By (9.13), taking the average with respect to the risk-neutral measure $\mathbf{P}^*$ gives us

$$V(t) = {_{T-t}p_{x+t}}\mathbf{E}^*(X \mid \mathcal{F}_t^1) - \int_t^T p(u)e^{-r(u-t)} \cdot {_{u-t}p_{x+t}}\,du. \tag{9.14}$$

Next, we let $\pi_t^*(T) = \mathbf{E}^*(X \mid \mathcal{F}_t)e^{-rt}$ and rewrite (9.14) in the form

$$V(t) = {}_{T-t}p_{x+t}\pi_t^*(T)e^{rt} - \int_t^T p(u)e^{-r(u-t)} \cdot {}_{u-t}p_{x+t}\, du. \tag{9.15}$$

Letting $\psi(t) = e^{-rt}/{}_{T-t}p_{x+t}$ and assuming that π_t^*, $V(t)$, and ${}_{u-t}p_{x+t}$ are sufficiently smooth, we get that

$$\begin{aligned}\frac{\partial\, {}_{u-t}p_{x+t}}{\partial t} &= \mu_{x+t} \cdot {}_{u-t}p_{x+t}, \quad u \geqslant t,\\ \frac{\partial}{\partial t}\psi(t) &= -(\mu_{x+t} + r)\psi(t),\\ \frac{\partial \pi_t^*}{\partial S} = \psi(t)\frac{\partial V}{\partial S}, &\quad \frac{\partial^2 \pi_t^*}{\partial S^2} = \psi(t)\frac{\partial^2 V}{\partial S^2},\\ \frac{\partial \pi_t^*}{\partial t} &= \psi(t)\left(\frac{\partial V}{\partial t} - (\mu_{x+t} + r)V(t) - p(t)\right).\end{aligned}$$

From (9.15),

$$\pi_t^*(T) = \psi(t)\left[V(t) + \int_t^T p(u)e^{-r(u-t)} \cdot {}_{u-t}p_{x+t}\, du\right]. \tag{9.16}$$

Using (9.16), the representation

$$dS_t = rS_t\, dt + \sigma S_t\, d\left(W_t + \frac{\mu - r}{\sigma}t\right) = rS_t\, dt + \sigma S_t\, dW_t^*,$$

and the Kolmogorov–Itô formula, we get that for any $s \geqslant t$

$$\begin{aligned}\pi_s^*(T) = \pi_t^*(T) &+ \int_t^s \psi(u)\frac{\partial V}{\partial S}\sigma S_u\, dW_u^*\\ &+ \int_t^s \psi(u)\left[\frac{\partial V}{\partial S}rS_u + \frac{1}{2}\sigma^2 S_u^2\frac{\partial^2 V}{\partial S^2} - (\mu_{x+u} + r)V(u) + \frac{\partial V}{\partial u} - p(u)\right] du.\end{aligned} \tag{9.17}$$

Since $\pi_t^*(T)$ is a martingale with respect to the martingale measure $\mathbf{P}^*$, we have that the third term on the right-hand side of (9.17) must be zero, and we arrive at a *generalization* of *Thiele's equation* for the reserves $V(t)$, which is well known in the economics of insurance:

$$\frac{\partial V}{\partial t} = p(t) + (\mu_{x+t} + r)V(t) - \frac{1}{2}\sigma^2 S_t^2\frac{\partial^2 V}{\partial S^2} - rS_t\frac{\partial V}{\partial S}. \tag{9.18}$$

When the reserves are independent of S_t, (9.18) passes into *Thiele's classical equation*:

$$\frac{\partial V}{\partial t} = p(t) + (\mu_{x+t} + r)V(t), \tag{9.19}$$

which describes the dynamics of the reserves of an insurance company in dependence on the premiums being received, the mortality rate, and the interest rate.

For a long time (and the equation (9.19) was discovered in 1875) this was sufficient for the problems that arose in life insurance. However, the intensive development of financial markets has had a strong influence on the structure of insurance products: among them have appeared a whole spectrum of equity-linked innovation proposals. Thus, the corresponding reserves have gotten another risk measurement connected with risky assets of the financial market, and the equation

(9.18) takes this into account. Furthermore, if we formally set the parameters $p(t)$ and μ_{x+t} (which are not characteristic for the "financial" theory) equal to zero, then (9.18) reduces to the Black–Scholes equation (3.27).

Thus, the above results on calculating premiums and reserves clearly attest to the unity of the methodological foundations of these calculations in finance and insurance.

§ 9.3. Estimation of the ruin probability

In the calculations of premiums and reserves it was noted that they depend on a parameter that had to be identified starting from various principles. In this section the *ruin probability* appears as a criterion for an identification of this kind. The ruin probability is a traditional characteristic of the solvency of an insurance company and is the main object of our investigation below.

There are several ways of estimating the ruin probability. Of all the approaches (the Cramér–Lundberg approach, renewal theory, and so on) the martingale approach, which lies at the basis of the general methodology expounded here for estimating the ruin probability, is beyond comparison from the viewpoint of the structural generality of the formation of capital of a company.

On a standard stochastic base $(\Omega, \mathcal{F}, \mathbb{F}, \mathbf{P})$ we consider the following collective risk model (see Chapter 2 and (9.1)):

$$V_t = v + B_t + M_t + X_t, \tag{9.20}$$

where v is the initial value, B is the predictable increasing process of premium payments, X is the pure jump process of claims, with the jump times *unpredictable* and the jumps ΔX_t themselves required to be negative, M is a continuous (local) martingale describing the influence of the "random" environment surrounding the company, and $B_0 = M_0 = X_0 = 0$ (a.s.).

It is technically convenient for us to assume (and this is in complete agreement with the "genealogy" of the process V) that on any time interval $(0, t]$ the series of its jumps converges (a.s.).

We represent the jump process of the risk X in the form

$$X_t = \int_0^t \int_{(-\infty,0)} x\mu(\omega, ds, dx),$$

where μ is the random jump measure of the semimartingale V, and we assume that there is a predictable random measure $\nu(\omega, dt, dx)$ with $\nu(\omega, \{t\}, (-\infty, 0)) = 0$ (a.s.) that compensates μ in the sense that

$$X_t - \int_0^t \int_{-\infty}^0 x\, d\nu$$

is a (local) martingale. It is natural to call such a measure ν the *intensity* of incoming claims.

We rewrite (9.20) in the form

$$V_t - v = \left(B_t + \int_0^t \int_{-\infty}^0 x\, d\nu\right) + M_t + \left(X_t - \int_0^t \int_{-\infty}^0 x\, d\nu\right). \tag{9.21}$$

The last two terms on the right-hand side of (9.21) are (local) martingales, and therefore can be assumed to be zero in the mean. Hence, the insurance company's attempt to form an amount of capital V_t exceeding the initial value v leads to the natural condition that

$$B_t > -\int_0^t \int_{-\infty}^0 x\, d\nu \tag{9.22}$$

for any $t > 0$ (**P**-a.s.). In the framework of the semimartingale model (9.20) under consideration for the collective risk, (9.22) is an analogue of the *net return condition* well known from the classical Cramér–Lundberg theory (see (9.1)), namely, $c \geqslant \lambda \cdot \mathbf{E}Y_1$, which means that the ratio of the intensity of accumulation of premiums to the intensity of the payments must exceed the mean size of a claim. Thus, the premium strategy of the company must be designed with the intensity of the claims taken into account.

Suppose that the sizes of the claims have kth exponential conditional moment for all $t > 0$ (**P**-a.s.):

$$\int_0^t \int_{-\infty}^{-1} e^{-kx}\, d\nu < \infty, \qquad 0 < k \leqslant k_0 < \infty. \tag{9.23}$$

Under the condition (9.23) the following characteristic of the semimartingale V, called its *cumulant*, is well defined:

$$G_t(k) = -kB_t + \frac{k^2}{2}\langle M\rangle_t + \int_0^t \int_{-\infty}^0 (e^{-kx} - 1)\, d\nu,$$

where $\langle M\rangle$ is the quadratic characteristic of M.

We form (assuming *continuity* of B_t for simplicity) the process

$$M_t(k) = e^{-k(V_t - v)}/e^{G_t(k)}, \qquad M_0(k) = 1, \tag{9.24}$$

which is a (local) martingale under our conditions.

To see this, we apply integration by parts to the ratio (9.24) (see Chapter 2) and get

(9.25)

$$\begin{aligned}
M_t(k) &= 1 + \int_0^t e^{-k(V_{s-} - v)}\, de^{-G_s(k)} + \int_0^t e^{-G_s(k)}\, de^{-k(V_s - v)} \\
&= 1 + \int_0^t e^{-k(V_{s-} - v) - G_s(k)} \\
&\qquad\qquad \times \left[k\, dB_s - \frac{k^2}{2}\, d\langle M\rangle_s - \int_{-\infty}^0 (e^{-kx} - 1)\, d\nu\right] \\
&\quad + \int_0^t e^{-G_s(k) - k(V_{s-} - v)} \\
&\qquad\qquad \times \left[-k\, dB_s - k\, dM_s + \frac{k^2}{2}\langle M\rangle_s + \int_{-\infty}^0 (e^{-kx} - 1)\, d\mu\right] \\
&= 1 - k\int_0^t e^{-k(V_{s-} - v) - G_s(k)}\, dM_s \\
&\quad + \int_0^t \int_{-\infty}^0 e^{-k(V_{s-} - v) - G_s(k)} (e^{-kx} - 1)\, d(\mu - \nu)
\end{aligned}$$

$$= 1 - k \int_0^t M_{s-}(k)\, dM_s$$
$$+ \int_0^t \int_{-\infty}^0 M_{s-}(k)(e^{-kx} - 1)\, d(\mu - \nu).$$

The equality (9.25) shows that $M_t(k)$ satisfies a linear stochastic equation with respect to (local) martingales and, according to the properties of stochastic exponentials (Chapter 2), is a (local) martingale.

We define next the stopping time

$$\tau = \inf\{t > 0 : V_t < 0\}$$

and call it (by actuarial tradition) the *ruin time* of the company.

Like every nonnegative (local) martingale, $M_t(k)$ is a supermartingale, and hence its expectations decrease as a function of time. Therefore,

$$1 = \mathbf{E}M_0(k) \geqslant \mathbf{E}M_{t\wedge\tau}(k) \geqslant \mathbf{E}(M_\tau(k) \mid \tau \leqslant t) \cdot \mathbf{P}(\tau \leqslant t)$$

for $t > 0$ and $k \leqslant k_0$. Further, we get from (9.24) and Jensen's inequality that

$$\mathbf{P}(\tau \leqslant t) \leqslant \frac{e^{-kv}}{\mathbf{E}(\exp\{-G_\tau(k)\} \mid \tau \leqslant t)}. \tag{9.26}$$

Note that $G_\tau(0) = 0$, and in view of (9.22)

$$-G'_\tau(0) = B_\tau + \int_0^\tau \int_{-\infty}^0 x\, d\nu > 0. \tag{9.27}$$

The strict convexity of the function $\mathbf{E}(\exp\{-kv + G_\tau(k)\} \mid \tau \leqslant t)$ with respect to k and the property (9.27) imply the existence of a unique minimum point $k^* \leqslant k_0$ such that

$$\mathbf{P}(\tau \leqslant t) \leqslant \frac{e^{-k^* v}}{\mathbf{E}(\exp\{-G_\tau(k^*)\} \mid \tau \leqslant t)}. \tag{9.28}$$

Then the following exponential estimate holds if there is a K_t for which the denominator on the right-hand side of (9.28) is equal to 1:

$$\mathbf{P}(\tau \leqslant t) \leqslant e^{-K_t v}. \tag{9.29}$$

From the estimate (9.29) of the ruin probability on a finite time interval $(0, t)$ we get an exponential estimate of the ruin probability on the infinite interval when there is a finite limit $K = \lim_{t\to\infty} K_t$:

$$\mathbf{P}(\tau < \infty) \leqslant e^{-Kv}. \tag{9.30}$$

We now consider a substantive example of the model (9.20) that is the classical Cramér–Lundberg model "perturbed" by a Wiener process W:

$$V_t = v + c \cdot t + \sigma W_t - \sum_{i=1}^{N_t} Y_i, \tag{9.31}$$

where c is the intensity of the premiums, N is a Poisson process with intensity of claims $\lambda > 0$, and the Y_i are independent identically distributed random variables determining the size of the claims and having distribution $F(dx)$. It is assumed that W, N, and Y are independent.

It is clear that $B_t = c \cdot t$, $M_t = \sigma W_t$, $\langle M \rangle_t = \sigma^2 \cdot t$, $\nu(\omega, dt, dx) = \lambda F^*(dx)\, dt$, $F^*(x) = 1 - F(-x)$, and the cumulant G_t in the model (9.31) takes the form

$$G_t(k) = g(k) \cdot t = \left(-kc + \frac{\sigma^2 k^2}{2} + \lambda[\mathbf{E}e^{kY_1} - 1]\right)t.$$

The process $M_t(k)$ constructed in the framework of the model (9.31) is a *martingale*, as follows from the independence of W, N, Y_i, the independence of the increments of W and N, and the chain of equalities (where we set $v = 0$ for simplicity)

$$\begin{aligned}
\mathbf{E}(M_t(k) \,|\, \mathcal{F}_s) &= \mathbf{E}\big(e^{-kV_t - g(k)t} \,|\, \mathcal{F}_s\big) = \mathbf{E}\big(e^{-k(V_t - V_s) - kV_s - g(k)t} \,|\, \mathcal{F}_s\big) \\
&= e^{-kV_s - g(k)s - g(k)(t-s)} \mathbf{E}\big(e^{-k(V_t - V_s)} \,|\, \mathcal{F}_s\big) \\
&= M_s(k)\, e^{-g(k)(t-s)} \mathbf{E}\big(e^{-kc(t-s) - k\sigma(W_t - W_s) + k\sum_{N_s+1}^{N_t} Y_i} \,|\, \mathcal{F}_s\big) \\
&= M_s(k)\, e^{-g(k)(t-s)} e^{-kc(t-s) + \frac{k^2\sigma^2}{2}(t-s) + \lambda[\mathbf{E}e^{kY_1} - 1](t-s)} \\
&= M_s(k)
\end{aligned}$$

for $s \leqslant t$.

The general methodology for estimating the ruin probability, which led to the estimates (9.29)–(9.30), admits a *concretization* for the model (9.31) such that the so-called *Lundberg constant* K is found as the positive solution of the equation $g(k) = 0$.

The indicated estimates can be derived immediately. For example, using the stopping theorem for martingales, we get that for $t > 0$

$$1 = \mathbf{E}M_0(K) = \mathbf{E}M_{t\wedge\tau}(K) = \mathbf{E}e^{-K(V_{t\wedge\tau} - v)},$$

and hence

$$\mathbf{E}e^{-KV_{t\wedge\tau}} = e^{-Kv}.$$

Further, since V_τ is negative,

$$\begin{aligned}
e^{-Kv} &= \mathbf{E}e^{-KV_{t\wedge\tau}} \geqslant \mathbf{E}e^{-KV_{t\wedge\tau}} I_{\{\tau \leqslant t\}} \\
&= \mathbf{E}e^{-KV_\tau} \cdot I_{\{\tau \leqslant t\}} \geqslant \mathbf{P}(\tau \leqslant t).
\end{aligned} \tag{9.32}$$

Passing to the limit as $t \to \infty$ in (9.32) leads to the desired exponential *Cramér–Lundberg estimate*:

$$\phi(v) = \mathbf{P}(\tau < \infty) \leqslant e^{-Kv}.$$

We turn now to a study of the ruin probability $\phi(v)$ of an insurance company whose capital evolves according to the Cramér–Lundberg model (9.1). In this case it turns out to be possible to get an integro-differential and even a differential equation for $\phi(v)$ or, what is equivalent, for the *non-ruin probability* $\psi(v) = 1 - \phi(v)$ under the assumption that the density $f(y)$ of the distribution function $F(y)$ of the claims is sufficiently smooth. Moreover, for exponential payments with $F(y) = 1 - \exp\{-\alpha^{-1}y\}$, $y > 0$, we find exact exponential formulas for $\phi(v)$ and $\psi(v)$ as functions of the initial value v.

By the formula for the total probability and by the properties of a Poisson process, we have that (as $\Delta t \to 0$)

$$\begin{aligned}
\psi(v) &= \psi(v + c\Delta t)\big(1 - \lambda\Delta t + o(\Delta t)\big) \\
&\quad + \lambda\Delta t \int_0^{v + c\Delta t} \psi(v + c\Delta t - y)\, dF(y) + o(\Delta t).
\end{aligned} \tag{9.33}$$

Using Taylor's formula, we can rewrite (9.33) in the form

$$\psi(v) = \big(\psi(v) + c\psi'(v)\Delta t\big)\big(1 - \lambda\Delta t + o(\Delta t)\big) + \lambda\Delta t\int_0^{v+c\Delta t} \psi(v + c\Delta t - y)\,dF(y) + o(\Delta t),$$

and hence

$$\psi(v)\big(\lambda\Delta t + o(\Delta t)\big) = c\psi'(v)\Delta t\big(1 - \lambda\Delta t + o(\Delta t)\big) + \lambda\Delta t\int_0^{v+c\Delta t} \psi(v + c\Delta t - y)\,dF(y) + o(\Delta t). \tag{9.34}$$

Next, we divide both sides of (9.34) by Δt and, passing to the limit as $\Delta t \to 0$, we get the *integro-differential equation*

$$\psi(v)\lambda = c\psi'(v) + \lambda\int_0^v \psi(v - y)\,dF(y). \tag{9.35}$$

For the exponential distribution $F(y)$ we differentiate the equation (9.35) and integrate by parts:

$$\begin{aligned}
\psi'(v)\lambda &= c\psi''(v) + \lambda\psi(0)\alpha^{-1}e^{-v/\alpha} + \lambda\int_0^v \psi'_v(v-y)\alpha^{-1}e^{-y/\alpha}\,dy \\
&= c\psi''(v) + \lambda\psi(0)\alpha^{-1}e^{-v/\alpha} - \lambda\int_0^v \alpha^{-1}e^{-y/\alpha}\,d\psi(v-y) \\
&= c\psi''(v) + \frac{\lambda}{\alpha}\psi(v) + \lambda\int_0^v \psi(v-y)\alpha^{-1}\,de^{-y/\alpha} \\
&= c\psi''(v) + \frac{\lambda}{\alpha}\psi(v) - \alpha^{-1}\left[\lambda\int_0^v \psi(v-y)e^{-y/\alpha}\,dy\right] \\
&= c\psi''(v) + \frac{\lambda}{\alpha}\psi(v) + \frac{c}{\alpha}\psi'(v) - \frac{\lambda}{\alpha}\psi(v) \\
&= c\psi''(v) + \frac{c}{\alpha}\psi'(v).
\end{aligned}$$

As a result we arrive at the *differential equation*

$$\psi''(v) + \psi'(v)\left(\alpha^{-1} - \frac{\lambda}{c}\right) = 0. \tag{9.36}$$

The general form of the solution of (9.36) is exponential:

$$\psi(v) = B + Ae^{v\left(\frac{\lambda}{c} - \alpha^{-1}\right)},$$

where $\dfrac{\lambda}{c} - \alpha^{-1} < 0$ in view of the *net return* condition. The constants A and B are found from the relations $\psi(\infty) = 1$ and $\psi(0)\lambda = c\psi'(0)$. Consequently,

$$\begin{aligned}
\psi(v) &= 1 - \frac{\alpha\lambda}{c}\exp\left\{\left(\frac{\lambda}{c} - \alpha^{-1}\right)v\right\}, \\
\phi(v) &= \frac{\alpha\lambda}{c}\exp\left\{\left(\frac{\lambda}{c} - \alpha^{-1}\right)v\right\}
\end{aligned} \tag{9.37}$$

are the exact exponential values of $\psi(v)$ and $\phi(v)$.

If we assume now that the company invests its capital in a bank account with interest rate $r \geqslant 0$, then analogous considerations lead to the integro-differential equation

$$\psi(v)\lambda = \psi(v)(rv + c) + \lambda \int_0^v \psi(v-y)\,dF(y),$$

which for exponential payments "turns into" the purely differential equation

$$\psi''(v)(rv+c) + \psi'(v)\left(\frac{rv+c}{\alpha} + r - \lambda\right) = 0. \tag{9.38}$$

The form of a solution of (9.38) is easy to find:

$$\psi(v) = A \int_0^v (rx+c)^{(\lambda/r-1)} e^{-x/\alpha}\,dx + B,$$

where A and B are determined from the same conditions as earlier for (9.36).

An uncomplicated analysis of the asymptotics of $\psi(v)$ and $\phi(v)$ as $r \to 0$ leads to the "natural" limits (9.37).

It is a more complicated matter to study the case when the company's capital is invested in stocks whose prices are governed by the stochastic equation

$$dS_t = S_t(\mu\,dt + \sigma\,dW_t)$$

(see the Black–Scholes model (3.19)).

Nonetheless, the non-ruin probability $\psi(v)$ again satisfies an integro-differential equation

$$\frac{\sigma^2}{2} v^2 \psi''(v) + (\mu v + c)\psi'(v) - \lambda\psi(v) + \lambda \int_0^v \psi(v-y)\,dF(y) = 0$$

and for exponential payments admits the asymptotic representation (as $v \to \infty$):

$$\psi(v) = \psi(\infty) + \text{const} \cdot v^{1-2\mu/\sigma^2} + o(v^{1-2\mu/\sigma^2}) \quad \text{for} \quad \mu > \frac{\sigma^2}{2},$$

$$\psi(v) = \psi(\infty) + \text{const} \cdot \int_v^\infty x^{-2} e^{-x/\alpha}\,dx + o\left(\int_v^\infty x^{-2} e^{-x/\alpha}\,dx\right) \quad \text{for} \quad \mu < \frac{\sigma^2}{2}.$$

Thus, *investing capital in a risky asset of the financial market violates the exponential asymptotics* (9.37), *making it power asymptotics.*

§ 9.4. Catastrophe risks and reinsurance of them on financial markets

Traditional insurance works well for insurance of events significant for the client but relatively minor for the company. However, more and more frequently in insurance practice there are cases which absolutely do not fit in the given scheme because of their significance for the solvency of the insurance companies themselves. There are *catastrophes*, or events affecting many people and causing damages exceeding 25 million dollars (5 million before 1997). The corresponding risks are difficult for an insurance business to accept, since the losses exceed ordinary losses by orders of magnitude, and hence the insurance claims are practically impossible to correlate with the premiums. For example, hurricane Andrew in the USA swallowed up $\frac{1}{5}$ of the *whole sum* of premiums during 1992. Nevertheless, catastrophic events are a reality which insurance encounters. There are several approaches to the insurance of risks of a catastrophic nature (*CAT-insurance*).

The *first* is *traditional reinsurance* with coverage of a percent of losses fixed in the contract, and so on. CAT-insurance has historically been the domain of reinsurance associations like the syndicate Lloyd's. However, the cost of CAT-reinsurance has substantially increased in recent times. For example, from 1989 to 1995 it increased twofold (from 8 to 15%) for "Guy Carpenter", the largest reinsurance brokerage firm in the USA. The scales of natural disasters can reach sizes comparable with the whole of the insurance and reinsurance business. For instance, as already mentioned, insurance and reinsurance in the USA is exhausted by roughly 270 billion dollars (230 and 40, respectively), while a computer simulation of possible California earthquakes led to damage estimates in the range of 50–100 billion dollars. Moreover, the statistics of expenditures on CAT-insurance also point to a clear "exhaustibility" of the resources of traditional insurance. For example, in 1992 it guaranteed 10 billion dollars for these purposes in the USA, while the international potential of reinsurance held fairly stable on a level of 15 billion dollars.

The *second* approach consists in the creation of *opportunities to exchange* such risks "inside" the insurance and reinsurance businesses themselves. This *institutional* approach was realized, for example, by the creation in 1996 of a specialized exchange in the USA: the CATEX, or Catastrophe Risk Exchange. Its activities, which are formally separate from the stock market, involve insurance companies, reinsurance companies, and insurance brokers.

The *third* approach has to do with the creation of a complex of *insurance securities* and diversification through them of catastrophe risks on the stock market. In principle such a possibility really does exist in view of the huge US "funds capacity" of 19 trillion dollars with its daily fluctuations of 130–140 billion. In essence, "risk securitization" is a new form of reinsurance. Here we distinguish two directions which have definitely been realized in practice:

- catastrophe bonds;
- catastrophe futures and options.

The best known example of a bond of this kind is the three-year Winterthur Insurance convertible bond with face value 4700 Swiss francs and 2.25% coupons. This convertible bond was issued at the beginning of 1997 by the Swiss insurance company "Winterthur Insurance". According to the provisions, it is convertible into five stock shares of the issuer, and the total amount of this bonded debt constitutes 300 million Swiss francs. The "catastrophe" qualities of the Winterthur Insurance convertible bond are determined as follows. The coupons for the bond are not paid off if there is a "catastrophe": more than 6000 motor vehicles insured by Winterthur Insurance being damaged by strong winds (more than 75 km/hr) or hail. The main question in pricing these bonds is the adequate valuation of the corresponding coupon payments. The number of such events is simulated with the help of a Poisson process with varying intensity, and the amount of the damage is simulated with the help of the Pareto distribution.

Let us dwell on an interesting aspect of the second of the indicated directions: *catastrophe derivative securities*—the futures and options introduced into the reinsurance business by the Chicago Board of Trade (CBOT) in 1992 and 1995, respectively.

The use of CAT-futures and CAT-options has the purpose of diversifying the scaled risks of events like natural disasters on the capital market. Here the problem unavoidably arises of what to take as the *underlying asset* for "writing" these derivative securities. In 1992 the *ISO-index* was taken in this capacity (named after the "Insurance Services Office", a statistical agency that accumulates data about losses of the most significant US insurance companies), and then the *PCS-index* in 1995 (named after the "Property Claims Service", a specialized noncommercial organization that has compiled losses in this special index since 1949).

CAT-futures based on the ISO-index were traded up to 1995 quarterly: March, June, September, December. Contracts were based on losses suffered in the current quarter and added by the participant companies to the end of the next quarter. The so-called reported period was composed in this way over six months. Moreover, an additional three month period was set aside for refinements, which is fairly typical for the practice of insurance compensations. The contract was fulfilled on the fifth day of the fourth month after the reported period.

Let $T_1 < T_2$ be the terminal times for the period of losses and reports, respectively. The cost of the contract was computed with the help of the ISO-index of losses, which was formed as follows.

The ISO agency gathered data from 100 insurance companies about quarterly losses. A representative pool was then singled out (from no fewer than 10 companies). On the basis of "weighted" data about losses and premiums of the members of this pool, the agency made up its index

$$I = \text{ISO-index} = \frac{\text{losses}}{\text{premiums}}.$$

The CBOT made public the list of companies and the number of premiums they earned before the start of trading, and the established cost of the contract was determined by the formula

$$F_{T_2} = 25000 \cdot \min\{I_{T_2}, 2\}, \tag{9.39}$$

where I_{T_2} is the value of the index at T_2.

It is clear from (9.39) that the exchange established a limit of 50 thousand dollars on losses per contract, and that one point of the index cost 250 dollars.

The determination of a futures price F_t in the contract period is based on the following considerations. The quantity $F_T = F_{T_2}$ is a contingent claim, and F_t is its price at time $t \leqslant T$. Hence, according to the general "financial" approach, this price should be taken to be

$$F_t = \mathbf{E}^*(F_T \mid \mathcal{F}_t)$$

(let the interest rate be zero, and let $\mathcal{F}_t$ be the market information up to time $t \leqslant T$), where $\mathbf{E}^*$ is the expectation with respect to a certain martingale measure to be determined, for example, with the help of the Esscher transform.

We will not spend any more time on this no longer traded instrument of CAT-reinsurance. Let us pass to instruments that are more interesting, from both a theoretical and practical point of view: PCS-futures and PCS-options.

A key deficiency of the statistics of the ISO agency is the *truth* of the declared losses and premiums supplied for ISO by *its own* insurance companies. This was clearly demonstrated in 1994 by the Northridge earthquake, when the ISO-index turned out to be essentially lower, and the corresponding payments were far from covering the actual losses. In 1995 the ISO-index was replaced by the PCS-index, which is based on the size of the *actual* losses from a catastrophic event during a definite period. It is usual to distinguish two periods: the *risk-period* (as a rule, the quarter when the PCS agency estimates the losses as catastrophic) and the *period of development of the losses* (usually half a year, when there is a refinement of the information from the risk-period).

At the beginning of the risk-period the index is set equal to zero, and then each point of it is equivalent to losses of 100 million dollars in the period being observed. The size of possible losses is limited to 50 billion dollars, and to get a monetary expression for options and futures, a monetary equivalent of each of its points, namely, 200 dollars, is introduced on the PCS-index. Accordingly, the PCS-index is defined by

$$L_t = \frac{X_t}{C},$$

where X_t is the loss process, and $C = 100$ million dollars.

Naturally, concretization of the process X_t is necessary in order to get calculation formulas for futures and options on this index as the underlying asset.

We consider $X_t = \sum_{i=1}^{N_t} Y_i$, where N_t is a Poisson process with intensity $\lambda > 0$, and (Y_i) is a sequence of independent (and independent of N_t) random variables having a gamma-distribution with density

$$g_{(n,\mu)}(x) = \frac{\mu^n}{\Gamma(n)} x^{n-1} e^{-\mu x}, \qquad x \geqslant 0,$$

where $\mu, n > 0$ and $\Gamma(\,\cdot\,)$ is the gamma-function.

Clearly, the price of a futures contract at time t is equal to the sum of two quantities: the losses up to the time t, divided by C (this is the value L_t), and the premium for the remaining risk, which is computed by the martingale method presented in § 9.1:

$$F_t = L_t + \frac{\lambda \mathbf{E}[Y_1 e^{\beta(Y_1)}](T-t)}{C}, \tag{9.40}$$

where the function $\beta(\,\cdot\,)$ determines the "insurance" martingale measure. In particular, by finding this measure from the exponential utility principle with $U(x) = 1 - \alpha^{-1} e^{-\alpha x}$, we arrive at $\beta(x) = \alpha x$, and (9.40) can be calculated exactly:

$$F_t = L_t + \frac{\lambda n \mu^n (T-t)}{(\mu - \alpha)^{n+1} C}, \qquad 0 \leqslant \alpha < \mu.$$

As for options on the PCS-index, an analogous simulation of the process X_t for the risk-period and the period of development, together with the Esscher technique for finding martingale measures, leads to the determination of an appropriate price.

References for Chapter 9

[**2**]–[**4**], [**12**], [**17**], [**24**], [**26**]–[**32**], [**35**], [**36**], [**39**], [**48**], [**52**]–[**54**], [**61**], [**63**]–[**68**], [**104**]–[**106**], [**109**]–[**112**], [**123**], [**124**], [**130**], [**131**], [**133**], [**136**]–[**138**], [**141**], [**146**], [**147**], [**150**], [**151**], [**159**]–[**161**].

Bibliographical Notes

Chapter 1

The descriptive nature of the survey in Chapter 1 reflects an attempt to give an idea about financial systems (§ 1.1), tendencies in their development (§ 1.3), models of financial markets, and pricing of financial instruments (§ 1.2). Here attention is focused on the classical Bachelier [**11**], Black–Scholes [**23**], Merton [**118**], and Cox–Ross–Rubinstein ([**33**], [**34**]) models, as well as a very enlightening example of an incomplete market with stochastic volatility in the form of a "telegraph signal" ([**43**], [**127**], [**158**]). The essentials of complete and incomplete markets are explained along with the main principles for hedging of contingent claims (derivative securities) and calculating the corresponding strategies and premiums, and sketches are given for the basic methodological approaches and mathematical techniques (methods of the theory of martingales and differential equations) ([**27**], [**40**], [**46**], [**49**], [**72**], [**73**], [**79**], [**81**], [**87**], [**90**], [**93**], [**94**], [**100**], [**106**], [**108**], [**126**], [**134**], [**141**]–[**143**], [**158**]). In the description of innovation tendencies in the evolution of a financial system ([**29**], [**122**]) attention is given to the extremely attractive idea of the "financial-innovation spiral" ([**118**], [**121**]). Also called the "Merton spiral" in this book, it acquires for the first time an intuitive graphical interpretation in terms of incompleteness parameters of financial markets ([**111**], [**112**]).

With regard to insurance (§ 1.4), which is intentionally treated in the general context of a financial system, we use the references [**17**], [**26**], [**35**], [**64**], [**68**]. Here we give the first facts about the connection observed between finance and insurance in terms of investment on financial markets ([**54**], [**106**]), calculating flexible insurance schemes, and catastrophe insurance ([**2**]–[**4**], [**28**], [**31**], [**32**], [**36**], [**112**], [**158**], [**161**]).

Chapter 2

The description of the necessary elements of the theory of random processes in § 2.1 uses mainly [**149**], [**154**], and that in § 2.2 uses [**98**], [**103**], [**154**]. The exposition in § 2.3 is based on [**106**], [**107**]. A more complete picture of the semimartingale calculus can be obtained from [**82**], [**104**]. Theorems on representation of martingales, on optional decomposition of semimartingales, and on Snell envelopes in the theory of controlled random processes can be found in [**49**], [**55**], [**81**], [**90**], [**94**], [**98**], [**141**], [**158**].

Chapter 3

The theory of hedging in complete markets presented in § 3.1 is given in semimartingale (and hence most general) form, and follows [**106**], [**107**], [**115**], [**158**] (see also [**27**], [**72**], [**73**], [**152**]). The methodology for finding martingale measures in § 3.2 is based on [**107**], [**115**]. This methodology is used to derive from general results in § 3.1 a spectrum of classical formulas for hedging strategies and call option prices for the Black–Scholes model ([**15**], [**16**], [**21**], [**44**], [**46**], [**51**], [**79**], [**83**], [**86**], [**87**], [**90**], [**100**], [**117**], [**118**], [**126**], [**141**], [**143**]), the Merton model [**118**], the Cox–Ross–Rubinstein binomial model ([**21**], [**33**], [**34**], [**44**], [**46**], [**51**], [**79**], [**100**], [**108**], [**113**], [**126**], [**141**], [**142**], [**158**]), and the complete jump-diffusion model ([**1**], [**13**], [**22**], [**100**], [**115**], [**116**], [**158**]). In the study in § 3.3 of optimal investment in the sense of maximalizing the mean utility function, the basic result is taken from [**95**], and from it formulas are obtained for optimal (with a logarithmic utility function) investment strategies in the case of the above (complete) models of a (B, S)-market ([**46**], [**51**], [**87**], [**90**], [**100**], [**113**], [**118**]–[**120**]).

Chapter 4

In § 4.1 a description is given (in semimartingale form) of financial markets in which there is more than one martingale measure, and this is identified with incompleteness of the market. To characterize a hedging methodology adequate for these conditions, we invoke the concept of superhedging with the use of theorems on optional decomposition and on Snell envelopes in the theory of controlled random processes (see Chapter 2) ([**38**], [**40**], [**49**], [**81**], [**90**], [**91**], [**94**], [**95**], [**107**], [**152**], [**158**]).

A fully representative example of an incomplete market is given [**20**] in which perfect hedging of a series of contingent claims is nevertheless possible.

In § 4.2 a systematic study is made of the Black–Scholes model with stochastic volatility ([**77**], [**80**], [**81**]). The stress is on the case when the volatility is a bounded process ([**8**]–[**10**], [**15**], [**50**], [**59**], [**60**], [**127**]). For the model with volatility in the form of a "telegraph signal" a corresponding methodology for hedging contingent claims is worked out that leads to an optimal control problem for diffusion processes and to the Bellman equation ([**98**], [**99**]). In this way the ask and bid prices of options are calculated by the small parameter method ([**43**], [**127**], [**158**]). An example is also given [**76**] of a complete diffusion market with stochastic volatility.

In § 4.3 a survey is given of methods for estimating volatility: point and interval estimation ([**5**], [**62**], [**153**]), along with estimation from the statistic of option prices with indication of the so-called smile effect [**77**].

Chapter 5

The basic theoretical propositions in § 5.1 on the topic of markets with constraints were taken from [**55**] and [**158**], with [**82**] as the main source on stochastic analysis. Concrete calculations for particular models with constraints like prohibition of short sales and with different credit and deposit interest rates are derived from general theoretical propositions, and this correlates with earlier particular results ([**14**], [**37**], [**90**], [**92**], [**144**], [**145**]).

In § 5.2 Leland's approach to hedging in the Black–Scholes model with transaction costs is presented ([**69**], [**102**], [**85**]). With regard to investment problems with small transaction costs the exposition follows [**157**], which in its formulation is related to [**7**], [**125**].

Concrete applications of the theory to the Russian financial market are presented in § 5.3.

Chapter 6

Chapter 6 concentrates on two types of imperfect hedging: mean-variance hedging in § 6.1, and quantile hedging in § 6.2.

For mean-variance hedging a general methodology is given ([**56**], [**101**], [**139**]), which is then interpreted in the concrete classical models: Black–Scholes, Merton, and Cox–Ross–Rubinstein ([**46**], [**47**], [**51**], [**93**], [**126**], [**141**]).

As for quantile hedging, which is to a certain degree a dynamic variant of the now popular VaR method in financial risk management ([**71**], [**84**]), this topic is scarcely represented at all in the monograph literature. The methodology used in treating this type of hedging is based on the papers [**57**] and [**58**], whose authors associate the origin of the approach first and foremost with the names D. Heath and M. Kulldorff. Concrete calculations, representing independent problems in themselves, are given as earlier for the Black–Scholes, Merton, Cox–Ross–Rubinstein, and jump-diffusion models ([**25**], [**97**], [**108**], [**148**]).

Chapter 7

The exposition in § 7.1 of problems involving American options and the general method of pricing them is based on optional decomposition (see Chapter 2) [**158**]. A complete solution of the problem is given there in the framework of the Black–Scholes model, by reducing it to the solution of the corresponding optimal stopping problem, based mainly on [**143**] (see also [**46**], [**51**], [**87**], [**90**], [**98**]). The method presented at the end of § 7.1 for solving the indicated problem is taken from [**19**].

In § 7.2 a whole spectrum of analytic formulas is given for various American options ([**96**], [**155**], [**156**]).

In § 7.3 a complete solution of the quantile hedging problem is given for American options in the Black–Scholes model, mostly following [**162**]. An analogous problem, in more general form but with less complete results, is treated in [**132**].

Chapter 8

The general idea of the term structure of bond prices is presented in § 8.1. This structure is made specific for the classical Ho–Lee and Heath–Jarrow–Morton models ([**22**], [**46**], [**74**], [**75**], [**79**], [**108**], [**126**], [**133**], [**135**], [**141**]).

In § 8.2 the problem is solved of hedging options on a bond in the framework of the Ho–Lee model and the Heath–Jarrow–Morton model ([**74**], [**75**], [**108**], [**126**], [**129**], [**135**], [**141**]).

In § 8.3 investment problems on a bond market are treated for the Ho–Lee and Heath–Jarrow–Morton models when an optimal strategy is found from the condition that the mean logarithmic utility function be maximized ([**46**], [**64**], [**108**], [**135**]).

Chapter 9

In § 9.1 a model of collective risk is studied, and a set of commonly accepted actuarial principles for calculating insurance premiums is presented ([**17**], [**26**], [**35**], [**53**], [**64**], [**131**], [**136**], [**151**]). It is shown in the framework of the Cramér–Lundberg model how the "financial" no-arbitrage principle works in insurance calculations, and from this the above actuarial principles for calculating insurance premiums are deduced ([**39**], [**146**]).

§ 9.2 is devoted to problems in life insurance ([**26**], [**65**], [**68**]). It is shown that the Black–Scholes formula arises in the structure of premiums when working out "flexible" insurance schemes (insurance payments are connected with the price of a stock in the Black–Scholes model), and a component of the Black–Scholes equation arises in Thiele's equation for reserves ([**3**], [**12**], [**24**], [**124**], [**130**], [**131**], [**133**], [**136**], [**150**]).

In § 9.3 questions are investigated concerning an insurance company's solvency. The ruin probability provides an adequate description of this. A study is made of the ruin probability in the framework of a semimartingale model for the capital of the insurance company [**147**], and then the results obtained are connected with more specific models ([**48**], [**137**]). In the case of the Cramér–Lundberg model, corresponding integro-differential equations and differential equations are obtained for the ruin probability as a function of the initial capital, equations leading to the Cramér–Lundberg exponential estimates. Further, it is indicated that the exponential asymptotics can be violated if the company invests capital on a Black–Scholes market ([**35**], [**61**], [**106**], [**136**]).

In § 9.4 questions involving the insurance of catastrophe risks are considered, and possible approaches are described in terms of traditional reinsurance, in terms of the creation of specialized exchanges, and in terms of the creation of new instruments: "catastrophe" bonds and "catastrophe" forwards, futures, and options ([**2**], [**4**], [**28**]–[**32**], [**36**], [**52**], [**105**], [**109**]–[**112**], [**123**], [**124**], [**138**], [**159**]–[**161**]).

Bibliography

1. K. K. Aase, *Contingent claim valuation when the security price is a combination of an Itô process and a random point process*, Stochastic Process. Appl. **28** (1988), 185–220.
2. K. K. Aase, *An equilibrium model of catastrophe insurance futures and spreads*, working paper, Norwegian School of Economics and Business Administration, Bergen, 1996.
3. K. K. Aase and S. A. Persson, *Pricing of unit-linked life insurance policies*, Scand. Actuar. J. **1** (1994), 26–52.
4. P. Albrecht, *An actuarial approach to risk management with CAT insurance contracts*, Proc. 5th AFIR Internat. Colloquium, vol. 3, Brussels, 1995, pp. 952–974.
5. T. Andersen, *Stochastic autoregression volatility: a framework for volatility modelling*, Math. Finance **4** (1994), 75–102.
6. P. Artzner, F. Delbaen, J.-M. Eber, and D. Heath, *Thinking coherently*, Risk **10** (1997), no. 11, 68–70.
7. C. Atkinson and P. Wilmott, *Portfolio management with transaction costs: an asymptotic analysis of the Morton and Pliska model*, Math. Finance **5** (1995), 357–367.
8. M. Avellaneda, A. Levy, and A. Paras, *Pricing and hedging of derivative securities in markets with uncertain volatilities*, Appl. Math. Finance **2** (1995), 73–88.
9. M. Avellaneda and A. Paras, *Dynamic hedging portfolios for derivative securities in the presence of large transaction costs*, Appl. Math. Finance **1** (1994), 165–193.
10. M. Avellaneda and A. Paras, *Managing the volatility risk of portfolios of derivative securities: the Lagrangian uncertainty volatility model*, Appl. Math. Finance **3** (1996), 21–52.
11. L. Bachelier, *Théorie de la speculation*, Ann. Sci. École Norm. Sup. **17** (1900), 21–86.
12. A. R. Bacinello and F. Ortu, *Pricing equity-linked life insurance with endogeneous minimum guarantees*, Insurance Math. Econom. **12** (1993), 245–257.
13. I. Bardhan and X. Chao, *Pricing options on securities with discontinuous returns*, Stochastic Process. Appl. **48** (1993), 123–137.
14. Ya. Yu. Bart, *Option hedging in the binomial model with differing interest rates*, Uspekhi Mat. Nauk **53** (1998), no. 5, 227–228; English transl., Russian Math. Surveys **53** (1998), 1084–1085.
15. M. Baxter, *Hedging in financial markets*, ASTIN Bulletin **28** (1998), no. 1, 5–16.
16. M. W. Baxter and A. J. O. Rennie, *Financial calculus: an introduction to derivative pricing*, Cambridge Univ. Press, Cambridge, 1996.
17. R. E. Beard, T. Pentikainen, and E. Pesonen, *Risk theory: the stochastic basis of insurance*, 3rd ed., Chapman and Hall, London, 1984.
18. W. H. Beaver and G. Parker, *Risk management: problems and solutions*, McGraw-Hill, New York, 1995.
19. M. Beibel and H. R. Lerche, *A new look at warrant pricing and related optimal stopping problems*, Discussion paper no. 12, Univ. Freiburg, Freiburg, 1995.
20. A. Bensoussan and R. J. Elliott, *Attainable claims in a Markov market*, Math. Finance **5** (1995), 121–131.
21. N. H. Bingham and R. Kiesel, *Risk-neutral valuation: pricing and hedging of financial derivatives*, Springer-Verlag, Berlin, 1998.
22. T. Björk, G. B. Di Masi, Yu. M. Kabanov, and W. J. Runggaldier, *Towards a general theory of bond markets*, Finance and Stochastics **1** (1997), 141–174.
23. F. Black and M. Scholes, *The pricing of options and corporate liabilities*, J. Political Economy **81** (1973), 637–654.
24. M. J. Brennan and E. S. Schwartz, *The pricing of equity-linked life insurance policies with an asset value guarantee*, J. Financial Economics **3** (1976), 195–213.

25. S. Browne, *Stochastic differential portfolio games*, J. Appl. Probab. **37** (2000), 126–127.
26. H. Bühlmann, *Mathematical methods in risk theory*, Springer-Verlag, New York, 1970.
27. H. Bühlmann, F. Delbaen, P. Embrechts, and A. N. Shiryaev, *No-arbitrage, change of measure, and conditional Esscher transforms*, Quarterly **9** (1996), no. 4, 291–317.
28. M. S. Canter, J. B. Cole, and R. L. Sandor, *Insurance derivatives: a new asset class for the capital markets and a new hedging tool in the insurance industry*, J. Appl. Corporate Finance **10** (1997), no. 3, 69–83.
29. H. Cavanna (ed.), *Financial innovations*, Routledge, London, 1992.
30. L. Chew, *Managing derivative risks: the use and abuse of leverage*, Wiley, New York, 1996.
31. *Catastrophe insurance: background report*, The Chicago Board of Trade, Chicago, 1995.
32. S. H. Cox and H. W. Pedersen, *Catastrophe risk bonds*, Trans. 26th Internat. Congr. Actuaries **7** (1996), 483–512.
33. J. C. Cox and S. A. Ross, *The valuation of options for alternative stochastic processes*, J. Financial Economics **3** (1976), 145–166.
34. J. C. Cox, R. A. Ross, and M. Rubinstein, *Option pricing: a simplified approach*, J. Financial Economics **7** (1979), no. 3, 229–263.
35. H. Cramér, *Collective risk theory*, Scandia Ins. Co. (Jubilee Volume), Stockholm, 1955.
36. J. D. Cummins and H. Geman, *Pricing catastrophe futures and call spreads: an arbitrage approach*, J. Fixed Income **4** (1995), 46–57.
37. J. Cvitanic, *Optimal trading under constraints*, Lecture Notes in Math., vol. 1656, Springer-Verlag, Berlin, 1997, pp. 123–190.
38. M. H. A. Davis and J. M. C. Clark, *A note on super-replicating strategies*, Philos. Trans. Roy. Soc. London Ser. A. **347** (1994), 485–494.
39. F. Delbaen and J. Haezendonck, *A martingale approach to premium calculation principles in an arbitrage free market*, Insurance Math. Econom. **8** (1989), 269–277.
40. F. Delbaen and W. Schachermayer, *A general version of the fundamental theorem of asset pricing*, Math. Ann. **300** (1994), 463–520.
41. F. Delbaen and W. Schachermayer, *The variance-optimal martingale measure for continuous processes*, Bernoulli **2** (1996), no. 1, 81–105.
42. E. Derman and I. Kani, *Riding on a smile*, Risk **7** (1994), 32–39.
43. G. B. Di Mazi, Yu. M. Kabanov, and W. J. Runggaldier, *Mean-variation hedging of options on stocks with Markov volatilities*, Teor. Veroyatnost. i Primenen. **39** (1994), 211–222; English transl., Theory Probab. Appl. **39** (1994), 172–182.
44. M. U. Dothan, *Prices in financial markets*, Oxford Univ. Press, Oxford, 1990.
45. J.-C. Duan, *The GARCH option pricing model*, Math. Finance **5** (1995), 13–22.
46. D. Duffie, *Dynamic asset pricing theory*, Princeton Univ. Press, Princeton, 1992.
47. D. Duffie and H. R. Richardson, *Mean-variance hedging in continuous time*, Ann. Appl. Probab. **1** (1991), 1–15.
48. F. Dufresne and H. U. Gerber, *Risk theory for the compound Poisson process that is perturbed by diffusion*, Insurance Math. Econom. **10** (1991), 51–59.
49. N. El Karoui and M. C. Quenez, *Dynamic programming and pricing of contingent claims in an incomplete market*, SIAM J. Control Optim. **33** (1995), 29–66.
50. N. El Karoui, M. J.-B. Picque, and S. E. Shreve, *Robustness of the Black–Scholes formula*, Math. Finance **8** (1998), 93–126.
51. R. Elliott and P. E. Kopp, *Mathematics of financial markets*, Springer-Verlag, Berlin, 1998.
52. P. Embrechts and S. Meister, *Pricing insurance derivatives, the case of CAT-futures*, Proc. 1st Bowles Symposium (Society of Actuaries) (S. Cox, ed.), Georgia State Univ., Atlanta, 1995.
53. P. Embrechts, C. Klüppelberg, and T. Mikosh, *Modelling extremal events for insurance and finance*, Springer-Verlag, Berlin, 1997.
54. P. Embrechts, R. Frey, and H. Furrer, *Stochastic processes in insurance and finance*, Handbook of Statistics (C. R. Rao and D. N. Shaubhag, eds.), vol. 18, Elsevier Science, Amsterdam, 2000.
55. H. Föllmer and D. O. Kramkov, *Optional decompositions under constraints*, Probab. Theory Related Fields **109** (1997), 1–25.
56. H. Föllmer and D. Sondermann, *Hedging of non-redundant contingent claims*, Contributions to Mathematical Economics (in Honor of Gerard Debreu) (W. Hilderbrand and A. Mas-Colell, eds.), North-Holland, Amsterdam, 1986, pp. 205–223.

57. H. Föllmer and P. Leukert, *Quantile hedging*, Finance and Stochastics **3** (1999), no. 3, 251–273.
58. H. Föllmer and P. Leukert, *Efficient hedging: cost versus shortfall risk*, Finance and Stochastics **4** (2000), no. 2, 117–146.
59. R. Frey, *Super-replication in stochastic volatility models and optimal stopping*, Finance and Stochastics **4** (2000), no. 2, 161–187.
60. R. Frey and C. A. Sin, *Bounds on European option prices under stochastic volatility*, Math. Finance **9** (1999), 97–116.
61. A. G. Frolova, *Some mathematical models of risk theory*, Abstracts, All-Russian School-Colloquium on Stochastic Methods in Geometry and Analysis, TVP, Moscow, 1994, pp. 117–118. (Russian)
62. H. Frydman, *Asymptotic inference for the parameters of a discrete-time square-root process*, Math. Finance **4** (1994), 169–181.
63. H. U. Gerber, *Martingales in risk theory*, Mitteilungen der Schweizerischen Vereinigung der Versicherungsmathematiker **1973**, 205–216.
64. H. U. Gerber, *An introduction to mathematical risk theory*, S. S. Huebner Foundation, Wharton School, Philadelphia, 1979.
65. H. U. Gerber, *Life insurance mathematics*, 2nd ed., Springer-Verlag, Zürich, 1995.
66. H. U. Gerber and E. S. W. Shiu, *Actuarial bridges to dynamic hedging and option pricing*, Insurance Math. Econom. **17** (1996), 152–163.
67. M. J. Goovaerts, F. De Vylder, and J. Haezendonck, *Insurance premiums*, North–Holland, Amsterdam, 1984.
68. J. Grandel, *Aspects of risk theory*, Springer-Verlag, New York, 1991.
69. P. Grandits and W. Schachinger, *Leland's approach to option pricing: the evolution of a discontinuity*, Preprint, Inst. für Stat., Univ. Wien, 1998.
70. E. R. Grannan and G. H. Swindle, *Minimizing transaction costs of option hedging strategies*, Math. Finance **6** (1996), 341–364.
71. S. Grayling (ed.), *VaR: understanding and applying Value-at-Risk*, Risk Books, London, 1997.
72. J. M. Harrison and D. M. Kreps, *Martingales and arbitrage in multiperiod securities markets*, J. Econom. Theory **20** (1979), 381–408.
73. J. M. Harrison and S. Pliska, *Martingales and stochastic integrals in the theory of continuous trading*, Stochastic Process. Appl. **11** (1981), no. 3, 215–260.
74. D. Heath, R. Jarrow, and A. Morton, *Bond pricing and the term structure of interest rates: a new methodology for contingent claims valuation*, Econometrica **60** (1992), no. 1, 77–105.
75. T. S. Y. Ho and S.-B. Lee, *Term structure movements and pricing interest rate contingent claims*, J. Finance **41** (1986), 1011–1029.
76. D. G. Hobson and L. C. G. Rogers, *Complete models with stochastic volatility*, Math. Finance **8** (1998), 27–48.
77. N. Hoffman, E. Platen, and M. Schweizer, *Option pricing under incompleteness and stochastic volatility*, Math. Finance **2** (1992), 189–196.
78. Huang Chi-Fu and R. H. Litzenberger, *Foundations for financial economics*, North–Holland, New York, 1988.
79. J. C. Hull, *Options, futures, and other derivative securities*, Prentice–Hall, Englewood Cliffs, NJ, 1999.
80. D. Hull and A. White, *The pricing of options on assets with stochastic volatilities*, J. Finance **42** (1987), 281–300.
81. S. D. Jacka, *A martingale representation result and an application to incomplete financial markets*, Math. Finance **2** (1992), 239–250.
82. J. Jacod and A. N. Shiryaev, *Limit theorems for stochastic processes*, Springer-Verlag, New York, 1994; Russian transl., vols. 1, 2, "Nauka", Moscow.
83. Jia-An Yan, *Introduction to martingale methods in option pricing*, City Univ. of Hong Kong, Centre for Math. Sci., Lecture Notes in Math., Hong Kong, 1998.
84. P. Jorion, *Value at Risk: the new benchmark for controlling market risk*, Irwin Professional Publ., Chicago, 1997.
85. Yu. M. Kabanov and M. M. Safarian, *On Leland's strategy of option pricing with transaction costs*, Finance and Stochastics **1** (1997), 239–250.

86. G. Kallianpur and R. L. Karandikar, *Introduction to option pricing theory*, Birkhäuser, Boston, 2000.
87. I. Karatzas, *Lectures in mathematical finance*, Amer. Math. Soc., Providence, RI, 1997.
88. I. Karatzas and S. G. Kou, *On the pricing of contingent claims with constrained portfolios*, Ann. Appl. Probab. **6** (1996), 321–369.
89. I. Karatzas, J. P. Lehoczky, S. E. Shreve, and G. L. Xu, *Martingales and duality methods for utility maximization in an incomplete market*, SIAM J. Control Optim. **29** (1991), 702–730.
90. I. Karatzas and S. E. Shreve, *Methods of mathematical finance*, Springer-Verlag, New York, 1998.
91. T. Kariya, *Quantitative methods for portfolio analysis*, Kluwer, Dordrecht, 1993.
92. R. Korn, *Contingent claim valuation in a market with different rates*, Math. Methods Oper. Res. **42** (1995), 255–274.
93. R. Korn, *Optimal portfolios*, World Scientific, Singapore, 1997.
94. D. O. Kramkov, *Optional decomposition of supermartingales and hedging contingent claims in incomplete security markets*, Probab. Theory Related Fields **105** (1996), 459–479.
95. D. O. Kramkov and W. Schachermayer, *The asymptotic elasticity of utility functions and optimal investment in incomplete markets*, Ann. Appl. Probab. **9** (1998), 904–950.
96. D. O. Kramkov and A. N. Vishnyakov, *Closed form representation for the minimal hedging portfolios of American type contingent claims*, Obozrenie Prikladnoi Promyshlennoi Matematiki **5** (1998), 387–409. (Russian)
97. R. N. Krutchenko and A. V. Mel′nikov, *Quantile hedging for a jump-diffusion financial market model*, Discussion paper, vol. 34, Humboldt Univ., Berlin, 2000.
98. N. V. Krylov, *Controlled processes of diffusion type*, "Nauka", Moscow, 1977; English transl., Springer-Verlag, New York, 1980.
99. N. V. Krylov, *Nonlinear elliptic and parabolic equations of the second order*, Reidel, Dordrecht, 1987.
100. D. Lamberton and B. Lapeyre, *Introduction to stochastic calculus applied to finance*, Chapman and Hall, London, 1996.
101. J. P. Laurent and H. Pham, *Dynamic programming and mean-variance hedging*, Finance and Stochastics **3** (1999), no. 1, 83–110.
102. H. E. Leland, *Option pricing and replication with transaction costs*, J. Finance **40** (1985), 1283–1301.
103. R. Sh. Liptser and A. N. Shiryaev, *Statistics of random processes*, "Nauka", Moscow, 1974; English transl., vols. I, II, Springer-Verlag, Berlin, 1977.
104. R. Sh. Liptser and A. N. Shiryaev, *Theory of martingales*, "Nauka", Moscow, 1986; English transl., Kluwer, Dordrecht, 1989.
105. R. H. Litzenberger, D. R. Beaglehole, and C. E. Reynolds, *Assessing catastrophe reinsurance-linked securities as a new asset class*, J. Portfolio Management, Special Issue (1996), 76–86.
106. A. V. Mel′nikov, *On stochastic analysis in the contemporary mathematics of finance and insurance*, Obozrenie Prikladnoi Promyshlennoi Matematiki **2** (1995), 514–526. (Russian)
107. A. V. Mel′nikov, *Stochastic differential equations: nonsmoothness of coefficients, regression models, and stochastic approximation*, Uspekhi Mat. Nauk **51** (1996), no. 5, 43–136; English transl., Russian Math. Surveys **51** (1996), 819–909.
108. A. V. Mel′nikov, *Financial markets: stochastic analysis and the pricing of derivative securities*, TVP, Moscow, 1997; English transl., Amer. Math. Soc., Providence, RI, 1999.
109. A. V. Mel′nikov, *Financial innovations and problems of controlling risk*, Upravlenie Riskom **4** (1997), 34–41. (Russian)
110. A. V. Mel′nikov, *On innovation and risk aspects of the evolution of a financial system*, Voprosy Analiza Riska **1** (1999), no. 1, 22–27. (Russian)
111. A. V. Mel′nikov, *On global innovation processes in financial markets*, Voprosy Analiza Riska **1** (1999), no. 2–3, 10–14. (Russian)
112. A. V. Mel′nikov, *Financial system: innovations and pricing of risks*, Univ. Karlsruhe, Technical Report, vol. 5, Karlsruhe, 1999.
113. A. V. Mel′nikov, *The binomial financial market in the context of stochastic exponents and martingales*, Theory Probab. Math. Statist. **62** (2000), 10.
114. A. V. Mel′nikov and M. L. Nechaev, *On mean-variance hedging of contingent claims*, Teor. Veroyatnost. i Primenen. **43** (1998), 672–691; English transl., Theory Probab. Appl. **43** (1998), 588–603.

115. A. V. Mel′nikov and A. N. Shiryaev, *Criteria for the absence of arbitrage in the financial market*, Frontiers in Pure and Applied Probability. II, TVP, Moscow, 1996, pp. 121–134.
116. F. Mercurio and W. J. Runggaldier, *Option pricing for jump diffusions: approximations and their interpretation*, Math. Finance **3** (1993), 191–200.
117. R. C. Merton, *Theory of rational option pricing*, Bell Journal of Economics and Management Science **4** (1973), 141–183.
118. R. C. Merton, *Continuous-time finance*, Basil-Blackwell, Oxford, 1990.
119. R. C. Merton, *Lifetime portfolio selection under uncertainty: the continuous time case*, Review of Economics & Statistics **51** (1969), 247–257.
120. R. C. Merton, *Optimum consumption and portfolio rules in a continuous-time model*, J. Econom. Theory **3** (1971), 373–413.
121. R. C. Merton, *Influence of mathematical models in finance on practice: past, present and future*, Mathematical Models in Finance (S. D. Howison et al., eds.), Chapman and Hall, London, 1995, pp. 1–12.
122. M. H. Miller, *Financial innovations and market volatility*, Blackwell, Oxford, 1991.
123. M. Möller, *Pricing PCS-options with the use of Esscher transforms*, Trans. 26th Internat. Congr. Actuaries (Nuremberg, 1996), Karlsruhe, 1998, pp. 1299–1310.
124. T. Moller, *Risk-minimizing hedging strategies for unit-linked life insurance contracts*, ASTIN Bulletin **28** (1998), 17–47.
125. A. J. Morton and S. R. Pliska, *Optimal portfolio management with fixed transaction costs*, Math. Finance **5** (1995), 337–356.
126. M. Musiela and M. Rutkowski, *Martingale methods in financial modelling*, Springer-Verlag, New York, 1997.
127. V. Naik, *Option valuation and hedging strategies with jumps in the volatility of asset returns*, J. Finance **48** (1993), 1969–1984.
128. E. H. Neave (ed.), *The economic organisation of a financial system*, Routledge, London, 1991.
129. M. L. Nechaev, *On mean-variance hedging in the Ho–Lee diffusion model*, Teor. Veroyatnost. i Primenen. **44** (1999), 115–119; English transl., Theory Probab. Appl. **44** (1999), 102–106.
130. R. Norberg, *Addendum to Hattendorff's theorem and Thiele's differential equation generalized*, Scand. Actuar. J. **1996**, no. 1, 50–53.
131. R. Norberg, *Methods of stochastic analysis in life insurance*, Obozrenie Prikladnoi Promyshlennoi Matematiki **5** (1998), 83–115. (Russian)
132. A. A. Novikov, *Hedging of options with a given probability*, Teor. Veroyatnost. i Primenen. **43** (1998), 152–161; English transl., Theory Probab. Appl. **43** (1998), 135–144.
133. H. W. Pedersen and E. S. W. Shiu, *Pricing of options on bonds by binomial lattices and by diffusion processes*, ARCH **2** (1988), 115–139.
134. S. Pliska, *Introduction to mathematical finance: discrete time models*, Basil-Blackwell, Oxford, 1997.
135. R. Rebonato, *Interest-rate option models*, Wiley, New York, 1995.
136. T. Rolski, H. Schmidli, V. Schmidt, and J. Teugels, *Stochastic processes for insurance and finance*, Wiley, Chichester, 1998.
137. H. Schmidli, *Cramér–Lundberg approximations for ruin probabilities of risk processes perturbed by diffusion*, Insurance Math. Econom. **16** (1995), 135–149.
138. U. Schmock, *Estimating the value of the Wincat coupons of the Winterthur insurance convertible bond*, ASTIN Bulletin **29** (1999), no. 1, 101–163.
139. M. Schweizer, *Approximation pricing and the variance-optimal martingale measure*, Ann. Probab. **24** (1996), no. 1, 206–236.
140. A. N. Shiryaev, *Statistical sequential analysis*, 2nd ed., "Nauka", Moscow, 1976; English transl. of 1st ed., Amer. Math. Soc., Providence, RI, 1972.
141. A. N. Shiryaev, *Essentials of stochastic finance: facts, models, theory*, vols. 1, 2, Fazis, Moscow, 1998; English transl., World Scientific, River Edge, NJ, 1999.
142. A. N. Shiryaev, Yu. M. Kabanov, D. O. Kramkov, and A. V. Mel′nikov, *Toward a theory of pricing options of European and American types*, I. *Discrete time*, Teor. Veroyatnost. i Primenen. **39** (1994), 23–79; English transl., Theory Probab. Appl. **39** (1994), 14–60.
143. A. N. Shiryaev, Yu. M. Kabanov, D. O. Kramkov, and A. V. Mel′nikov, *Toward a theory of pricing options of European and American types*, II. *Continuous time*, Teor. Veroyatnost. i Primenen. **39** (1994), 80–129; English transl., Theory Probab. Appl. **39** (1994), 61–102.

144. S. E. Shreve, H. M. Soner, and G.-L. Xu, *Optimal investment and consumption with two bonds and transaction costs*, Math. Finance **13** (1991), 53–84.
145. S. E. Shreve and G.-L. Xu, *A duality method for optimal consumption and investment under short-selling prohibition*: I. *General market coefficients*, II. *Constant market coefficients*, Ann. Appl. Probab. **2** (1992), 87–112, 314–328.
146. D. Sondermann, *Reinsurance in arbitrage-free markets*, Insurance Math. Econom. **10** (1991), 191–202.
147. M. Sorensen, *A semimartingale approach to some problems in risk theory*, ASTIN Bulletin **26** (1996), no. 1, 15–23.
148. G. Spivak and J. Cvitanic, *Maximizing the probability of a perfect hedge*, Ann. Appl. Probab. **9** (1999), 1303–1328.
149. P. Spreij, *A crash course in stochastic calculus and application to mathematical finance*, Quarterly **9** (1996), no. 4, 357–388.
150. M. Steffensen, *A no-arbitrage approach to Thiele's differential equation*, Insurance Math. Econom. **27** (2000), 201–214.
151. E. Straub, *Non-life insurance mathematics*, Springer-Verlag, Zürich, 1988.
152. C. Stricker, *Arbitrage et lois de martingale*, Ann. Inst. H. Poincaré Probab. Statist. **26** (1990), 451–460.
153. S. J. Taylor, *Modelling stochastic volatility: a review and comparative study*, Math. Finance **4** (1994), 183–204.
154. A. D. Venttsel′ [Wentzell], *A course in the theory of stochastic processes*, "Nauka", Moscow, 1975; English transl., McGraw-Hill, New York, 1981.
155. S. N. Volkov, *A generalization of an integral option*, Vestnik Moskov. Gos. Univ. Ser. I. Mat. Mekh. **1995**, no. 6, 51–55; English transl., Moscow Univ. Math. Bull. **50** (1995), no. 6, 10–13.
156. S. N. Volkov, *Computation of the cost of one American-type option*, Uspekhi Mat. Nauk **50** (1995), no. 6, 173–174; English transl., Russian Math. Surveys **50** (1995), 1318–1320.
157. S. N. Volkov, *Optimal investment with operating costs proportional to the value of the portfolio*, Uspekhi Mat. Nauk **51** (1996), no. 6, 193–194; English transl., Russian Math. Surveys **51** (1996), 1227–1228.
158. S. N. Volkov and D. O. Kramkov, *On the methodology of hedging options*, Obozrenie Prikladnoi Promyshlennoi Matematiki **4** (1997), 18–65. (Russian)
159. C. Vorm Christensen and H. Schmidli, *Pricing catastrophe insurance products based on actually reported claims*, Insurance Math. Econom. **27** (2000), 189–200.
160. C. Vorm Christensen, *A new model for pricing catastrophe insurance derivatives*, working paper, vol. 28, Centre for Analytical Finance, Aarhus University, 1999.
161. F. Wagner, *Risk securitization: an alternative of risk transfer of insurance companies*, The Geneva Papers on Risk and Insurance, vol. 23, 1998, pp. 540–607.
162. E. V. Yagnyatinskiĭ, *A problem of optimal sequential investment*, Uspekhi Mat. Nauk **52** (1997), no. 4, 193–194; English transl., Russian Math. Surveys **52** (1997), 850–851.

Subject Index

Titles in This Series

212 **A. V. Mel′nikov, S. N. Volkov, and M. L. Nechaev,** Mathematics of financial obligations, 2002

211 **Takeo Ohsawa,** Analysis of several complex variables, 2002

210 **Toshitake Kohno,** Conformal field theory and topology, 2002

209 **Yasumasa Nishiura,** Far-from-equilibrium dynamics, 2002

208 **Yukio Matsumoto,** An introduction to Morse theory, 2002

207 **Ken'ichi Ohshika,** Discrete groups, 2002

206 **Yuji Shimizu and Kenji Ueno,** Advances in moduli theory, 2002

205 **Seiki Nishikawa,** Variational problems in geometry, 2001

204 **A. M. Vinogradov,** Cohomological analysis of partial differential equations and Secondary Calculus, 2001

203 **Te Sun Han and Kingo Kobayashi,** Mathematics of information and coding, 2002

202 **V. P. Maslov and G. A. Omel′yanov,** Geometric asymptotics for nonlinear PDE. I, 2001

201 **Shigeyuki Morita,** Geometry of differential forms, 2001

200 **V. V. Prasolov and V. M. Tikhomirov,** Geometry, 2001

199 **Shigeyuki Morita,** Geometry of characteristic classes, 2001

198 **V. A. Smirnov,** Simplicial and operad methods in algebraic topology, 2001

197 **Kenji Ueno,** Algebraic geometry 2: Sheaves and cohomology, 2001

196 **Yu. N. Lin′kov,** Asymptotic statistical methods for stochastic processes, 2001

195 **Minoru Wakimoto,** Infinite-dimensional Lie algebras, 2001

194 **Valery B. Nevzorov,** Records: Mathematical theory, 2001

193 **Toshio Nishino,** Function theory in several complex variables, 2001

192 **Yu. P. Solovyov and E. V. Troitsky,** C^*-algebras and elliptic operators in differential topology, 2001

191 **Shun-ichi Amari and Hiroshi Nagaoka,** Methods of information geometry, 2000

190 **Alexander N. Starkov,** Dynamical systems on homogeneous spaces, 2000

189 **Mitsuru Ikawa,** Hyperbolic partial differential equations and wave phenomena, 2000

188 **V. V. Buldygin and Yu. V. Kozachenko,** Metric characterization of random variables and random processes, 2000

187 **A. V. Fursikov,** Optimal control of distributed systems. Theory and applications, 2000

186 **Kazuya Kato, Nobushige Kurokawa, and Takeshi Saito,** Number theory 1: Fermat's dream, 2000

185 **Kenji Ueno,** Algebraic Geometry 1: From algebraic varieties to schemes, 1999

184 **A. V. Mel′nikov,** Financial markets, 1999

183 **Hajime Sato,** Algebraic topology: an intuitive approach, 1999

182 **I. S. Krasil′shchik and A. M. Vinogradov, Editors,** Symmetries and conservation laws for differential equations of mathematical physics, 1999

181 **Ya. G. Berkovich and E. M. Zhmud′,** Characters of finite groups. Part 2, 1999

180 **A. A. Milyutin and N. P. Osmolovskii,** Calculus of variations and optimal control, 1998

179 **V. E. Voskresenskiĭ,** Algebraic groups and their birational invariants, 1998

178 **Mitsuo Morimoto,** Analytic functionals on the sphere, 1998

177 **Satoru Igari,** Real analysis—with an introduction to wavelet theory, 1998

176 **L. M. Lerman and Ya. L. Umanskiy,** Four-dimensional integrable Hamiltonian systems with simple singular points (topological aspects), 1998

175 **S. K. Godunov,** Modern aspects of linear algebra, 1998

174 **Ya-Zhe Chen and Lan-Cheng Wu,** Second order elliptic equations and elliptic systems, 1998

173 **Yu. A. Davydov, M. A. Lifshits, and N. V. Smorodina,** Local properties of distributions of stochastic functionals, 1998

172 **Ya. G. Berkovich and E. M. Zhmud′,** Characters of finite groups. Part 1, 1998

171 **E. M. Landis,** Second order equations of elliptic and parabolic type, 1998

170 **Viktor Prasolov and Yuri Solovyev,** Elliptic functions and elliptic integrals, 1997

169 **S. K. Godunov,** Ordinary differential equations with constant coefficient, 1997

168 **Junjiro Noguchi,** Introduction to complex analysis, 1998

167 **Masaya Yamaguti, Masayoshi Hata, and Jun Kigami,** Mathematics of fractals, 1997

166 **Kenji Ueno,** An introduction to algebraic geometry, 1997

165 **V. V. Ishkhanov, B. B. Lur′e, and D. K. Faddeev,** The embedding problem in Galois theory, 1997

164 **E. I. Gordon,** Nonstandard methods in commutative harmonic analysis, 1997

163 **A. Ya. Dorogovtsev, D. S. Silvestrov, A. V. Skorokhod, and M. I. Yadrenko,** Probability theory: Collection of problems, 1997

162 **M. V. Boldin, G. I. Simonova, and Yu. N. Tyurin,** Sign-based methods in linear statistical models, 1997

161 **Michael Blank,** Discreteness and continuity in problems of chaotic dynamics, 1997

160 **V. G. Osmolovskiĭ,** Linear and nonlinear perturbations of the operator div, 1997

159 **S. Ya. Khavinson,** Best approximation by linear superpositions (approximate nomography), 1997

158 **Hideki Omori,** Infinite-dimensional Lie groups, 1997

157 **V. B. Kolmanovskiĭ and L. E. Shaĭkhet,** Control of systems with aftereffect, 1996

156 **V. N. Shevchenko,** Qualitative topics in integer linear programming, 1997

155 **Yu. Safarov and D. Vassiliev,** The asymptotic distribution of eigenvalues of partial differential operators, 1997

154 **V. V. Prasolov and A. B. Sossinsky,** Knots, links, braids and 3-manifolds. An introduction to the new invariants in low-dimensional topology, 1997

153 **S. Kh. Aranson, G. R. Belitsky, and E. V. Zhuzhoma,** Introduction to the qualitative theory of dynamical systems on surfaces, 1996

152 **R. S. Ismagilov,** Representations of infinite-dimensional groups, 1996

151 **S. Yu. Slavyanov,** Asymptotic solutions of the one-dimensional Schrödinger equation, 1996

150 **B. Ya. Levin,** Lectures on entire functions, 1996

149 **Takashi Sakai,** Riemannian geometry, 1996

148 **Vladimir I. Piterbarg,** Asymptotic methods in the theory of Gaussian processes and fields, 1996

147 **S. G. Gindikin and L. R. Volevich,** Mixed problem for partial differential equations with quasihomogeneous principal part, 1996

146 **L. Ya. Adrianova,** Introduction to linear systems of differential equations, 1995

145 **A. N. Andrianov and V. G. Zhuravlev,** Modular forms and Hecke operators, 1995

144 **O. V. Troshkin,** Nontraditional methods in mathematical hydrodynamics, 1995

143 **V. A. Malyshev and R. A. Minlos,** Linear infinite-particle operators, 1995

142 **N. V. Krylov,** Introduction to the theory of diffusion processes, 1995

141 **A. A. Davydov,** Qualitative theory of control systems, 1994

140 **Aizik I. Volpert, Vitaly A. Volpert, and Vladimir A. Volpert,** Traveling wave solutions of parabolic systems, 1994

139 **I. V. Skrypnik,** Methods for analysis of nonlinear elliptic boundary value problems, 1994
138 **Yu. P. Razmyslov,** Identities of algebras and their representations, 1994
137 **F. I. Karpelevich and A. Ya. Kreinin,** Heavy traffic limits for multiphase queues, 1994
136 **Masayoshi Miyanishi,** Algebraic geometry, 1994
135 **Masaru Takeuchi,** Modern spherical functions, 1994
134 **V. V. Prasolov,** Problems and theorems in linear algebra, 1994
133 **P. I. Naumkin and I. A. Shishmarev,** Nonlinear nonlocal equations in the theory of waves, 1994
132 **Hajime Urakawa,** Calculus of variations and harmonic maps, 1993
131 **V. V. Sharko,** Functions on manifolds: Algebraic and topological aspects, 1993
130 **V. V. Vershinin,** Cobordisms and spectral sequences, 1993
129 **Mitsuo Morimoto,** An introduction to Sato's hyperfunctions, 1993
128 **V. P. Orevkov,** Complexity of proofs and their transformations in axiomatic theories, 1993
127 **F. L. Zak,** Tangents and secants of algebraic varieties, 1993
126 **M. L. Agranovskiĭ,** Invariant function spaces on homogeneous manifolds of Lie groups and applications, 1993
125 **Masayoshi Nagata,** Theory of commutative fields, 1993
124 **Masahisa Adachi,** Embeddings and immersions, 1993
123 **M. A. Akivis and B. A. Rosenfeld,** Élie Cartan (1869–1951), 1993
122 **Zhang Guan-Hou,** Theory of entire and meromorphic functions: deficient and asymptotic values and singular directions, 1993
121 **I. B. Fesenko and S. V. Vostokov,** Local fields and their extensions: A constructive approach, 1993
120 **Takeyuki Hida and Masuyuki Hitsuda,** Gaussian processes, 1993
119 **M. V. Karasev and V. P. Maslov,** Nonlinear Poisson brackets. Geometry and quantization, 1993
118 **Kenkichi Iwasawa,** Algebraic functions, 1993
117 **Boris Zilber,** Uncountably categorical theories, 1993
116 **G. M. Fel′dman,** Arithmetic of probability distributions, and characterization problems on abelian groups, 1993
115 **Nikolai V. Ivanov,** Subgroups of Teichmüller modular groups, 1992
114 **Seizô Itô,** Diffusion equations, 1992
113 **Michail Zhitomirskiĭ,** Typical singularities of differential 1-forms and Pfaffian equations, 1992
112 **S. A. Lomov,** Introduction to the general theory of singular perturbations, 1992
111 **Simon Gindikin,** Tube domains and the Cauchy problem, 1992
110 **B. V. Shabat,** Introduction to complex analysis Part II. Functions of several variables, 1992
109 **Isao Miyadera,** Nonlinear semigroups, 1992
108 **Takeo Yokonuma,** Tensor spaces and exterior algebra, 1992
107 **B. M. Makarov, M. G. Goluzina, A. A. Lodkin, and A. N. Podkorytov,** Selected problems in real analysis, 1992
106 **G.-C. Wen,** Conformal mappings and boundary value problems, 1992
105 **D. R. Yafaev,** Mathematical scattering theory: General theory, 1992

For a complete list of titles in this series, visit the
AMS Bookstore at **www.ams.org/bookstore/**.